Cold Regions Ice/Snow Actions in Hydrology, Ecology and Engineering

Cold Regions Ice/Snow Actions in Hydrology, Ecology and Engineering

Editors

Zhijun Li
Fang Li
Sasan Tavakoli
Xuemei Liu
Changlei Dai

MDPI

Basel • Beijing • Wuhan • Barcelona • Belgrade • Novi Sad • Cluj • Manchester

Editors

Zhijun Li
State Key Laboratory of Coastal and Offshore Engineering
Dalian University of Technology
Dalian
China

Fang Li
School of Ocean and Civil Engineering
Shanghai Jiao Tong University
Shanghai
China

Sasan Tavakoli
Department of Infrastructure Engineering
University of Melbourne
Melbourne
Australia

Xuemei Liu
Northeast Institute of Geography and Agroecology
Chinese Academy of Sciences
Changchun
China

Changlei Dai
School of Hydraulic and Electric-power
Heilongjiang University
Harbin
China

Editorial Office
MDPI
St. Alban-Anlage 66
4052 Basel, Switzerland

This is a reprint of articles from the Special Issue published online in the open access journal *Water* (ISSN 2073-4441) (available at: www.mdpi.com/journal/water/special_issues/4IIA93XQZY).

For citation purposes, cite each article independently as indicated on the article page online and as indicated below:

Lastname, A.A.; Lastname, B.B. Article Title. *Journal Name* **Year**, *Volume Number*, Page Range.

ISBN 978-3-7258-0628-7 (Hbk)
ISBN 978-3-7258-0627-0 (PDF)
doi.org/10.3390/books978-3-7258-0627-0

Contents

Preface

Ice and snow are important elements of the earth's hydrosphere, playing significant roles in the climate and in our daily lives through hydrology, ecology, and engineering. This is a reprint of the articles gathered through a Special Issue on the topic of "Cold Regions Ice/Snow Actions in Hydrology, Ecology, and Engineering". The aim of the Special Issue is to collect articles on the physical, thermal, mechanical, optical, and electrical properties of crystal ice/snow, the melting water from ice/snow, as well as permafrost. The scope also includes theoretical studies and practical applications in remote sensing, experiments and numerical modeling in cold regions, snow/ice forming and melting processes in water bodies and permafrost, contributions to ecosystems, and behaviors in engineering and entertainment. In total, 12 articles have been published in the Special Issue, covering topics including identification of snow and ice through image analysis, snow and sea ice formation processes, physical and mechanical properties of snow and ice, ice flood disasters, changes in sea ice characteristics, and ecosystems under ice. The authors would like to thank all the authors for their high-quality contribution to this Special Issue, which promotes our understanding of ice and snow and our capability to cope with problems related to ice and snow. This reprint could be taken as a textbook or training material for postgraduate students and researchers. We hope to motivate readers to explore more about ice and snow in cold regions.

Zhijun Li, Fang Li, Sasan Tavakoli, Xuemei Liu, and Changlei Dai
Editors

water MDPI

Editorial

Cold Region Ice/Snow Actions in Hydrology, Ecology and Engineering

Zhijun Li [1,*], Fang Li [2], Sasan Tavakoli [3], Xuemei Liu [4] and Changlei Dai [5]

1 State Key Laboratory of Coastal and Offshore Engineering, Dalian University of Technology, Dalian 116024, China
2 School of Ocean and Civil Engineering, Shanghai Jiao Tong University, Shanghai 200240, China; fang.li@sjtu.edu.cn
3 Department of Infrastructure Engineering, University of Melbourne, Melbourne 3052, Australia; s.tavakoli@unimelb.edu.au
4 Northeast Institute of Geography and Agroecology, Chinese Academy of Sciences, Changchun 130102, China; liuxuemei@iga.ac.cn
5 School of Hydraulic and Electric-Power, Heilongjiang University, Harbin 150080, China; daichanglei@126.com
* Correspondence: lizhijun@dlut.edu.cn

1. Introduction

In the Earth's hydrosphere, 96.50% of the total mass is composed of seawater, while the remaining 3.50% is freshwater. Much of the sea water freezes into sea ice, either seasonally as first-year ice, or over the long term as multi-year ice. Among fresh water, 77.44% exists in the form of ice/snow at the polar regions [1]; comparatively little ice/snow is found in other inland cold regions. Throughout the course of human development, the ice/snow from the Quaternary glacial period has played an essential role in changing and evolving human life, involving handling and accommodating with ice/snow via multiple means. Some scholars consider the climatic changes at the end of the Quaternary ice age as essential prerequisites for the development of Homo sapiens civilization [2]. Scholars have also connected the fluctuations in grain production caused by modern climate variations with dynastic changes [3]. This is because ice/snow is an integral part of human activities, closely linked to water sources, food, transportation, entertainment, and eventually human development [4]. With the progress of science and technology, human understanding of nature and the ability to handle ice/snow has improved, prompting research into the various phenomena and essence of ice/snow that serve human activities. These indicate that ice and snow research has been standing as a critical frontier in the context of hydrology, ecology, and engineering.

According to IPCC WGI AR4 (2007), the main components of the cryosphere are snow, river and lake ice, sea ice, glaciers and ice caps, ice shelves, ice sheets, and frozen ground, which are all changing under global warming [5]. As the freezing point of ice/snow (−12.8–0 °C) falls into the variation range of atmospheric temperature, phase changes happen along with the temperature varying between minus and plus degrees Celsius, exhibiting sensitivity to temperature. During recent years, as global warming continues, challenges have emerged regarding frozen landscapes following the change in snow/ice. Currently, global warming profoundly alters ice and snow dynamics worldwide, accelerating the melting of glaciers, diminishing polar ice caps, and reducing snow cover. Ice in the polar region becomes more dynamic and more multiyear ice drifts to lower altitude due to melting. The ice cover in the cold regions becomes warmer and its mechanical properties significantly change [6].

In the past, humans primarily used glaciers for agriculture, domestic water supply, and, as an exclusive form of luxury consumption, for tourism and exploration. As human activities extended to Greenland and the Antarctic interior, the role of glaciers have become further elevated, connected to records of climate change. Notably, ice core research has

Citation: Li, Z.; Li, F.; Tavakoli, S.; Liu, X.; Dai, C. Cold Region Ice/Snow Actions in Hydrology, Ecology and Engineering. *Water* **2024**, *16*, 689. https://doi.org/10.3390/w16050689

Received: 1 February 2024
Accepted: 19 February 2024
Published: 27 February 2024

become a crucial aspect of this [7]. These polar expeditions and studies require ice structures, roads, airports, and other logistical support closely related to supply chains [8]. The instability of ice caps caused by climate warming has motivated scientists to link ice caps, ice shelves, and icebergs to ocean circulation, forming a larger-scale atmosphere–ice–water system. Atmospheric circulation, ocean circulation, biological migration, and more are all encompassed within this system. Scientifically subdividing the atmosphere–ice–water system into different scales has given rise to emerging fields such as ice meteorology, oceanography, and sea ice science. Meanwhile, humanity's desire to rationally utilize fisheries, mining, and transportation in ice-covered regions has sparked enthusiasm for ice engineering. New research topics, such as offshore wind power generation and solar panels in ice-covered areas, have emerged in addition to traditional navigation and drilling in ice. Moreover, the ocean, covering two-thirds of the Earth, is considered a crucial carbon sink, resulting in great potential in polar fishery.

The extensive expansion of human activities and resource utilization needs has reached more ice/snow areas. Scientific understanding and technology development are necessary to cope with the challenges posed by ice/snow. Engineers working on projects ranging from buildings and bridges to power lines and transportation systems must stand with the mechanical properties of ice, ensuring structures are resilient in the face of evolving climates. The ice cover above polar waters exerts a profound influence on the underwater ecosystem, shaping the conditions for marine life. Climate change-induced alterations to ice cover disrupt these dynamics, affecting species' behavior, migration patterns, and overall ecosystem health. The delicate balance between ice cover and the underwater environment underscores the vulnerability of polar ecosystems in the face of global climate change. The threat of ice flood disasters is big in regions prone to freezing conditions. Understanding the triggers and patterns of ice-related flooding is crucial for engineering effective preventive measures and emergency response strategies. Balancing the need for sustainable development with the risks posed by changing ice dynamics requires a holistic approach that integrates climate science, ecology, and cutting-edge engineering solutions.

To understand the implications of climate change to our ecosystem to maintain sustainability, and its effect on the engineering world to guarantee safety and mitigate risks, it is important to understand the change in the physical, thermal, mechanical, optical, and electrical properties of crystal ice/snow and the melting water from ice/snow, as well as permafrost. After our previous Special Issue, "Sea, River, Lake Ice Properties and Their Applications in Practice" [9], this Special Issue, entitled "Cold Regions Ice/Snow Actions in Hydrology, Ecology and Engineering", intends to continue research under this context. It invites researchers from different fields to investigate ice/snow-related problems in hydrology, ecology and engineering and publish their results. In this Special Issue, we focus on the physical and mechanical properties of ice and snow, as well as their impact on hydrology, ecology, and engineering. This Special Issue can guide future ice science and engineering in polar regions under climate change. The scope also includes theoretical studies and practical applications of various snow/ice properties in remote sensing, investigation, experiments, and numerical modeling in cold region snow/ice formation and melting processes in water bodies and permafrost, its contributions to the ecosystem, and behaviors in engineering and entertainment.

2. List and Summaries of the Contributions

This Special Issue received 15 manuscripts and all of them were subject to the rigorous *Water* review process. In total, 12 papers were finally accepted for publication and inclusion in this Special Issue. The contributions are listed in List of Contribution.

As shown in Table 1, the contributions covered wide perspectives concerned with ice. The contents of these published papers are snow and ice identification, snow and ice growth, the mechanical properties of ice and snow, the thermal diffusivity of ice, ice condition monitoring, ice flood disaster, and ecosystems under ice.

Table 1. Analysis of the published contributions in the Special Issue.

Number of Contribution	Research Area	Focus	Research Methods	Potential Applications
1	River ice characteristics	Real-time monitoring of river surface ice dynamics	Image processing	Hydrology
2	Sea ice characteristics	Polar objects, e.g., ice floe, identification from remote sensing and onboard images	Image processing	Sea ice engineering
3	Snow characteristics	Cloud removal to recover snow coverage by machine learning from remote sensing image	Image processing	Remote sensing and new technology
4	Snow formation	Snow Crystal Necks and the Effect on Hardness	Experiment	Snow engineering
5	Sea ice formation	Temperature and ice growth	Data collection and statistics	Sea ice engineering
6	Ice thermal properties	Review of research on thermal diffusivity of ice	Literature review	Ice engineering
7	Model ice mechanical properties	Mechanical property of model ice	Experiment and simulation	Ice engineering
8	Snow mechanical properties	Mechanical property of snow	Experiment	Snow engineering
9	Snow/ice mechanical properties	Property and topology of brash ice and its effect on mechanics	Experiment	Ice engineering
10	River ice engineering conditions	Risks of ice flood disaster	Catastrophe theory	River ice engineering
11	Sea ice engineering design conditions	Change of ice characteristics in the Bohai Bay	Data statistics	Sea ice engineering
12	Ecosystem under ice	Productivity of the ecosystem under ice cover	Simulation	Ecological service in cold regions

3. An Overview of Published Articles

As listed in Section 2, research topics related to ice/snow in the field of hydrology, ecology and engineering are all covered by the published articles. An overview of these articles is provided here.

Ice/snow engineering problems are often investigated through experimental or numerical modeling methods, e.g., [10,11], which usually assume an idealized ice field. To support engineering applications, it is of equal importance to identify and understand complex ice/snow fields in nature. Image technology has been developed a lot over recent years for real ice/snow conditions [12]. In-depth understanding of measurement data, in addition to theoretical analysis, has seen a rise in artificial intelligence (AI) and machine learning as new technologies. AI and machine learning are now widely applied to analyze the patterns, mechanisms, and trends in various natural phenomena with a certain level of randomness. As temperature-sensitive materials, ice/snow can benefit from AI and machine learning to assist in extracting valuable insights from available data resources, exploring the relationships between ice behavior and the main controlling factors, as well as numerous secondary controlling factors [13]. Three papers were published on the topic of snow and ice identification using machine learning or artificial intelligence. Yang et al. (Contribution 1) presented a comprehensive approach to the real-time monitoring of river surface ice dynamics using deep learning methods and camera imagery. Focusing on the Nenjiang River in China, the study introduces a four-step methodology. First, image preprocessing involves calibrating camera images to real-world coordinates. Subsequently, a lightweight semantic segmentation network is employed to identify ice and water pixels, facilitating the calculation of ice concentration and area. The motion detection process is enhanced using the segmentation results, and the particle video tracking algorithm is adapted for ice velocity measurement through a novel tracking point generation strategy. The research, encapsulated in the IPC_RI_IDS dataset, contributes to our understanding of ice break-up processes. The goal is to support real-time short-term forecasts of ice floods by analyzing surface ice data and predicting the stages of the ice break-up process. The innovation of this approach is to address previous limitations in tracking river ice velocity, providing valuable insights for mitigating risks associated with ice blockages in water routes. In addressing the safety concerns of ship navigation in polar regions, Ding et al. (Contribution 2) emphasize the need for the prompt detection of sea ice, icebergs, and passing ships. Recognizing the limitations of individual data sources, they construct a comprehensive polar multi-target local-scale dataset with categories such as sea ice, icebergs, ice melt ponds, icebreakers, and inter-ice channels. Using a single-shot detector (SSD), they achieved a final mean average precision (mAP) value of 70.19%. The study further presents a remote sensing sea ice dataset with 15,948 labels, enhancing the You Only Look Once (YOLOv5) model with advanced features. Through ablation experiments, the improved YOLOv5 demonstrated notable enhancements, surpassing other models like YOLOv3 in performance. This facilitates the detection of multiple targets on various scales in polar regions, promoting data fusion and providing valuable support for polar ship path planning. Zhu et al. (Contribution 3) presented a novel approach for fine-spatiotemporal-resolution snow monitoring at the watershed scale, crucial for effective snow water resource management. They proposed a cloud removal algorithm based on snow grain size (SGS) gap-filling using a space–time extra tree. This addresses the limitation of cloud occlusion in long-time-series snow products, enhancing coverage and time resolution. The study focused on the Kaidu River Basin (KRB) and incorporated spatiotemporal information into dimensional data to characterize geomorphic features and snow duration. By training a spatiotemporal extreme tree model, the nonlinear mapping relationship between multidimensional inputs and SGS was simulated. This method demonstrated effectiveness, particularly when cloud cover was below 70%, achieving satisfactory SGS estimation and successful snow cover reconstruction. Compared to traditional methods, the proposed approach exhibited superior detail characterization and performance in complex mountainous environments, resulting in a significant reduction in cloud coverage and an increase in snow coverage from 2000 to 2020. This advancement in cloud removal significantly improved the time resolution of snow cover data without compromising accuracy.

The growth of ice/snow is an important topic relating meteorology to engineering. Ice/snow is a product of a cold environment. Its formation is closely related to hydrological and meteorological conditions. New research results have continuously emerged on the freezing processes of different types of water bodies in different environments, and even beyond natural environmental conditions [14,15]. This Special Issue published two relevant articles, focusing on the current natural environmental conditions of snow morphology and the processes of ice formation and melting. Two papers focus on the topic of snow and ice growth. Wei et al. (Contribution 4) conducted a constant-density (200 kg/m^3) snow metamorphism experiment to investigate the snow microstructure at different metamorphism times and isolate the snow neck area. Their findings highlighted the significant influence of temperature, solar radiation, snow density, and specific humidity on the neck region, with wind speed having a minimal effect. In response to atmospheric forcing, the authors developed a multiple linear regression equation, "$S = 288T + 2E + 189\rho + 12{,}194V - 20{,}443RH - 42{,}729$", where solar radiation (E), temperature (T), snow density (ρ), specific humidity (RH), and wind speed (V) collectively explained 84% of the neck area variability. Notably, the correlation between snow hardness and the neck area reached 71%, potentially rising to 91% in later metamorphic stages. A predictive model for hardness based on the neck area, "$H = 0.002764S + 67.922837$", was established. This study provides insights into the growth variations in the metamorphic snow cover's neck region, illustrating how external factors impact both microstructure and macroscopic physical characteristics. During the winters of 2009/2010 and 2020/2021, Ma et al. (Contribution 5) conducted observations at an eastern port of Liaodong Bay to investigate variations in sea ice thickness and atmospheric conditions. Two main observation items, daily ice thickness (DIT) and cumulative ice thickness (CIT), were studied. For DIT, sea ice thickness gradually decreased with rising temperatures, with a freezing rate of $1.48\ cm/(°C\cdot d)^{1/2}$. Concerning CIT, at -12 °C, the maximum growth rate decreased from 3.5 cm/d to 1.5 cm/d as ice thickness increased from 0 to 20 cm. The study applied the residual method to calculate oceanic heat flux, a crucial parameter in ice modeling. Both analytic (Stefan's law) and numerical models (high-resolution thermodynamic snow-and-ice model) were employed. Results indicated a high simulation accuracy with a growth coefficient in the analytic model. The numerical model, with an oceanic heat flux of 2 $W{\cdot}m^{-2}$, showed a maximum error of 60% in 2010 and 3.7% in 2021. Using the calculated oceanic heat flux, errors were significantly reduced to 4.2% in the winter of 2009/2010 and 1.5% in 2020/2021. Furthermore, the oceanic heat flux in Liaodong Bay exhibited a decreasing trend with increasing ice thickness and air temperature.

In ice/snow engineering, the mechanical properties and response of materials form the basis of engineering mechanics. The thermodynamics of ice are the first factor determining its mechanical properties. As a material formed through natural process, the mechanical properties of ice and snow have long been a challenging problem yet to be solved. Exploring the properties and behaviors of multiphase ice/snow with phase transitions is crucial, especially under the influence of climate change, particularly in the case of global warming [6]. Thorough research on the thermodynamics and mechanical behaviors of ice/snow in these contexts is still necessary [16]. Ongoing studies in this area continue to explore and advance our understanding. In this Special Issue, one paper focuses on the physical properties of ice, specifically the thermal conductivity, and three papers relate to the mechanical properties of snow and ice. Li et al. (Contribution 6) presented a review of ice thermal parameters, essential for accurately simulating ice phenology, distribution, and thickness—a facet considered a "vulnerable group" in ice research. Despite the perceived technical simplicity, the authors emphasized the complexity of obtaining accurate ice thermal property parameters, necessitating a rigorous research process. While progress in understanding the thermal conductivity of ice in China stagnated after explorations in the 1980s, the current century introduced mathematical methods. In this study, inversion identification and analysis utilized time-series data from in situ testing of vertical temperature profiles in ice layers to derive thermal diffusivities for different natural ices. Cross-validation demonstrated variations in thermal diffusivity due to impurities within unfrozen water among ice crystals. The paper highlighted the importance of parameterizing thermal diffusivity in the phase transition zone of ice under the influence of global warming trends. Future research directions were envisioned, encompassing the physical mechanisms, application value, and parameterization schemes for the thermal diffusivity of natural ice. Tian et al. (Contribution 7) presented a study on the flexural strength of columnar saline model ice, crucial for designing structures in ice-infested waters. Conducted at the China Ship Scientific Research Center, circular plate center loading tests explored varied loading rates and ice temperatures. Using FEM and LS-DYNA, a numerical model validated and compared the results, unveiling crack propagation, stress distribution, and failure modes. The model ice exhibited typical brittle failure, with flexural strength linearly linked to temperature and no significant correlation with loading rate. Porosity influenced load response and failure time, but not the failure mode. Specifically, 7% porosity resulted in a 7.8% reduction in load response compared to nonporous ice. This research establishes a method for analyzing model ice flexural strength, laying a foundation for the further exploration of structure–ice sheet interactions. Han et al. (Contribution 8) addressed the limited understanding of snow's mechanical properties, crucial for polar infrastructure construction. Uniaxial compression tests under step loading were conducted to investigate snow behavior in cold regions. Using the Maxwell model, constitutive equations were developed, incorporating different temperatures, densities, and loading rates. Findings revealed that loading rates had no significant impact on snow's elastic modulus and viscosity coefficient. Both parameters exhibited an exponential relationship with density, increasing with higher density. As temperature decreased, the elastic modulus and viscosity coefficient initially declined and then increased, though no specific functional relationship was identified. Notably, a novel constitutive equation, accounting for snow density, was derived based on the Maxwell model, contributing to a more comprehensive understanding of snow mechanics in polar environments. Zhaka et al. (Contribution 9) conducted a comprehensive study to address the critical differences between brash ice and surrounding level ice, crucial for understanding full-scale brash ice channel development. This research, spanning the winters of 2020–2021, 2021–2022, and 2023, focused on channels near the Bay of Bothnia, Luleå, Sweden. Measurements included snow, slush, and total ice thickness, along with analyses of ice microstructure and strength. Notably, this is the first paper reporting on the influence of snow in brash ice channels. The study revealed that snow covered the channels between ship passages, submerging and forming slush-filled voids after each passage. These voids then transformed into snow ice (SI) clusters frozen with columnar ice. Image analyses estimated SI content in brash ice, highlighting varying percentages in different locations, providing valuable insights into the complex dynamics of brash ice channels.

To cope with the impact of ice and snow on human activities, structural or non-structural measures have been adopted. This is the common practice in ice and snow engineering. Many research outcomes have been achieved for various engineering types, covering aspects related to both ice and snow and the use of structures, such as ice navigation, offshore wind farms, ports, reservoir dams, and more [17–19]. These studies encompass research on both ice conditions and the types of structures involved. This Special Issue published two articles with a specific focus on ice conditions. In the field of hydrology, one research article focuses on the ice flood disaster. Li et al. (Contribution 10) established a risk evaluation system for ice flood disasters, prevalent in frigid high-latitude and high-altitude regions, posing threats to personal and property safety through ice dam or ice jam flooding. Utilizing catastrophe theory and the Pearson correlation coefficient, the study aims to provide a comprehensive and necessary risk assessment for preventing ice flood disasters. The system incorporates hierarchical cluster analysis to simplify indicators and select typical years based on correlations. Results reveal catastrophe membership values in the Mohe, Tahe, and Huma regions from 2000 to 2020 ranging from 0.86 to 0.93. The evaluation system, coupled with actual disaster situations, yielded a four-level classification of risk ratings. A comparison with the fuzzy comprehensive evaluation method demonstrates similar risk levels, affirming the effectiveness and practicality of applying catastrophe theory to ice flood risk evaluation. This study introduces a novel method for studying and understanding ice floods. One research article focuses on the change in ice characteristics in the Bohai Bay. Li et al. (Contribution 11) investigated the safety of winter activities in the Bohai Sea, emphasizing the need for more detailed information on ice characteristics and a refined ice zone division. Utilizing a $1/12^{\circ}$-resolution sea ice characteristic dataset derived from the NEMO-LIM2 ice-ocean coupling model, which assimilated MODIS satellite sea ice observations spanning 2005 to 2022, new sea ice hindcasting data was obtained. These data facilitated the analysis of ice period, thickness, concentration, temperature, salinity, and design ice thickness for various return periods in a $1/4^{\circ}$-resolution refined zoning. The findings, compared with the previous 21-ice-zone standard, revealed a significant reduction in ice conditions along the west coast of the Bohai Sea, emphasizing the importance of updated information for enhancing safety measures in winter maritime activities.

Under meteorological conditions, especially driven by solar radiation flux penetrating through the ice, understanding how the ecological environment under ice responds and how changes in the ice cover, influenced by climate change, correlate with the water ecological environment under ice has become an important issue in the context of recent climate change [20,21]. This has received lots of attention, and this Special Issue also published a relevant article. In the field of ecology, one research article focuses on the ecosystem under ice cover. Zhang et al. (Contribution 12) employed a Vertically Generalized Production Model (VGPM) suitable for ice-covered periods to investigate the complete change process of primary productivity in a temperate lake, shedding light on the connection between ice physical characteristics and biological production. Despite the significance of primary productivity in understanding the impact of global warming on temperate lake ecosystems, few studies have delved into the entire change process during the ice-covered period. The study focused on Hanzhang Lake, revealing a substantial primary productivity level (189.1 ± 112.6 mg $C \cdot m^{-2} \cdot d^{-1}$) under the ice. Contrary to common assumptions, phytoplankton production under the ice was not severely restricted by light; instead, water temperature emerged as a more crucial factor influencing primary productivity changes compared to light conditions. The research provided insights into the variability of primary productivity throughout the entire ice-covered age, enhancing understanding of how warmer temperatures affect the aquatic environment of lakes in seasonal ice-covered areas.

Author Contributions: Writing—original draft preparation, Z.L.; writing—review and editing, F.L., S.T., X.L. and C.D. All authors have read and agreed to the published version of the manuscript.

Funding: This organization of this Special Issue was supported by different projects from National Natural Science Foundation of China (grant numbers U23A2012, 52301331, 42207088, 51979024).

Conflicts of Interest: The authors declare no conflicts of interest.

List of Contributions:

1. Yang, Z.; Zong, J.; Zhu, Y.; Liu, X.; Tao, R.; Yu, Y. River Ice Regime Recognition Based on Deep Learning: Ice Concentration, Area, and Velocity. *Water* **2024**, *16*, 58. https://doi.org/10.3390/w16010058.
2. Ding, S.; Zeng, D.; Zhou, L.; Han, S.; Li, F.; Wang, Q. Multi-Scale Polar Object Detection Based on Computer Vision. *Water* **2023**, *15*, 3431. https://doi.org/10.3390/w15193431.
3. Zhu, L.; Ma, G.; Zhang, Y.; Wang, J.; Kan, X. Reconstruction of Snow Cover in Kaidu River Basin via Snow Grain Size Gap-Filling Based on Machine Learning. *Water* **2023**, *15*, 3726. https://doi.org/10.3390/w15213726.
4. Wei, J.; Lu, P.; Hu, S.; Zhao, Q.; Yuan, S.; Huo, P.; Wang, Q. Research on the Evolution of Snow Crystal Necks and the Effect on Hardness during Snowpack Metamorphism. *Water* **2024**, *16*, 48. https://doi.org/10.3390/w16010048.
5. Ma, Y.; Ding, D.; Xu, N.; Yuan, S.; Shi, W. Ice Mass Balance in Liaodong Bay: Modeling and Observations. *Water* **2023**, *15*, 943. https://doi.org/10.3390/w15050943.
6. Li, Z.; Fu, X.; Shi, L.; Huang, W.; Li, C. Recent Advances and Challenges in the Inverse Identification of Thermal Diffusivity of Natural Ice in China. *Water* **2023**, *15*, 1041. https://doi.org/10.3390/w15061041.
7. Tian, Y.; Zhao, W.; Yu, C.; Gang, X.; Lu, P.; Yue, Q. Investigations on Flexural Strength of a Columnar Saline Model Ice under Circular Plate Central Loading. *Water* **2023**, *15*, 3371. https://doi.org/10.3390/w15193371.
8. Han, H.; Yang, M.; Liu, X.; Li, Y.; Gao, G.; Wang, E. Study on the Constitutive Equation and Mechanical Properties of Natural Snow under Step Loading. *Water* **2023**, *15*, 3271. https://doi.org/10.3390/w15183271.
9. Zhaka, V.; Bridges, R.; Riska, K.; Nilimaa, J.; Cwirzen, A. Observations of Snow–Slush–Snow Ice Transformation and Properties of Brash Ice in Ship Channels. *Water* **2023**, *15*, 2360. https://doi.org/10.3390/w15132360.
10. Li, Y.; Han, H.; Sun, Y.; Xiao, X.; Liao, H.; Liu, X.; Wang, E. Risk Evaluation of Ice Flood Disaster in the Upper Heilongjiang River Based on Catastrophe Theory. *Water* **2023**, *15*, 2724. https://doi.org/10.3390/w15152724.
11. Li, G.; Jiao, Y.; Chen, X.; Zhao, Y.; Li, R.; Guo, D.; Ge, L.; Hou, Q.; Wang, Q. Investigation of the Recent Ice Characteristics in the Bohai Sea in the Winters of 2005–2022 Using Multi-Source Data. *Water* **2024**, *16*, 290. https://doi.org/10.3390/w16020290.
12. Zhang, J.; Xie, F.; Song, H.; Meng, J.; Zhang, Y. Simulation and Key Physical Drivers of Primary Productivity in a Temperate Lake during the Ice-Covered Period: Based on the VGPM Model. *Water* **2023**, *15*, 918. https://doi.org/10.3390/w15050918.

References

1. Oki, T. Hydrosphere–The water realm which supports human life. In *Human Geoscience*, 1st ed.; Himiyama, Y., Satake, K., Oki, T., Eds.; Springer: Singapore, 2020; pp. 39–46. [CrossRef]
2. Hays, J.D. A review of the Late Quaternary climatic history of Antarctic Seas. In *Antarctic Glacial History and World Palaeoenvironments*, 1st ed.; Bakker, E.M., Ed.; Cambridge University Press: Leiden, The Netherlands, 1978; pp. 57–71. [CrossRef]
3. Su, Y.; Fang, X.Q.; Yin, J. Impact of climate change on fluctuations of grain harvests in China from the Western Han Dynasty to the Five Dynasties (206 BC–960 AD). *Sci. China Earth Sci.* **2014**, *57*, 1701–1712. [CrossRef]
4. Bokhorst, S.; Pedersen, S.H.; Brucker, L.; Anisimov, O.; Bjerke, J.W.; Brown, R.D.; Ehrich, D.; Essery, R.L.; Heilig, A.; Ingvander, S.; et al. Changing Arctic snow cover: A review of recent developments and assessment of future needs for observations, modelling, and impacts. *Ambio* **2016**, *45*, 516–537. [CrossRef] [PubMed]

5. IPCC. Summary for Policymakers. In *Climate Change 2007-Synthesis Report*; Bernstein, L., Bosch, P., Canziani, O., Eds.; Cambridge University Press: New York, NY, USA, 2007; pp. 2–22. Available online: https://www.ipcc.ch/site/assets/uploads/2018/02/ar4_syr_full_report.pdf (accessed on 1 January 2024).
6. Gharamti, I.E.; Dempsey, J.P.; Polojärvi, A.; Tuhkuri, J. Creep and fracture of warm columnar freshwater ice. *Cryosphere* **2021**, *15*, 2401–2413. [CrossRef]
7. Thompson, L.G. Ice core evidence for climate change in the Tropics: Implications for our future. *Quat. Sci. Rev.* **2000**, *19*, 19–35. [CrossRef]
8. Gurtu, A.; Afshari, H.; Jaber, M.Y. Managing supply chains and transportation in the Arctic–challenges and opportunities: A literature review and research directions. In *Supply Chain Operations in the Arctic*, 1st ed.; Tsvetkova, A., Timoshenko, K., Eds.; Routledge: New York, NY, USA, 2023; pp. 23–48. [CrossRef]
9. Li, Z.J.; Kolerski, T.; Zhou, L.; Shi, X.H.; Zhang, Z.Y.; Li, F. Sea, river, lake ice properties and their applications in practices. *Water* **2023**, *15*, 899. [CrossRef]
10. Lemström, I.; Polojärvi, A.; Puolakka, O.; Tuhkuri, J. Load, pressure, rubble pile geometry and video data from model-scale tests on shallow water ice-structure interaction. *Data Brief* **2022**, *45*, 108580. [CrossRef] [PubMed]
11. Bjørnø, J.; Berg, M.V.D.; Lu, W.J.; Skjetne, R.; Lubbad, R.; Løset, S. Performance quantification of icebreaker operations in ice management by numerical simulations. *Cold Reg. Sci. Technol.* **2022**, *194*, 103435. [CrossRef]
12. Doering, J.C.; Morris, M.P. A digital image processing system to characterize frazil ice. *Can. J. Civ. Eng.* **2003**, *30*, 1–10. [CrossRef]
13. Pant, P.; Rajawat, A.S.; Goyal, S.B.; Chakrabarti, P.; Bedi, P.; Salau, A.O. Machine learning-based approach to predict ice meltdown in glaciers due to climate change and solutions. *Environ. Sci. Pollut. Res.* **2023**, *30*, 125176–125187. [CrossRef]
14. Knack, I.M.; Shen, H.T. A numerical model study on Saint John River ice breakup. *Can. J. Civ. Eng.* **2018**, *45*, 817–826. [CrossRef]
15. Xie, F.; Lu, P.; Li, Z.J.; Wang, Q.K.; Zhang, H. A floating remote observation system (FROS) for lake ice evolution studies. *Cold Reg. Sci. Technol.* **2022**, *199*, 103557. [CrossRef]
16. Voermans, J.J.; Rabault, J.; Marchenko, A.; Nose, T.; Waseda, T.; Babanin, A.V. Estimating the elastic modulus of landfast ice from wave observations. *J. Glaciol.* **2023**, 1–11. [CrossRef]
17. Charlebois, L.; Frederking, R.; Timco, G.W.; Watson, D.; Richard, M. Evaluation of pack ice pressure approaches and engineering implications for offshore structure design. *Cold Reg. Sci. Technol.* **2018**, *149*, 71–82. [CrossRef]
18. Xie, C.; Zhou, L.; Ding, S.F.; Lu, M.F.; Zhou, X. Research on self-propulsion simulation of a polar ship in a brash ice channel based on body force model. *Int. J. Nav. Arch. Ocean Eng.* **2023**, *15*, 100557. [CrossRef]
19. Dong, W.B.; Zhou, L.; Ding, S.F.; Wang, A.; Cai, J.Y. Two-staged method for ice channel identification based on image seg-mentation and corner point regression. *China Ocean Eng.* **2024**, *38*, 1–13.
20. Jewson, D.H.; Granin, N.G.; Zhdanov, A.A.; Gnatovsky, R.Y. Effect of snow depth on under-ice irradiance and growth of *Aulacoseira baicalensis* in Lake Baikal. *Aquat. Ecol.* **2009**, *43*, 673–679. [CrossRef]
21. Sharma, S.; Richardson, D.C.; Woolway, R.I.; Imrit, M.A.; Bouffard, D.; Blagrave, K.; Daly, J.; Filazzola, A.; Granin, N.; Korhonen, J.; et al. Loss of ice cover, shifting phenology, and more extreme events in northern hemisphere Lakes. *J. Geophys. Res. Biogeosci.* **2021**, *126*, e2021JG006348. [CrossRef]

water

MDPI

Article

River Ice Regime Recognition Based on Deep Learning: Ice Concentration, Area, and Velocity

Zhiyong Yang [1], Jun Zong [2], Yuelong Zhu [1], Xiuheng Liu [1], Ran Tao [2] and Yufeng Yu [1,*]

[1] College of Computer and Information, Hohai University, Nanjing 211100, China; yzy_hohai@hhu.edu.cn (Z.Y.)

[2] Nanjing Research Institute of Hydrology and Water Conservation Automation, Ministry of Water Resources, Nanjing 210012, China

* Correspondence: yfyu@hhu.edu.cn; Tel.: +86-139-5167-0067

Abstract: The real-time derivation of the concentration, area, and velocity of river surface ice based on camera imagery is essential for predicting the potential risks related to ice blockages in water routes. The key lies in the continuous tracking and velocity measuring of river ice, and reliable ice motion detection is a prerequisite for the dynamic perception of tracking targets. Previous studies did not utilize motion tracking for measuring ice velocity, and particle image velocimetry and feature point matching were used. This study aimed to use deep learning methods to address the challenging problems of deriving the ice concentration, area, and velocity based on camera imagery, and the focus was on measuring the ice velocity and drawing trajectories using the particle video tracking algorithm. We built a dataset named IPC_RI_IDS and collected information during the ice cover break-up process in the Nenjiang River (China). Our suggested approach was divided into four steps: (1) image preprocessing, where the camera image was calibrated to real-world coordinates; (2) determining the ice and water pixels in the camera image using the lightweight semantic segmentation network and then calculating the ice concentration and area; (3) enhancing and optimizing motion detection using the semantic segmentation results; and (4) adapting the particle video tracking algorithm to measure ice velocity using the proposed tracking points generation strategy. Finally, we analyzed the surface ice data in the study area and attempted to predict the stage of the ice break-up process to provide support for the real-time short-term forecasts of ice floods.

Keywords: river ice; ice regime; ice velocity; ice tracking; particle video; deep learning

Citation: Yang, Z.; Zong, J.; Zhu, Y.; Liu, X.; Tao, R.; Yu, Y. River Ice Regime Recognition Based on Deep Learning: Ice Concentration, Area, and Velocity. *Water* **2024**, *16*, 58. https://doi.org/10.3390/w16010058

Academic Editors: Xuemei Liu, Zhijun Li, Sasan Tavakoli, Changlei Dai and Fang Li

Received: 3 November 2023
Revised: 8 December 2023
Accepted: 9 December 2023
Published: 22 December 2023

1. Introduction

Large-scale floating ice collisions cause significant damage to hydraulic structures and inland transportation along rivers, and the accumulation of floating ice can quickly raise water levels, leading to ice-jam floods [1]. River ice hazards cause substantial economic losses. In 2017, during the spring melt, ice-jam floods cost approximately USD 300 million in North America alone [2]. In 2021, a large floating ice mass hit and destroyed the Xinxing Bridge in the Ant River, Fangzheng County, Harbin, China [3]. Real-time river ice regime recognition can provide practical information and support for the early warning of ice floods to reduce disaster losses. Based on computer vision technology with deep learning techniques, the pixel distributions of ice and water in a camera image were identified to extract additional high semantic information, such as ice concentration, area, velocity, distribution, and change process, which provided essential data support for the analysis and prediction of ice floods. This paper studies river ice regime recognition based on camera images, as shown in Figure 1.

River ice break-up forming an ice run is a river's short-term natural behavior, and it often occurs in one day or over several days. It is necessary to study this change process by recognizing river ice regimes through real-time monitoring. Researchers mainly use three types of river ice images. The first is a high-spatial-resolution remote sensing image (satellite imagery) [4–6]. The advantage is that it can observe and calculate the overall river ice regime over an extensive range. However, only coarse-grained river ice changes can be captured due to the long image-shooting interval. The second is unmanned aerial vehicle imaging (UAV imagery) [7–9], which has the advantage of its relatively low cost and ability to capture hourly data for river ice changes occurring anywhere. However, it is challenging to capture the long-term data for river ice changes. The third is a fixed-position camera image with an oblique perspective (camera imagery) [10–12], which has the advantage of monitoring the long-term continuous changes in a river ice regime from a fixed perspective, and it more accurately captures the details of river ice in a river section and is convenient for locking the measurement condition variables to analyze the changes in a river ice regime. Aiming to address the suddenness of river ice hazards, a real-time monitoring camera presents more advantages for short-term forecasting.

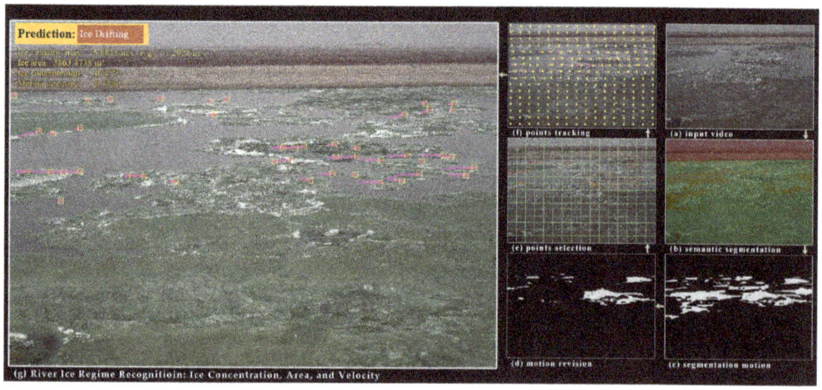

Figure 1. Process overview of river ice regime recognition, including ice concentration, area, and velocity.

Previous studies have provided some methods for river ice regime recognition. Related to this paper, Daigle et al. (2013) [13] used an artificial neural network and a particle image velocimetry method to measure the concentration and velocity of river ice on a camera, and this was a relatively early and comprehensive study on river ice recognition. The shortcoming was that it could not continuously track ice and measure ice velocity. Wang et al. (2022) [7] took UAV images from a high-altitude, overlooking a river, during ice flood periods in the Heilongjiang Basin in China, and they selected two images with an interval of one second, extracted and matched similar feature points of the river ice using the scale-invariant feature transform (SIFT) algorithm and the brute force (BF) algorithm, and then measured the river ice velocity according to the displacement difference. A shortcoming was the inability to monitor the river ice for a long time by UAV. Zhang et al. (2020–2023) [8,14,15] conducted a series of studies on the semantic segmentation of river ice on oblique UAV images. Finally, they achieved real-time semantic segmentation while ensuring high accuracy, and they further calculated and analyzed the concentration of the river ice. This method required a lot of image calibration work to recognize the river ice parameters. Xin et al. (2023) [10] used the boundary rectangle method and Harris corner detection method to measure a river's ice surface area and velocity on camera images from the Huma River Basin in the Daxing'an Mountains. The author mentioned that ice velocity measurements required manual operation. Li et al. (2023) [11] collected river ice images based on a fixed camera at the Yellow River. Their main work was to estimate the Gaussian distribution of the sizes and shapes of the river ice and establish the relationship function between the river ice concentration and the ice drift velocity, which helped to understand and analyze the freezing and thawing mode of the river ice. The semantic segmentation of the river ice was manually completed, and their work was not automated.

In summary, camera imagery is of more practical significance for the real-time monitoring of and risk early warnings for river ice. Research on multi-target continuous tracking and velocity measurements of river ice is still lacking. Under these motivations, this paper studies river ice regime recognition using camera imagery, including the ice concentration, area, and velocity. The difficulties were as follows: (1) there was a need for a river ice dataset of camera imagery; (2) in practice, we found that ice motion detection results could be more precise; (3) the continuous and accurate river ice tracking and velocimetry; and (4) forecasting the possibility of ice hazards through ice concentrations and velocity was not enough, and it also required the parameter of ice motion intensity (i.e., a visual scale of the surface ice motion, similar to the volume of ice flow in three-dimensional space). This work began with building the dataset, enhancing motion detection, and improving the ice velocimetry to solve the above problems.

The main contributions of this study are as follows:

1. It addressed the motion detection problem caused by the color similarity of river ice to river water. Compared with traditional methods, motion detection using the results of the semantic segmentation of river ice can extract a more significant and precise binary map of river ice motion, and then it can use the river ice mask to further modify the binary map of motion, which can be used to obtain a reliable binary map of the motion of the river ice.
2. We proposed a novel continuous ice velocity measurement method based on particle video tracking. The difference between the velocity measurement method and previous works (e.g., the PIV-based method from 2013 [13] and the SIFT-based matching method from 2023 [7]) is the continuous tracking, and the features of the regions adjacent to the points were extracted using the feature method and the optical flow method without global image matching.
3. The relationship between river ice concentration, area, velocity, and motion intensity in the ice cover break-up process was analyzed. We proposed the calculation of the motion intensity of the ice run and designed a feed-forward neural network to predict the stage of the ice cover break-up process using the above ice parameters.
4. We built a dataset named IPC_RI_IDS of river ice regime recognition that contained the complete ice cover break-up process. We annotated 113 dataset images with semantic segmentation and provided preliminary numerical information, such as ice concentration, area, velocity, and motion intensity, for each image. Subsequent research on river ice regime recognition will be supported by this research.

2. Study Area and Materials

2.1. Study Area

The Nenjiang River, located in Northeast China, is a tributary of the Songhua River. The river is 1370 km long, with an average flow of 823.4 m^3/s. At medium and high water levels, the maximum water surface width is 450–8000 m, and the maximum water depth is 6–13 m. At low water levels, the maximum water surface width is 170–180 m, the maximum water depth is 1.6–7.2 m, and 300–500 ton ships can navigate the middle and lower reaches. The freezing period is from mid-November to mid-April of the following year.

The observation location in this study was located in Baishatan Village, Dandai Township, Zhenlai County, Jilin Province, at the entrance of the Nenjiang River in Jilin Province (as shown in Figure 2). The river's surface was 150 m in width. Monitoring and early warning points were set here to ensure the safety of the downstream river. We set up a nine-meter-high network camera on the right bank facing the opposite river to monitor the ice cover break-up in real time, and it captured the complete ice cover break-up process on 31 March 2023. The ice cover break-up started at 10 a.m. Beijing time. The ice run began at 2 p.m. and ended at 5 p.m. The ice cover, frozen for several months, broke up in one day and flowed downstream. The data collection of the ice cover break-up process is significant for the real-time short-term forecasting of ice floods.

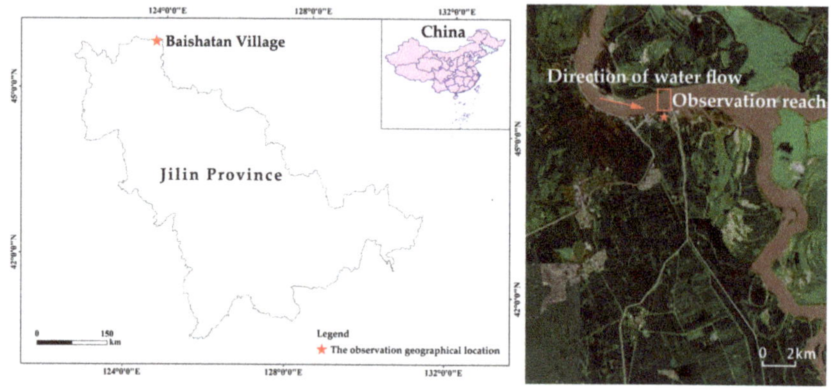

Figure 2. Schematic diagram of the observation geographical location. The base map is from Tianditu.com [16] accessed on 1 November 2023.

2.2. Materials

A dataset of the ice cover break-up process of the Nenjiang River or other similar rivers was needed. Therefore, we collected a monitoring video of the ice cover break-up process of the Nenjiang River in the study area on 31 March 2023, and we saved 43 video clips (data size of 43 GB). We divided the ice cover break-up process into five stages according to the morphology and intensity of the ice cover and ice run, namely, (1) the ice frozen stage, (2) the ice break-up beginning stage, (3) the ice drifting stage, (4) the ice break-up ending stage, and (5) the ice-free stage. According to each stage's characteristics and river ice morphologies, we extracted 26 one-minute short video clips from the five stages (1, 6, 10, 6, and 3), each with a frame rate of 10 fps, and we were able to extract 600 sequence images. Therefore, there were 15,600 sequence images, as shown in Table 1. This dataset was named IPC_RI_IDS, as shown in Figure 3, and we annotated it with refined semantics, as shown in Figure 4.

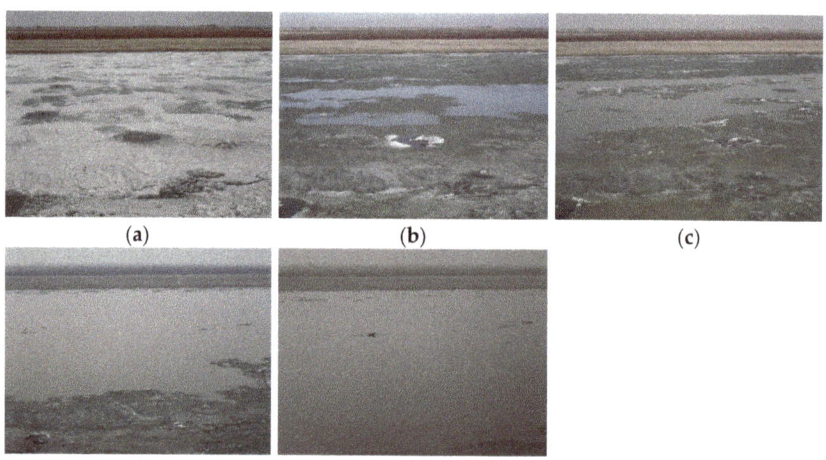

Figure 3. The five stages of our river ice dataset: (**a**) ice frozen stage; (**b**) ice break-up beginning stage; (**c**) ice drifting stage; (**d**) ice break-up ending stage; and (**e**) ice-free stage.

Table 1. The number of images in each stage.

No.	The Stage of Ice Break-Up	The Number of Videos	The Number of IMAGES
1	ice frozen	1	600
2	ice break-up beginning	6	3600
3	ice drifting	10	60,000
4	ice break-up ending	6	3600
5	ice-free	3	1800
	Total	26	15,600

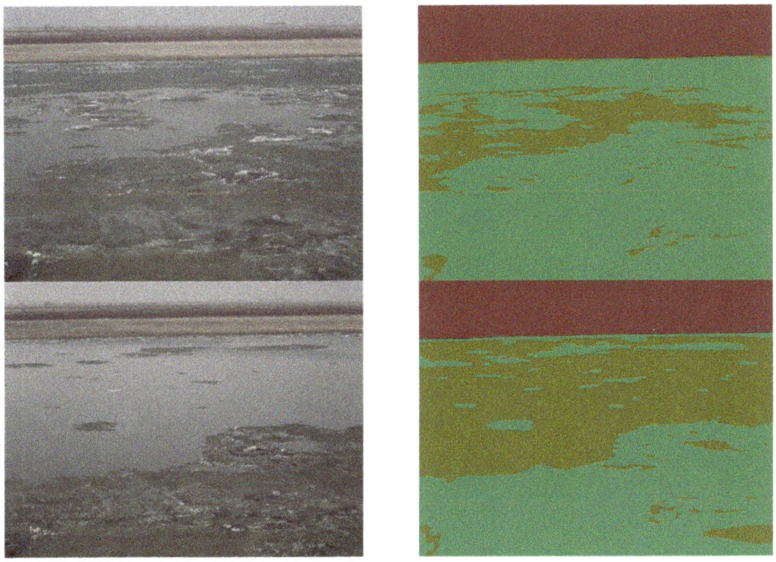

(**a**) (**b**)

Figure 4. Partial samples from our dataset: (**a**) the original images and (**b**) the labels for the original image. The green color represents the ice, the yellow color represents the water, and the red color represents the background.

In the first stage, one video clip was selected where the river ice had no change during the freezing stage, and each video clip was the same. In other stages, only the ice-drifting stage had a significant image change. We extracted 10 video clips, and we also made refined annotations on them in terms of semantic segmentation. A total of 113 images were annotated; on average, each image annotation took 2 h.

By observing the river ice video, it was found that in the ice drifting stage, the shape of the floating ice was complex, and the mixture of fragmented ice residue and water brought difficulties to semantic segmentation and motion detection. The drift velocity of the floating ice was too fast, and the shapes changed too fast, bringing significant challenges to the subsequent ice velocity measurement task.

3. Methods

In this study, the suggested approach was divided into six main steps. The goal was to extract the river ice concentration, area, velocity, and motion intensity to predict the stage of the ice cover break-up process, as shown in Figure 5.

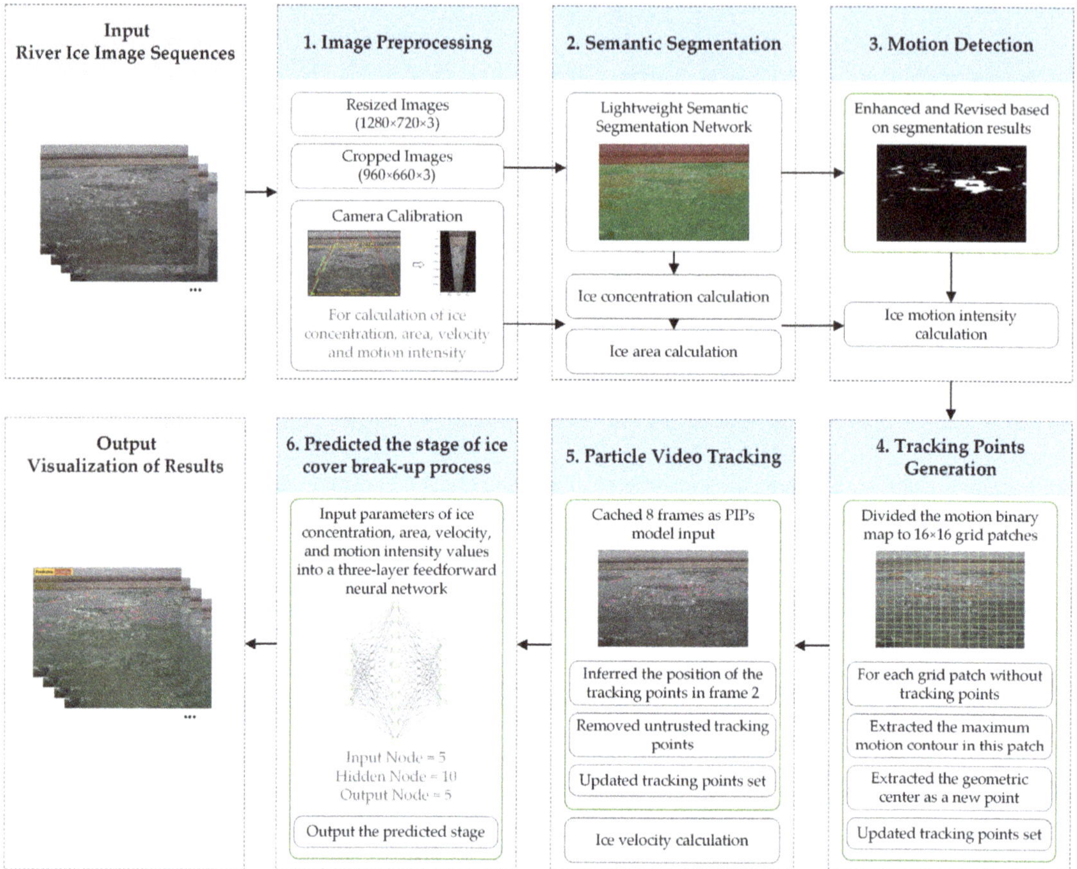

Figure 5. Process overview of the proposed approach. The contents in the green boxes are the main contributions of this study.

First, the camera image was calibrated to real-world coordinates during the image preprocessing step. Second, the lightweight depth convolution neural network was used to perform semantic segmentation on the river surface ice, extracting the ice pixels from the camera imagery to calculate the ice concentration and area. Third, we innovatively used the semantic segmentation results to improve the motion detection to enhance the significance of the motion binary map and optimize it to calculate the motion intensity. Fourth, the tracking point generation strategy was proposed, in which the tracking points were dynamically controlled by dividing the 16 × 16 grid patches of the motion binary map of the river ice. Fifth, the particle video tracking method was adjusted to adapt to the dynamic tracking of the river ice, and when measuring the ice velocity, the maximum velocity and average velocity were recorded. Finally, the five parameters of concentration, area, maximum velocity, average velocity, and motion intensity were input into the feed-forward neural network to predict the stage of the ice cover break-up process. The processing procedures in each step are introduced individually in subsequent sections.

3.1. Image Preprocessing and Calibration

In the image preprocessing step, the input camera images were resized to 1280 pixels wide × 720 pixels high × 3 RGB channels and then cropped to 960 × 660 × 3 by aligning the bottoms and centers.

Projective Transform

The camera images were calibrated to obtain the vertical-looking non-deformed images to calculate more accurate river ice parameter values. We used the projective transform method [17] to convert the camera images to real-world coordinates and used the actual widths and heights of the pixels to calculate the ice concentration, area, velocity, and motion intensity (see Figure 6).

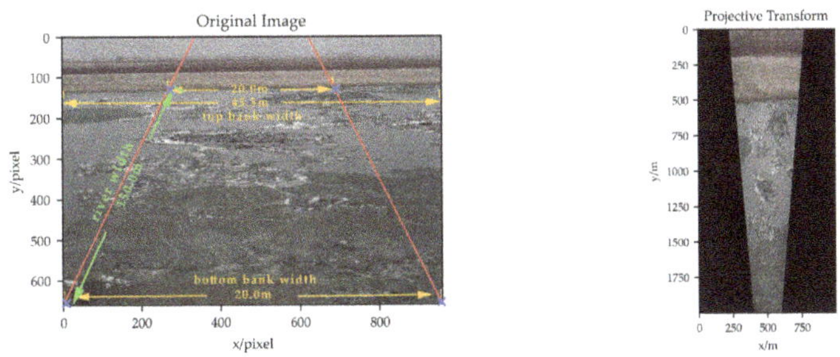

(**a**) Distance parameters (**b**) After conversion

Figure 6. Distance parameters of the camera images. (**a**) Distance parameters. The bottom bank width was 20.0 m, the top bank width was 45.5 m, and the river width was 150.0 m. (**b**) After the image coordinate system conversion.

If we let x and y represent the pixel coordinates of the original images, x' and y' represent the pixel coordinates of the converted images, and h represent the transformation coefficient, then the formulas for computing the transformation of x' and y' are as follows:

$$\begin{bmatrix} x' \\ y' \\ 1 \end{bmatrix} = \begin{bmatrix} h_{11} & h_{12} & h_{13} \\ h_{21} & h_{22} & h_{23} \\ h_{31} & h_{32} & 1 \end{bmatrix} \begin{bmatrix} x \\ y \\ 1 \end{bmatrix} \text{ and} \tag{1}$$

$$x' = \frac{h_{11}x + h_{12}y + h_{13}}{h_{31}x + h_{32}y + 1}, \ y' = \frac{h_{21}x + h_{22}y + h_{23}}{h_{31}x + h_{32}y + 1}. \tag{2}$$

We selected four point-pairs according to the distance parameters (the blue '×' in Figure 6a) to calculate the transformation coefficient, as set out in Equations (3) and (4) below:

$$[X, Y] = \begin{bmatrix} 270 & 130 \\ 0 & 660 \\ 685 & 125 \\ 960 & 660 \end{bmatrix}, \ [X', Y'] = \begin{bmatrix} 400 & 500 \\ 400 & 2000 \\ 600 & 500 \\ 600 & 2000 \end{bmatrix} \text{ and} \tag{3}$$

$$\begin{bmatrix} h_{11} & h_{12} & h_{13} \\ h_{21} & h_{22} & h_{23} \\ h_{31} & h_{32} & 1 \end{bmatrix} = \begin{bmatrix} 0.7692 & 1.8690 & 152.0306 \\ 0.1585 & 11.7309 & -814.4399 \\ 0.00008 & 0.0038 & 1 \end{bmatrix}. \tag{4}$$

We proposed PW (the actual pixel width) and HW (the actual pixel height) to easily calculate the pixel actual area. For the pixel P_{xy}, the coordinates were x and y in the image. We let PW represent the actual width of the pixel P_{xy} in the projection coordinates, and PH represented the actual height of the pixel P_{xy} in the projection coordinates, as set out in Equations (5) and (6) below:

$$PW = (x+1)' - x' \text{ and} \tag{5}$$

$$PH = (y+1)' - y'. \tag{6}$$

3.2. River Ice Semantic Segmentation

Semantic segmentation is a popular and mature image recognition task. The goal is to classify each pixel in an image. As shown in Figure 4b, the green color represents the ice, the yellow color represents the water, and the red color represents the background. Since Long et al. [18] proposed using a full convolutional neural network, the effect of the semantic segmentation task has been qualitatively improved. Excellent examples of this method include DeepLabV3+ [19], K-Net [20], and Mask2Former [21]. In our previous work [22], we conducted a special study on the zero-shot semantic segmentation of river ice in this scene.

This work was more concerned with the processing efficiency of semantic segmentation. We trained several lightweight semantic segmentation methods on IPC_RI_IDS. The selected models were FastScnn [23], MobileSeg [24], PPLiteSeg [25], and PPMobileSeg [26]. The codes were from the open-source project PaddleSeg [27] repository. The model with the best efficiency and precision would be the final semantic segmentation model.

After the semantic segmentation step, the ice concentration and area were calculated as follows:

Ice Concentration. We calculated the *ice concentration* according to the category of each pixel. We let P_{ice} represent the pixel classified as river ice and P_{water} represent the pixel classified as water, as set out in Equation (7) below:

$$Ice\ Concentration = \frac{Count(P_{ice})}{Count(P_{ice}) + Count(P_{water})}. \tag{7}$$

Ice Area. We calculated the *ice area* according to the actual area corresponding to each ice pixel P_{ice}, and then we summed them. *IS_ICE* meant 1 when P_{xy} was the P_{ice} category; otherwise, it meant 0, as set out in Equation (8) below:

$$Ice\ Area = \sum_{y=0}^{IMG\_H} \sum_{x=0}^{IMG\_W} PW \times PH \times IS\_ICE. \tag{8}$$

3.3. Motion Detection

A motion detection algorithm identifies pixels with motion changes in continuous images, and it usually includes background subtraction, temporary differences, optical flows, and so on [28].

This work used the classic and efficient ViBe [29] algorithm to obtain the motion binary map of the surface ice. The traditional strategy of a motion detection algorithm cannot achieve accurate results when directly applied to an original image. Therefore, we proposed a novel strategy for improving motion detection based on the semantic segmentation map, and the obtained motion binary map was more prominent. To distinguish the motion of the ice water, we used the river ice region in the semantic segmentation map to trim the motion binary image, and we modified the river ice motion binary map.

The purpose of obtaining the motion binary maps was to calculate the motion intensity parameters of the surface ice, and the scale of the surface ice movement was crucial for forecasting the possibility of an ice hazard.

Motion Intensity. To express the scale of the surface ice movement, we used the concentration of the motion binary map multiplied by the standard deviation of the motion pixels to obtain the *Motion Intensity*. P_{motion} represents the moving pixel, and *std ()* represents the "*numpy·std ()*" method in Python to calculate the standard deviation, as set out in Equation (9) below:

$$Motion\ Intensity = \frac{count(P_{motion})}{IMG\_W \times IMG\_H} \times \frac{std(P_{motion})}{max(IMG\_W, IMG\_H)}. \tag{9}$$

3.4. Tracking Points Generation

To dynamically control and generate new river ice tracking points, we proposed a tracking point generation strategy based on grid patches. The strategy was to divide the image into $n \times n$ ($n = 16$ in this paper) grids, and then the tracking points were taken from the geometric center of the maximum motion contour in each patch. The goal was to generate, at most, one tracking point per grid patch each time the tracking points were generated, as shown in Figure 7. For the steps, see Algorithm 1.

Algorithm 1: Strategy for the tracking point generation

```
Data: motion binary map, tracking point set
Result: tracking point set
n ← 16;
contour area threshold ← 50;
grid patches ← divide the motion binary map into n × n grid patches
for patch in grid patches, do
    if does not contain tracking point in the patch, then
        maximum contour ← extract the maximum motion contour in this patch
        if maximum contour area > contour area threshold, then
            tracking point ← extract the geometric center point
            update tracking point into tracking point set
        end if
    end if
end for
return tracking point set
```

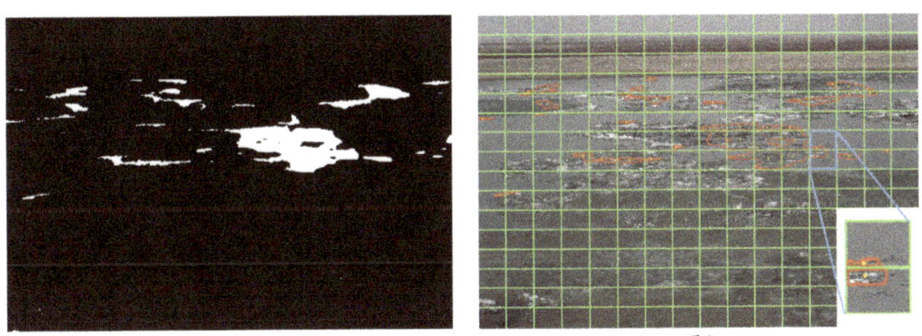

(**a**) (**b**)

Figure 7. Strategy of the tracking point generation. (**a**) Binary motion detection map. The white color represents the moving ice. (**b**) Tracking point generation extracted the geometric center point of the motion contour from each grid patch. The green color represents the grid patch; the red color represents the maximum motion contour; the yellow color represents the geometric center.

3.5. Particle Video Tracking

The standard global image motion estimation method is PIV (particle image velocimetry) [30]. By calculating the velocity field between two images, a velocity field can describe the motion mode of an image's content. Since Sand and Teller [31] proposed the particle video, the particle tracking effect has been more accurate and smooth. Harley et al. [32] proposed the persistent independent particles (PIPs) method based on deep learning, and this method makes the similarity template more reliable and further improves the point-tracking performance. In this study, the PIPs method was adjusted to adapt the river ice tracking, and the advantage was that it could obtain accurate river ice trajectories and velocities, as shown in Figure 8. For the ice tracking and velocimetry steps, see Algorithm 2.

Algorithm 2: The algorithm steps for the particle video tracking and velocimetry

Data: original images, semantic segmentation maps, tracking point set
Result: tracking point set and velocities
Image buffer ← empty;
Segmentation buffer ← empty;

```
1:  while inputting original image and semantic segmentation map, do
2:      put original image into image buffer
3:      put semantic segmentation map into segmentation buffer
4:      if len (image buffer) < 8, then
5:          continue
6:      otherwise
7:          if next frame of original image exists, then
8:              tracking point set ← infer second frame points from image buffer
9:              remove points where confidence < 0.5
10:             remove points where distance to image border < 30
11:             remove points where tracked frame count > 50
12:             remove points where not in ice mask from segmentation buffer
13:             calculate actual distance as velocity between buffer frames 1 and 2
14:         otherwise
15:             tracking point set ← infer the next seven frame's points
16:             repeat steps 9 to 13 to remove discarded points
17:         end if
18:         pop the first item of image buffer and segmentation buffer
19:     end if
20: end while
```

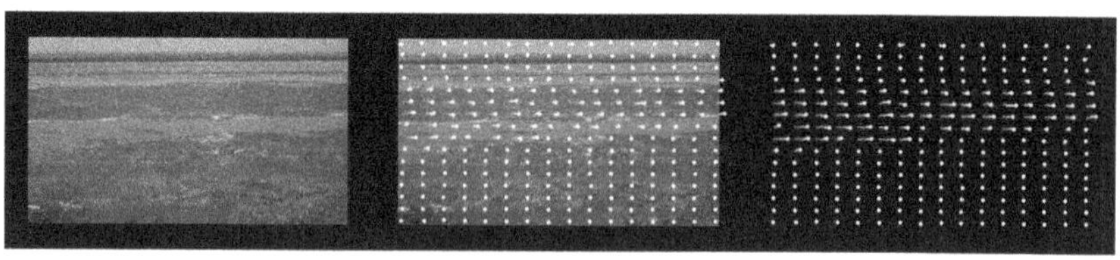

Figure 8. Example of the persistent independent particle method's global particle video tracking. The gradient red line represents the tracking trajectory.

Ice Velocity. If we let P_{xy}' represent the current real-world coordinate of the tracking point, P_{x1y1}' represents the real-world coordinate of the next frame of the tracking point, and duration represents the frame time interval (100 ms). Then, in each frame, only the velocity of tracking points tracked for more than five consecutive frames was included in the statistics, where the maximum velocity was recorded as *MAX velocity*, and the average velocity was recorded as *AVG velocity*, as set out in Equation (10) below:

$$Ice\ Velocity = \frac{\sqrt{(P_x{}' - P_{x1}{}')^2 - (P_y{}' - P_{y1}{}')^2}}{duration \times 1000}. \quad (10)$$

3.6. Prediction of the Ice Cover Break-Up Stages

In this study, the process of ice cover break-up was divided into the following five stages: (1) ice frozen stage, (2) ice break-up beginning stage, (3) ice drifting stage, (4) ice break-up ending stage, and (5) ice-free stage. The real-time short-term warning of ice floods would be realized by predicting the current stage of the ice cover break-up process. A three-layer feed-forward neural network was designed to predict the stage. The first layer was the input layer, which inputs the following five values of river surface ice: concentration, area, MAX velocity, AVG velocity, and motion intensity; so there were five input neurons in total. The second layer was the hidden layer with ten neurons, followed by a Sigmoid activation function. The third layer was the output layer, with five neurons representing the five stages of the ice cover break-up process, as shown in Figure 9. The loss function adopted cross-entropy loss, and the optimizer adopted the Adam (adaptive moment estimation) method.

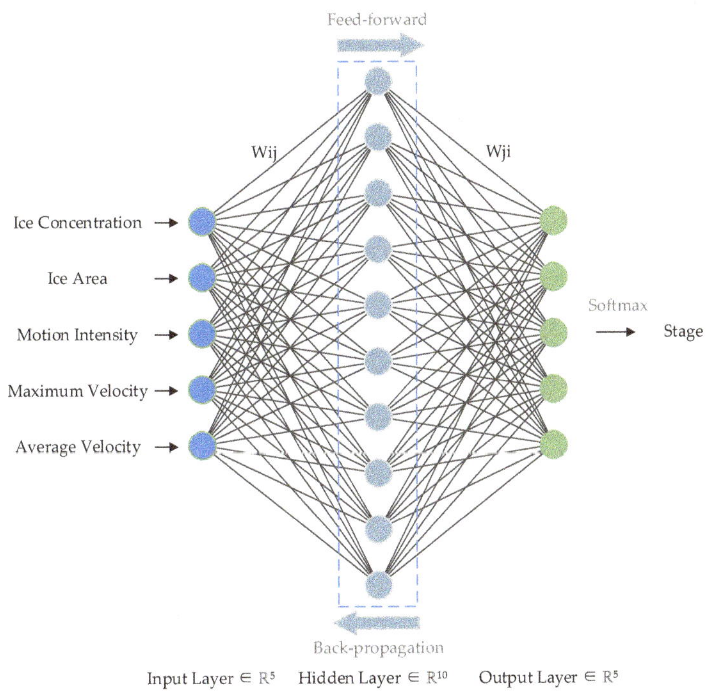

Figure 9. The architecture of the feed-forward neural network.

Data Preprocessing. We analyzed and handled missing or abnormal values in the dataset to ensure that the data could be correctly read by the model. We visualized the data distribution through box plots and observed any abnormal values in the data. Records with missing values were removed, and abnormal values were modified to limit values. Then, data normalization was deemed necessary since neural network models are sensitive to data scales. The ice area and velocity values were normalized to [0, 1] based on min–max normalization. The min–max normalization was calculated using the equation established in reference [33], as follows:

$$Wi = \frac{X_i - \min(X)}{\max(X) - \min(X)}, \tag{11}$$

where X_i is the number of features required to be normalized, and Wi represents the normalized features [33].

4. Results and Analysis

This experiment used the deep learning frameworks PaddlePaddle 2.5.1, PaddleSeg 2.8, Python 3.8.17, CUDA 11.7, and CUDNN 8.4.1.50. The experimental equipment included an NVIDIA GPU GeForce RTX 3070 laptop (NVIDIA Corporation., Santa Clara, CA, USA) with 8 GB of VRAM, 32 GB of RAM, an AMD Ryzen 7 5800 H CPU (Advanced Micro Devices, Inc., Santa Clara, CA, USA), and the Windows 11 operating system.

Because of the discontinuity of each video clip, it was not easy to obtain data on the river ice motion and velocity in the 20 frames at the beginning of the video clip. Therefore, the data from the first 20 frames were removed during the data extraction of the experimental results, and the total data were reduced from 15,600 (23 × 600) frames to 15,080 (26 × 580) frames.

4.1. Ice Concentration and Area

The ice concentration and area were calculated based on the semantic segmentation results. We tested four lightweight semantic segmentation methods on our IPC_RI_IDS dataset, and PPMobileSeg [26] had the best effect, as shown in Table 2.

Table 2. Comparison of the different methods on the IPC_RI_IDS dataset.

Methods	mIoU	Acc	Time
FastScnn [23]	0.9687	0.9821	112 ms
MobileSeg [24]	0.9666	0.9810	115 ms
PPLiteSeg [25]	0.9672	0.9813	121 ms
PPMobileSeg [26]	0.9762	0.9865	121 ms

Through the pixel area calculation of the semantic segmentation results for each frame, the curves of the ice concentration and area were obtained, as shown in Figures 10 and 11. It can be seen that as the ice breaking progressed, the ice concentration and ice area gradually decreased, and the ice concentration and ice area were linearly correlated.

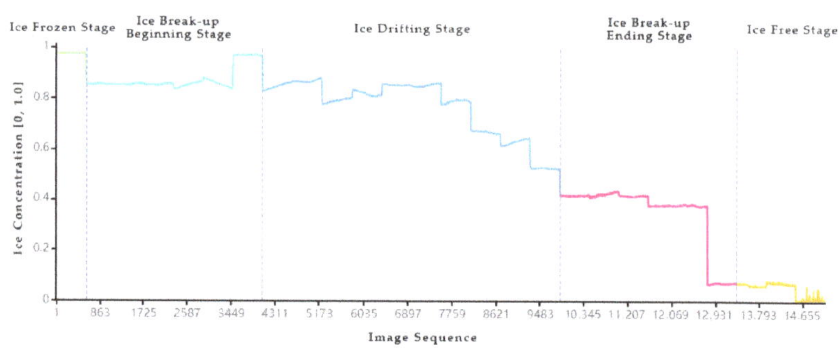

Figure 10. Ice concentration curve. The different colors represent the different stages. The ice concentration decreased continuously over time, and the floating ice from upstream increased the concentration.

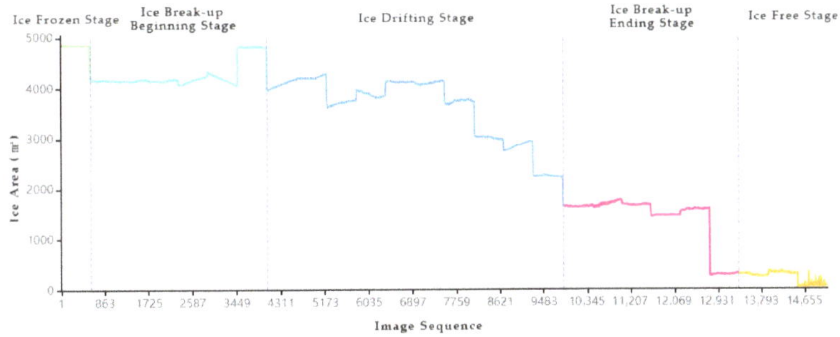

Figure 11. Ice area curve. The different colors represent the different stages. Similar to the ice concentration, the ice area decreased continuously over time, and the floating ice from upstream increased the area.

4.2. Motion Intensity

The motion intensity of the surface ice was calculated based on the results of the motion detection. The experiment demonstrated that the motion detection method directly applied to the original image could not obtain effective results, as shown in Figure 12b. The motion detection on the semantic segmentation maps achieved more prominent outcomes, as shown in Figure 12c. After revision by the semantic segmentation maps, only the motion of the river ice was retained in the motion binary map. The white part represents the moving ice, as shown in Figure 12d.

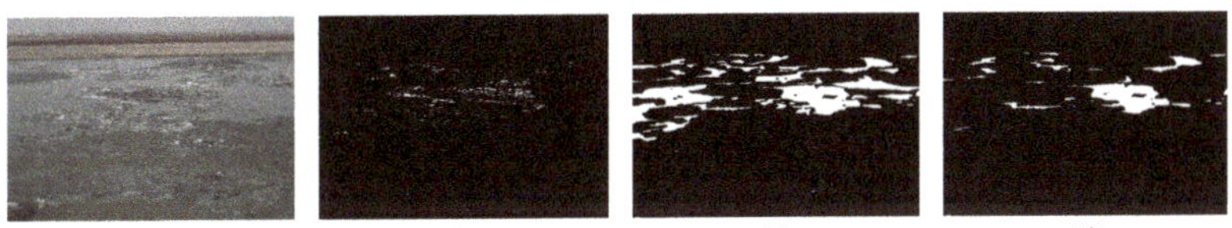

Figure 12. The motion detection based on the semantic segmentation map was more prominent than the original image motion detection. (**a**) Original image. (**b**) Motion detection on the original image. (**c**) Motion detection on the segmentation map. (**d**) Motion detection revision by the segmentation map.

The motion intensity of the river ice was calculated according to the ice concentration and dispersion of the motion binary map. Because the values were relatively small, they were enlarged 50 times to [0, 1]. As shown in Figure 13, the ice drifting stage had the highest motion intensity, and there was also a short section with high motion intensity at the ice break-up ending stage. The zigzag curve was because each video clip was not continuous, resulting in constant changes in motion intensity from low to high.

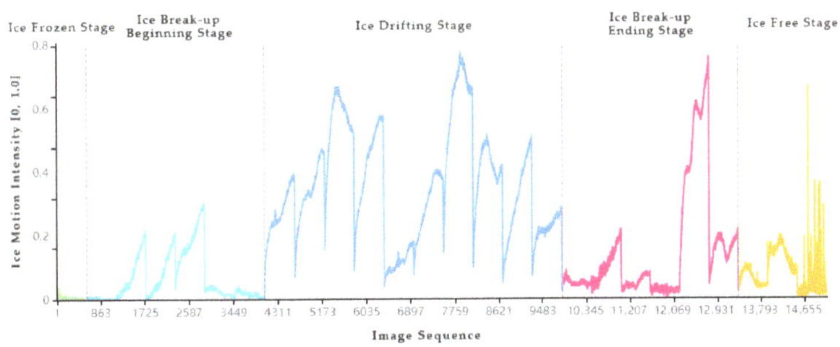

Figure 13. Ice motion intensity curve. The different colors represent the different stages. The motion intensity in the ice drifting stage was significantly higher than it was in other stages.

4.3. *Ice Velocity*

The ice velocity was calculated using particle video velocimetry, and the maximum velocity (MAX velocity) and average velocity (AVG velocity) were counted for each frame. As shown in Figures 14 and 15, the maximum velocity was approximately 3 m/s, and the average velocity was approximately 0.5 m/s. The ice velocity in the ice break-up ending stage was higher than it was in other stages, which was related to the absence of ice jams after river dredging.

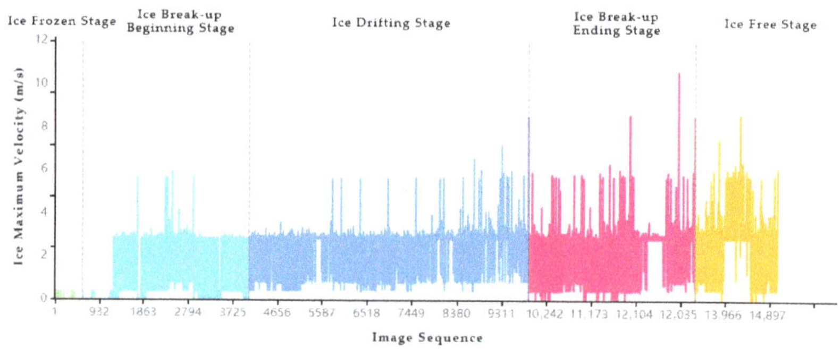

Figure 14. River ice max velocity curve. The x-axis is the image sequence number. The y-axis is the max velocity. The different colors represent the different stages.

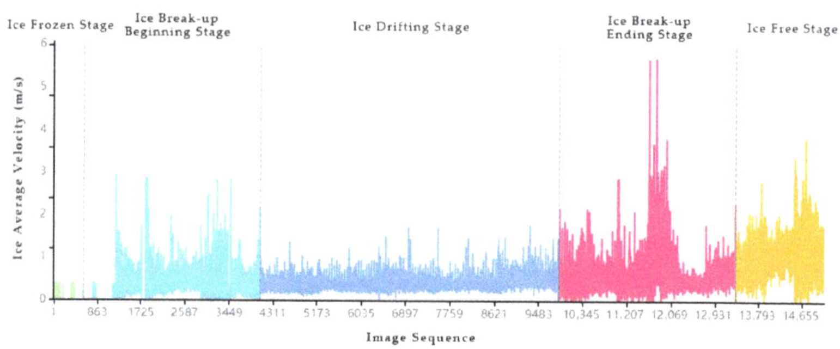

Figure 15. River ice average velocity curve. The x-axis is the image sequence number. The y-axis is the average velocity. The different colors represent the different stages.

4.4. *Prediction of the Ice Cover Break-Up Stages*

4.4.1. Data Preprocessing

A total of 15,080 frames in the dataset were preprocessed using the following steps: (1) missing values processing, (2) abnormal values processing, and (3) data normalization. The data before preprocessing are shown in Table 3. All data were correctly read, and the Python interface's 'pandas.isna ()' method was used to check for missing values. The box plot was used to analyze the abnormal values, as shown in Figure 16. It was found that there were abnormal values greater than 10 m/s and 5 m/s in the maximum velocity and average velocity, respectively, of the ice. In this work, the abnormal values were processed by modifying them to the nearest normal value, and then we normalized the ice area and velocity to [0, 1], as shown in Figure 17.

Table 3. Partial data for the ice regime parameters in the IPC_RI_IDS dataset.

No.	Stage	Ice Concentration	Ice Area	Motion Intensity	Maximum Velocity	Average Velocity
2	1	0.9761	4857.9570	0.0100	0.0	0.0
4	1	0.9760	4857.3160	0.0050	0.0	0.0
5	1	0.9775	4865.0730	0.0200	0.0	0.0
6	1	0.9764	4860.0120	0.0050	0.0	0.0
			…[1]			
3037	2	0.8705	4264.3650	0.0400	0.3904	0.1946
3040	2	0.8698	4259.4620	0.0350	0.3884	0.0971
3041	2	0.8707	4263.4260	0.0400	0.4136	0.2005
3042	2	0.8707	4264.1170	0.0350	0.4136	0.2010
			…			
6121	3	0.8171	3857.9760	0.4800	2.8409	0.4689
6122	3	0.8165	3854.6580	0.4750	3.3266	0.5130
6124	3	0.8159	3850.7420	0.4800	2.9883	0.2939
6125	3	0.8165	3854.5810	0.4800	2.8676	0.4193
			…			
10,531	4	0.4162	1655.6450	0.0550	1.2755	0.4717
10,533	4	0.4150	1652.1780	0.0450	1.2076	0.3242
10,536	4	0.4136	1638.5600	0.0400	2.8678	1.1598
10,537	4	0.4140	1645.8530	0.0400	2.9536	0.5697
			…			
14,547	5	0.0074	51.8180	0.0100	0.4564	0.4551
14,548	5	0.0062	43.3105	0.0050	0.4564	0.2282
14,549	5	0.0071	49.8568	0.0150	0.9077	0.6820
14,550	5	0.0085	59.5084	0.0300	0.0	0.0

Note: [1] Omit data.

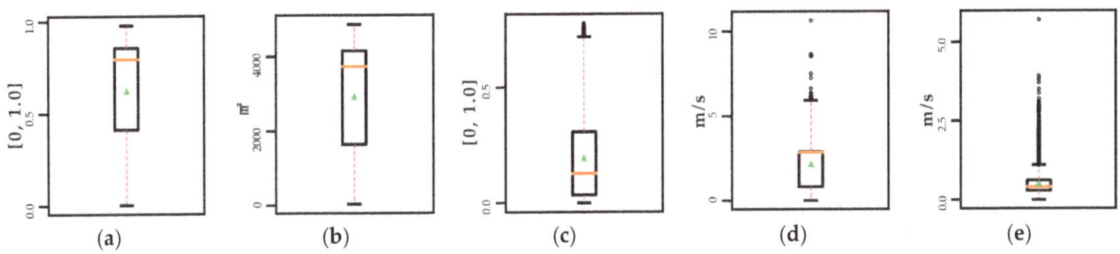

Figure 16. Box plot of the dataset before data preprocessing. (**a**) Ice concentration. (**b**) Ice area. (**c**) Ice motion intensity. (**d**) Ice maximum velocity. We found abnormal values greater than 10 m/s. (**e**) The ice average velocity. We found abnormal values greater than 5 m/s.

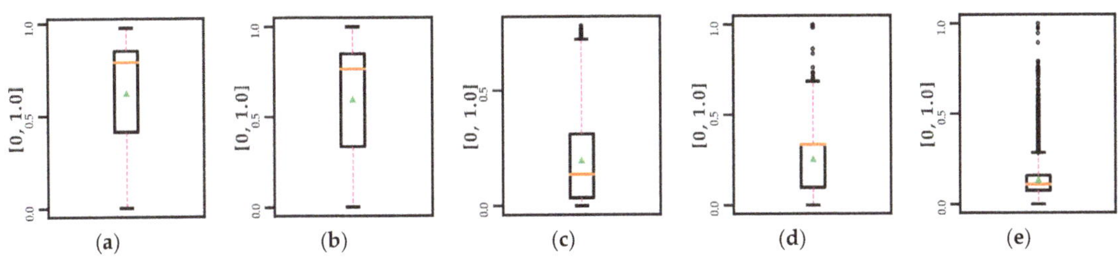

Figure 17. Box plot of the dataset after data preprocessing, with the modified abnormal values and the ice area and velocity normalized to [0, 1]. (**a**) Ice concentration. (**b**) Ice area. (**c**) Ice motion intensity. (**d**) Ice maximum velocity. (**e**) Ice average velocity.

4.4.2. Training Network

To balance the data distribution on the 15,080 frames dataset, the training set, verification set, and test set were extracted from each video clip in the ratio of 6:2:2. A total of 9047 frames were divided into the training set, 2985 into the verification set, and 3048 into the test set. The Adam optimizer was adopted, the learning rate was set to 0.01, cross-entropy was adopted as the loss function, and the iterations were set to 2000 times. The curves of the training accuracy and loss are shown in Figure 18.

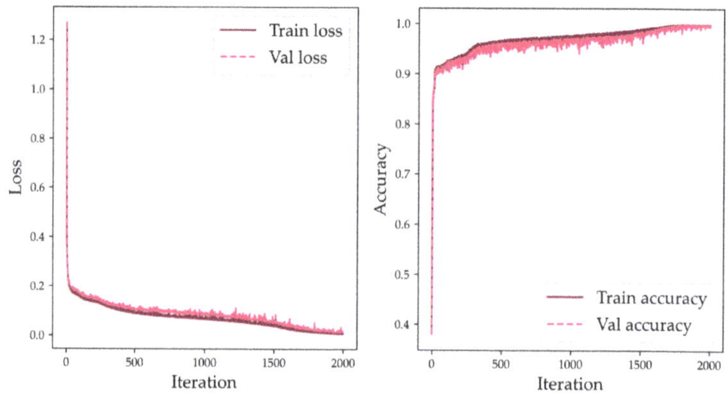

Figure 18. Training log curve. Accuracy and loss on the training set and validation set, respectively.

The optimization of hyperparameters in neural networks has a significant impact on the performance of a model. We conducted experiments on the network at three learning rates, with learning rates of 0.1, 0.01, and 0.001, respectively. It was found that a larger learning rate converged faster but was unstable on the curve, while a smaller one converged slower but was stable. Both the 0.01 and 0.1 learning rates ultimately achieved the highest accuracy of 0.9990 on the validation set, as shown in Figure 19. We saved the optimal model parameters with the highest accuracy of 0.9990 on the 1871st iteration of the validation set. The accuracy of the optimal model was 0.9984 on the test set.

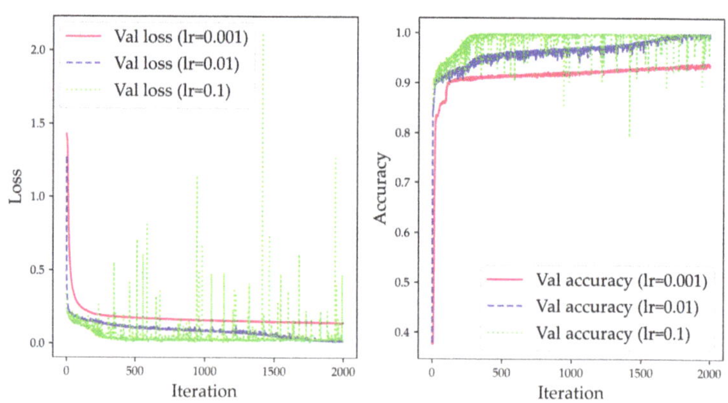

Figure 19. Loss curve on the validation set with the different learning rates. A larger learning rate converged faster but was unstable, while a smaller one converged slower but was stable. Both the 0.01 and 0.1 learning rates ultimately achieved the highest accuracy of 0.9990 on the validation set.

4.4.3. Experimentation Comparison

The softmax regression [34] and support vector machine (SVM) [35] methods were used for the comparison. Logistic regression is a commonly used linear model for handling binary classification problems, and softmax regression is an extension of logistic regression in multi-classification problems. SVM handles binary classification problems through hyperplanes, and combining multiple SVM classifiers can be extended to multi-classification problems.

The softmax regression method used five-dimensional feature inputs and five-dimensional category outputs in this experiment. The Gradient Descent method was used as the optimizer. Cross entropy was the loss function, and the iterations were set to 2000 times. It was found that setting the learning rate to five would achieve faster convergence. The best model was saved at the 1720th iteration, with the highest accuracy (0.9005) of the validation set. The SVM method used the implementation from the sklearn model of the Python package, set the hyperparameter C to 1.0, and tested four kernel functions on the dataset.

The comparison results of the three methods on the test set are shown in Table 4. The feed-forward neural network outperformed the other methods in terms of overall performance. The future prediction of the ice break-up process would introduce more hydrological features such as water level, water flow velocity, discharge, temperature, wind speed, etc., and the feed-forward neural network would be more suitable for this task.

Table 4. Comparison results of the three methods.

Methods	Kernel	Accuracy	Loss
Softmax regression	-	0.9008	0.2782
SVM	Linear	0.8967	-
	Poly (degree = 5)	0.9646	-
	RBF	0.9190	-
	Sigmoid	0.2684	-
Ours	-	0.9813	0.0173

4.5. Real-Time Monitoring of the River Ice Regime

Combining all the above steps as a single pipeline achieved the real-time monitoring of the river ice regime, and then the short-term early warning of ice floods was realized through the prediction of the stage of the ice cover break-up process. An early warning prompt could be issued when the ice cover break-up process entered the ice break-up beginning stage. As shown in Figure 20 and Video S1, our method accurately predicted different ice cover breaking stages and real-time displays of river ice regime information. The significance of the stage prediction was the short-term risk early warning prompts. In this work, the river ice-related risk warning level of the five stages is shown in Table 5.

Table 5. River ice-related risk warning level of the five stages.

No.	The Stage of Ice Break-Up	Warning Level	Note
1	Ice frozen	Medium	Observe if there is an ice jam
2	Ice break-up beginning	High	Ice run is about to occur
3	Ice drifting	High	Pay attention to blockage and collisions
4	Ice break-up ending	Low	The risk is minimal
5	Ice-free	None	The river has been opened

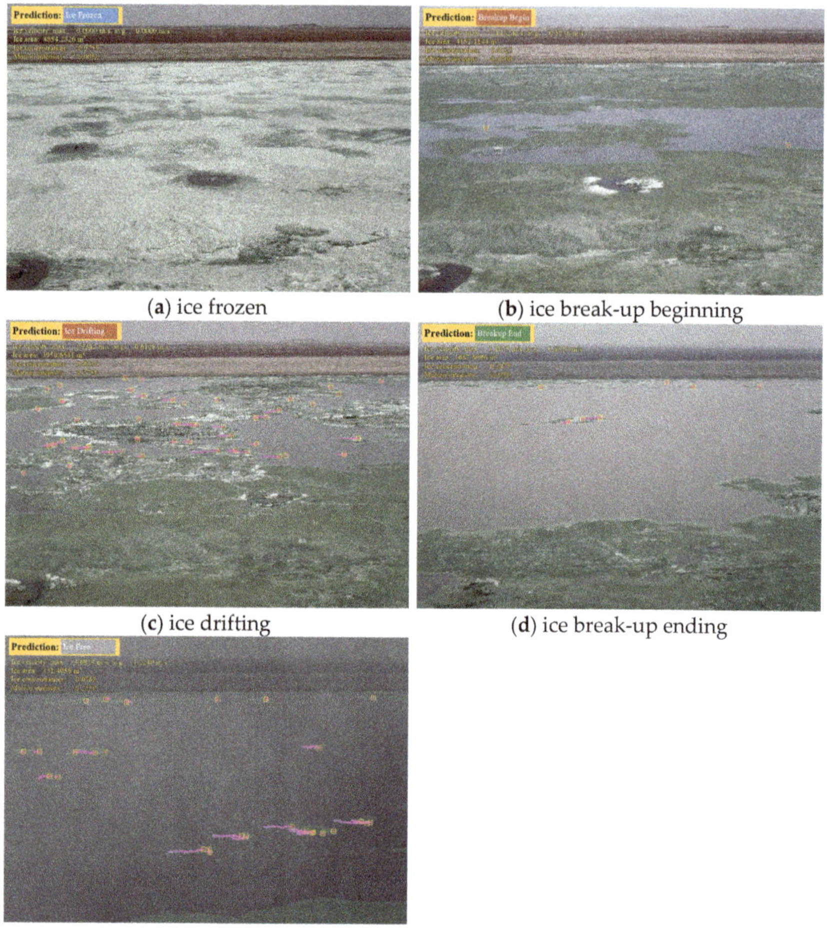

(**a**) ice frozen (**b**) ice break-up beginning (**c**) ice drifting (**d**) ice break-up ending (**e**) ice-free

Figure 20. Real-time monitoring of the river ice regime in each break-up stage. The images were captured from the tenth-second frame of the corresponding video clip. The blue color represents the medium warning level; the red color represents the high warning level; the green color represents the low warning level; the grey color represents the none warning level.

5. Discussion

This work addressed the ice tracking and velocimetry problem on camera imagery and used the derivation of the concentration, area, and motion intensity to realize river ice regime monitoring and short-term ice flood warnings. There are still many shortcomings.

5.1. Uncertainty Quantification

Uncertainty quantification has been proven to effectively assist decision-makers in understanding the uncertainty associated with the prediction results of neural networks and taking appropriate action [36]. This study was no exception; the neural network mechanically predicted any input image into five stages without creating uncertainty. Sometimes, highly uncertain prediction results can mislead decision-makers into making incorrect decisions.

In future work, we will estimate the uncertainty of neural networks from two aspects. Meanwhile, uncertainty quantification can help us improve the network.

Aleatoric uncertainty is the uncertainty caused by noisy data. Therefore, the network must estimate the error of the samples during inference to obtain uncertainty. We can modify a neural network's estimation of probability distribution instead of using a simple prediction.

Epistemic uncertainty is the uncertainty caused by a noisy model. Uncertainty can be estimated through multiple model results. In this work, we needed to estimate the uncertainty for the classification using the probability distribution of the softmax regression to estimate the uncertainty.

5.2. Capability Expansion

The task of river ice regime recognition involves not only the monitoring of surface ice but also the monitoring of underwater ice. It is significant to realize the three-dimensional monitoring of river ice. Xin et al. [10] proposed the estimation of river ice thickness. River ice regime identification is still facing many challenges that need to be further addressed, as listed below:

1. The estimation of river ice thickness;
2. The integration of hydrological monitoring elements such as temperature, water level, water flow velocity, discharge, wind speed, and evaporation;
3. Joint monitoring and predictions using multiple cameras;
4. Joint monitoring by fusing satellite remote sensing and ground camera images.

By solving the above four problems, systematic river ice regime recognition can be realized, and then comprehensive river ice flood warning and prediction can be realized.

6. Conclusions

The success of the river ice recognition method based on camera imagery in this paper lies in the following: first, it addressed the motion detection problem caused by the color similarity of the river ice to the water to extract a more accurate and precise motion binary map of the river ice. Second, a novel ice velocity measurement method was proposed. By dividing the 16 $\times$ 16 grid patches of the river ice motion binary map into dynamic control points, the particle video tracking method could be based on deep learning to adapt the continuous tracking scene of the river ice, which made the river ice velocity measurement more accurate. Finally, the ice concentration, area, motion intensity, and velocity were extracted to predict the stage of the ice cover break-up process to realize the short-term early warning of ice floods.

Supplementary Materials: The following supporting information can be downloaded at: https://www.mdpi.com/article/10.3390/w16010058/s1, Video S1: River ice monitoring based on web cameras.

Author Contributions: Conceptualization, Z.Y., J.Z. and Y.Y.; data curation, R.T.; formal analysis, J.Z.; funding acquisition, J.Z.; investigation, R.T.; methodology, Z.Y., Y.Z. and Y.Y.; project administration, Y.Y.; resources, J.Z. and R.T.; software, Z.Y. and X.L.; supervision, Y.Z.; validation, Z.Y. and X.L.; visualization, Z.Y.; writing—original draft, Z.Y.; writing—review and editing, J.Z. and Y.Y. All authors have read and agreed to the published version of the manuscript.

Funding: This research was funded by the National Key R&D Program of China (no. 2022YFC3204500) and the Central Public-interest Scientific Institution Basal Research Fund (no. Y522011).

Data Availability Statement: The data used in this study are available upon request to the corresponding author. The code is available at https://github.com/PL23K/RiverIceRegime, accessed on 15 October 2023.

Acknowledgments: Thanks to the PaddlePaddle team for providing the code base.

Conflicts of Interest: The authors declare no potential conflicts of interest with respect to the research, authorship, and/or publication of this article.

References

1. Lindenschmidt, K.-E. *River Ice Processes and Ice Flood Forecasting: A Guide for Practitioners and Students*; Springer International Publishing: Cham, Switzerland, 2020; ISBN 978-3-030-28678-1.
2. Yang, X.; Pavelsky, T.M.; Allen, G.H. The Past and Future of Global River Ice. *Nature* **2020**, *577*, 69–73. [CrossRef] [PubMed]
3. Baidu Encyclopedia. Fangzheng County Bridge Collapse Accident on 29 March 2021. Available online: https://baike.baidu.com/item/3%C2%B729%E6%96%B9%E6%AD%A3%E5%8E%BF%E5%A4%A7%E6%A1%A5%E5%9D%8D%E5%A1%8C%E4%BA%8B%E6%95%85 (accessed on 6 November 2023).
4. Priya, M.G.; Krishnaveni, D.; Bahuguna, I.M. Glacier Ice Surface Velocity Using Interferometry. In *Futuristic Communication and Network Technologies*; Subhashini, N., Ezra, M.A.G., Liaw, S.-K., Eds.; Lecture Notes in Electrical Engineering; Springer Nature: Singapore, 2023; Volume 966, pp. 67–75. ISBN 978-981-19833-7-5.
5. Li, G.; Mao, Y.; Feng, X.; Chen, Z.; Yang, Z.; Cheng, X. Monitoring Ice Flow Velocity of Petermann Glacier Combined with Sentinel-1 and −2 Imagery. *Int. J. Appl. Earth Obs. Geoinf.* **2023**, *121*, 103374. [CrossRef]
6. Lee, S.; Kim, S.; An, H.; Han, H. Ice Velocity Variations of the Cook Ice Shelf, East Antarctica, from 2017 to 2022 from Sentinel-1 SAR Time-Series Offset Tracking. *Remote Sens.* **2023**, *15*, 3079. [CrossRef]
7. Wang, E.; Hu, S.; Han, H.; Li, Y.; Ren, Z.; Du, S. Ice Velocity in Upstream of Heilongjiang Based on UAV Low-Altitude Remote Sensing and the SIFT Algorithm. *Water* **2022**, *14*, 1957. [CrossRef]
8. Zhang, X.; Zhao, Z.; Ran, L.; Xing, Y.; Wang, W.; Lan, Z.; Yin, H.; He, H.; Liu, Q.; Zhang, B. FastICENet: A Real-Time and Accurate Semantic Segmentation Model for Aerial Remote Sensing River Ice Image. *Signal Process.* **2023**, *212*, 109150. [CrossRef]
9. Iqbal, U.; Riaz, M.Z.B.; Zhao, J.; Barthelemy, J.; Perez, P. Drones for Flood Monitoring, Mapping and Detection: A Bibliometric Review. *Drones* **2023**, *7*, 32. [CrossRef]
10. Xin, D.; Tian, F.; Zhao, Y. Parameter Identification of Ice Drift toward Cross-River Bridges in Cold Regions. *J. Cold Reg. Eng.* **2023**, *37*, 04023008. [CrossRef]
11. Li, C.; Li, Z.; Zhang, B.; Deng, Y.; Zhang, H.; Wu, S. A Survey Method for Drift Ice Characteristics of the Yellow River Based on Shore-Based Oblique Images. *Water* **2023**, *15*, 2923. [CrossRef]
12. Pei, C.; She, Y.; Loewen, M. Deep Learning Based River Surface Ice Quantification Using a Distant and Oblique-Viewed Public Camera. *Cold Regions Sci. Technol.* **2023**, *206*, 103736. [CrossRef]
13. Daigle, A.; Bérubé, F.; Bergeron, N.; Matte, P. A Methodology Based on Particle Image Velocimetry for River Ice Velocity Measurement. *Cold Reg. Sci. Technol.* **2013**, *89*, 36–47. [CrossRef]
14. Zhang, X.; Jin, J.; Lan, Z.; Li, C.; Fan, M.; Wang, Y.; Yu, X.; Zhang, Y. ICENET: A Semantic Segmentation Deep Network for River Ice by Fusing Positional and Channel-Wise Attentive Features. *Remote Sens.* **2020**, *12*, 221. [CrossRef]
15. Zhang, X.; Zhou, Y.; Jin, J.; Wang, Y.; Fan, M.; Wang, N.; Zhang, Y. ICENETv2: A Fine-Grained River Ice Semantic Segmentation Network Based on UAV Images. *Remote Sens.* **2021**, *13*, 633. [CrossRef]
16. Ministry of Natural Resources, People's Republic of China Tianditu Online Map. Available online: https://map.tianditu.gov.cn/ (accessed on 28 October 2023).
17. Zhang, Z. A Flexible New Technique for Camera Calibration. *IEEE Trans. Pattern Anal. Mach. Intell.* **2000**, *22*, 1330–1334. [CrossRef]
18. Long, J.; Shelhamer, E.; Darrell, T. Fully Convolutional Networks for Semantic Segmentation. *IEEE Trans. Pattern Anal. Mach. Intell.* **2017**, *39*, 640–651.
19. Chen, L.-C.; Zhu, Y.; Papandreou, G.; Schroff, F.; Adam, H. Encoder-Decoder with Atrous Separable Convolution for Semantic Image Segmentation. In *Computer Vision–ECCV 2018*; Ferrari, V., Hebert, M., Sminchisescu, C., Weiss, Y., Eds.; Lecture Notes in Computer Science; Springer International Publishing: Cham, Switzerland, 2018; Volume 11211, pp. 833–851. ISBN 978-3-030-01233-5.
20. Zhang, W.; Pang, J.; Chen, K.; Loy, C.C. K-Net: Towards Unified Image Segmentation. *Adv. Neural Inf. Process. Syst.* **2021**, *34*, 10326–10338.
21. Cheng, B.; Misra, I.; Schwing, A.G.; Kirillov, A.; Girdhar, R. Masked-Attention Mask Transformer for Universal Image Segmentation. In Proceedings of the 2022 IEEE/CVF Conference on Computer Vision and Pattern Recognition, New Orleans, LA, USA, 18–24 June 2022; pp. 1290–1299.
22. Yang, Z.; Zhu, Y.; Zeng, X.; Zong, J.; Liu, X.; Tao, R.; Cong, X.; Yu, Y. An Easy Zero-Shot Learning Combination: Texture Sensitive Semantic Segmentation IceHrNet and Advanced Style Transfer Learning Strategy. *arXiv* **2023**, arXiv:2310.00310.
23. Poudel, R.P.K.; Liwicki, S.; Cipolla, R. Fast-SCNN: Fast Semantic Segmentation Network. *arXiv* **2019**, arXiv:1902.04502.
24. Sandler, M.; Howard, A.; Zhu, M.; Zhmoginov, A.; Chen, L.-C. MobileNetV2: Inverted Residuals and Linear Bottlenecks. In Proceedings of the 2018 IEEE Conference on Computer Vision and Pattern Recognition, Salt Lake City, UT, USA, 18–23 June 2018; pp. 4510–4520.
25. Peng, J.; Liu, Y.; Tang, S.; Hao, Y.; Chu, L.; Chen, G.; Wu, Z.; Chen, Z.; Yu, Z.; Du, Y.; et al. PP-LiteSeg: A Superior Real-Time Semantic Segmentation Model. *arXiv* **2022**, arXiv:2204.02681.
26. Tang, S.; Sun, T.; Peng, J.; Chen, G.; Hao, Y.; Lin, M.; Xiao, Z.; You, J.; Liu, Y. PP-MobileSeg: Explore the Fast and Accurate Semantic Segmentation Model on Mobile Devices. *arXiv* **2023**, arXiv:2304.05152.
27. PaddlePaddle Authors PaddleSeg, End-to-End Image Segmentation Kit Based on PaddlePaddle. Available online: https://github.com/PaddlePaddle/PaddleSeg (accessed on 31 October 2023).

28. Singh, S.; Prasad, A.; Srivastava, K.; Bhattacharya, S. Object Motion Detection Methods for Real-Time Video Surveillance: A Survey with Empirical Evaluation. In *Smart Systems and IoT: Innovations in Computing*; Somani, A.K., Shekhawat, R.S., Mundra, A., Srivastava, S., Verma, V.K., Eds.; Smart Innovation, Systems and Technologies; Springer: Singapore, 2020; Volume 141, pp. 663–679. ISBN 9789811384059.
29. Barnich, O.; Van Droogenbroeck, M. ViBe: A Universal Background Subtraction Algorithm for Video Sequences. *IEEE Trans. Image Process.* **2011**, *20*, 1709–1724. [CrossRef]
30. Adrian, R.J. Twenty Years of Particle Image Velocimetry. *Exp. Fluids* **2005**, *39*, 159–169. [CrossRef]
31. Sand, P.; Teller, S. Particle Video: Long-Range Motion Estimation Using Point Trajectories. *Int. J. Comput. Vis.* **2008**, *80*, 72–91. [CrossRef]
32. Harley, A.W.; Fang, Z.; Fragkiadaki, K. Particle Video Revisited: Tracking Through Occlusions Using Point Trajectories. In *Computer Vision–ECCV 2022*; Avidan, S., Brostow, G., Cissé, M., Farinella, G.M., Hassner, T., Eds.; Lecture Notes in Computer Science; Springer Nature Switzerland: Cham, Switzerland, 2022; Volume 13682, pp. 59–75. ISBN 978-3-031-20046-5.
33. Phan, D.T.; Tran, V.N.; Tran, L.H.; Park, S.; Choi, J.; Kang, H.W.; Oh, J. Enhanced Precision of Real-Time Control Photothermal Therapy Using Cost-Effective Infrared Sensor Array and Artificial Neural Network. *Comput. Biol. Med.* **2022**, *141*, 104960. [CrossRef]
34. Bisong, E.; Bisong, E. Logistic Regression. In *Building Machine Learning and Deep Learning Models on Google Cloud Platform: A Comprehensive Guide for Beginners*; Apress: Berkeley, CA, USA, 2019; pp. 243–250.
35. Cortes, C.; Vapnik, V. Support-Vector Networks. *Mach. Learn.* **1995**, *20*, 273–297. [CrossRef]
36. Begoli, E.; Bhattacharya, T.; Kusnezov, D. The Need for Uncertainty Quantification in Machine-Assisted Medical Decision Making. *Nat. Mach. Intell.* **2019**, *1*, 20–23. [CrossRef]

water

MDPI

Article

Multi-Scale Polar Object Detection Based on Computer Vision

Shifeng Ding [1], Dinghan Zeng [1], Li Zhou [2,*], Sen Han [1], Fang Li [2] and Qingkai Wang [3]

1 School of Naval Architecture and Ocean Engineering, Jiangsu University of Science and Technology, Zhenjiang 212100, China; 15001945469@163.com (S.D.); 211210101417@stu.just.edu.cn (D.Z.); hansen@just.edu.cn (S.H.)
2 School of Naval Architecture, Ocean & Civil Engineering, Shanghai Jiao Tong University, Shanghai 200030, China; fang.li@sjtu.edu.cn
3 State Key Laboratory of Coastal and Offshore Engineering, Dalian University of Technology, Dalian 116024, China; wangqingkai@dlut.edu.cn
* Correspondence: zhouli209@sjtu.edu.cn

Abstract: When ships navigate in polar regions, they may collide with ice masses, which may cause structural damage and endanger the safety of their occupants. Therefore, it is essential to promptly detect sea ice, icebergs, and passing ships. However, individual data sources have limits and should be combined and integrated to obtain more thorough information. A polar multi-target local-scale dataset with five categories was constructed. Sea ice, icebergs, ice melt ponds, icebreakers, and inter-ice channels were identified by a single-shot detector (SSD), with a final mAP value of 70.19%. A remote sensing sea ice dataset with 15,948 labels was constructed. The You Only Look Once (YOLOv5) model was improved with Squeeze-and-Excitation Networks (SE), Funnel Activation (FReLU), Fast Spatial Pyramid Pooling, and Cross Stage Partial Network (SPPCSPC-F). In the detection stage, a slicing operation was performed on remote sensing images to detect small targets. Simulated sea ice data were included to verify the model's generalization ability. Then, the improved model was trained and evaluated in an ablation experiment. The mAP, recall (R), and precision (P) values of the improved YOLOv5 were 75.3%, 70.3, and 75.4%, with value increases of 3.5%, 3.4%, and 1.9%, respectively, compared to the original model. The improved YOLOv5 was also compared with other models such as YOLOv3, Faster-RCNN, and YOLOv4-tiny. The results indicated that the performance of the proposed model surpassed those of the other conventional models. This study achieved the detection of multiple targets on different scales in a polar region and realized data fusion, avoiding the limitations of using a single data source, and provides a method to support polar ship path planning.

Keywords: computer vision; single-shot detector (SSD); You Only Look Once (YOLOv5); multi-source data; polar object; remote sensing image; sea ice

Citation: Ding, S.; Zeng, D.; Zhou, L.; Han, S.; Li, F.; Wang, Q. Multi-Scale Polar Object Detection Based on Computer Vision. *Water* **2023**, *15*, 3431. https://doi.org/10.3390/w15193431

Academic Editor: Juraj Parajka

Received: 16 August 2023
Revised: 26 September 2023
Accepted: 27 September 2023
Published: 29 September 2023

1. Introduction

Ice along the Arctic shipping waterways is gradually thawing under the influence of global warming, and new shipping routes to polar areas are becoming available [1]. This could greatly reduce the navigation time and increase safety [2]. Glacial surges, fog, and ice flow will affect the navigation safety and may result in collisions with ice and ship damage. Therefore, it is important to promptly detect sea ice, icebergs, and passing ships to avoid ship–ice and ship–ship collisions [3]. A detection system should provide information about the position and size of the objects on navigation routes, so as to support polar ship path planning and make ship navigation safer and more energy efficient.

Field observation focuses mostly on ships and buoys. As described, visual observation was combined with field measurements [4], determining for instance, ice thickness through the on-site drilling of ice samples. However, on-site detection in the harsh polar environment is challenging, and data collection is limited [5,6]. In recent years, image processing and remote sensing technology have been applied to the acquisition of polar information, and indirect detection techniques have been developed [7]. Methods such as ship walk observation, shipborne radar observation, and unmanned aircraft observation are used for local-scale detection, while active and passive microwave remote sensing is mainly used for large-scale observations [8].

For local-scale environmental information, shipboard cameras are commonly used to acquire and analyze optical images. Weissling et al. [9] developed a ship-based, ice condition imagery acquisition, processing, and analysis system. Worby et al. [10] evaluated the ice distribution characteristics in the Antarctic based on 20,000 images acquired during Antarctic ship voyages. In addition, researchers are studying how to apply machine learning and deep learning to polar target detection. Li et al. [11] proposed a two-stream radiative transfer model for ponded sea ice. The upwelling irradiance from the pond surface was determined and then its spectrum was transformed into RGB color space. Cai et al. [12] employed convolutional neural networks to detect sea ice by instance segmentation using a simulation ice pool dataset and estimated ice size and concentration.

For large-scale ice detection, passive and active microwave remote sensing images are mostly used. Some algorithms for calculating ice concentration were proposed, including NASA Team (National Aeronautics and Space Administration), Bootstrap, and ASI (ARTIST Sea Ice) [13]. For the identification and classification of sea ice, techniques such as the maximum likelihood method, SVM (support vector machines), Markov random field model, and neural networks have been utilized. Belchansky et al. [14] used SSM/I (Special Sensor Microwave/Image) bright temperature data and remote sensing ice images acquired by the ERS and Okean satellites as inputs to train neural networks. Karvoven et al. [15] segmented and classified six types of ice from Synthetic Aperture Radar (SAR) images using an impulse-coupled neural network. Ressel et al. [16] utilized an artificial neural network to classify ice, and the results demonstrated that the method was resistant to image noise. However, generally, the models used were not modified and improved according to the characteristics of the remote sensing ice images to be analyzed.

The detection based on shipboard optical images is characterized by high resolution, rapidity, and the ability to provide rich information [17], but it cannot allow a continuous monitoring of the environment and is affected by adverse weather conditions. The detection based on remote sensing images can be applied to wide polar regions and is independent of the weather conditions, but its spatial distribution is relatively low, and it is not sufficiently accurate to distinguish small targets. Most studies focused on ship detection rather than on ice detection, and those that investigated ice detection systems mainly used a single data source consisting of remote sensing or optical images.

In this paper, we combined data of local-scale optical images and remote sensing images to integrate their specific strengths. Polar datasets at different scales were constructed. The SSD model was used for polar target detection at the local scale. For remote sensing detection, the YOLOv5 model was improved according to the characteristics of the sea ice, and ablation and comparison experiments were conducted to verify the model. We performed a slicing operation on the images to ensure that small sea ice targets could be detected and we constructed hybrid datasets to verify the proposed model.

2. Polar Multi-Target Detection at the Local Scale

2.1. Target Detection

The region proposal method and the end-to-end method are based on two primary detection deep learning algorithms. Overfeat, R-CNN (Region-CNN), Faster R-CNN [18], etc., are involved in the region proposal-based method while YOLO and SSD are part of the end-to-end-based method [19]. The region proposal-based method has a significant advantage in detection accuracy with respect to the end-to-end-based method because it includes "two steps" and is more accurate for target localization and classification. On the other hand, it has a significant disadvantage in the detection of speed because it requires a long time to generate the region proposal. The end to end based detection method directly extracts features for object localization and classification using convolution. The SSD relies on the RPN (Region Proposal Network) mechanism of the Faster R-CNN, which combines the detection speed of the end-to-end method with the detection accuracy of the region nomination method. Therefore, in this paper, we chose the SSD model for polar multi-target detection on the local scale.

2.2. *SSD Model*

The SSD model consists of two major components, a base network and additional network layers, as shown in Figure 1. The base network uses the structure of Visual Geometry Group (VGG 16) and converts the last two fully connected layers into convolutional layers, Conv4_3 and Fc7. The additional network layers include four sets of convolutional layers: Conv6_2, Conv7_2, Conv8_2, and Conv9_2. The SSD detection model operations are as reported below.

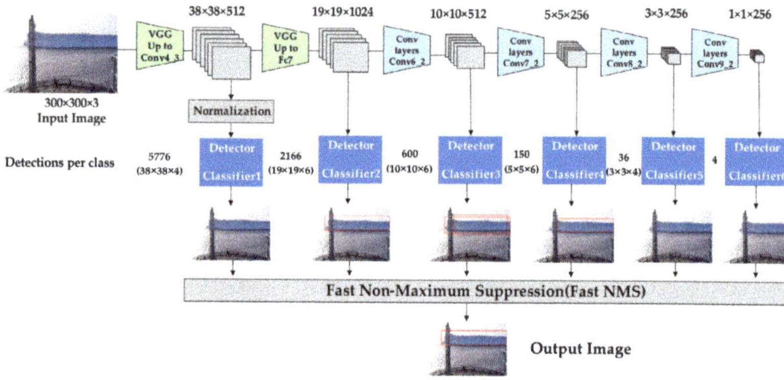

Figure 1. Structure of the SSD model.

Firstly, the input image is converted to a three-channel RGB (Red Green Blue) image with a resolution of 300 × 300 or 500 × 500. The image is fed into the network to extract multi-scale feature information, and the scales of each feature layer are 38 × 38, 19 × 19, 10 × 10, 5 × 5, 3 × 3 and 1 × 1.

Then, target feature extraction is performed through six feature layers of different scales. Default boxes are generated for each point of the feature map, and the number of default boxes is different for each layer.

Finally, all the generated default boxes are integrated, analyzed by non-maximum suppression (NMS) and filtered with an intersection over union (IOU) higher than 0.5. The final output contains information about the location, category, and confidence level of the target.

The SSD is characterized by its efficiency as a single-stage detector, performing detection directly in a single forward pass without the need for region proposals, which results in a faster detection compared to other models. It leverages multi-scale features and default boxes and can detect objects of various sizes. These advantages make SSD an effective model widely applied in practical scenarios.

2.3. *Construction of a Local-Scale Polar Multi-Target Dataset*

Due to the lack of a publicly accessible dataset for polar targets, constructing a new dataset is an important step. A total of 650 images were obtained through searching, deweighting, annotation, and review to create a local-scale polar multi-objective dataset. Some of the images were downloaded from The Norwegian Institute (https://icewatch.met.no, accessed on 19 August 2022). The dataset was divided into 5 categories, namely, sea ice (first-year ice), icebreakers, icebergs, inter-ice waterways, and melting pools on ice. Labellmg, an image annotation tool, was used to label the images as fy, icebreaker, iceberg, channel, and pool, respectively [20,21]. Finally, the dataset is randomly divided into training and testing sets at the ratio of 8:2. The details are shown in Table 1.

Table 1. Local-scale polar multi-target dataset.

Category Name	Label Name	Image Number	Label Number
Sea ice	fy	150	2446
Icebreakers	icebreaker	150	160
Icebergs	iceberg	150	167
Inter-Ice Waterways	channel	100	100
Melting pools on ice	pool	100	558

The majority of the images were captured by shipboard cameras and UAVs (Unmanned Aerial Vehicles), and the photographed scenes corresponded to polar ship navigation scenarios. Some of the sample images are shown in Figure 2.

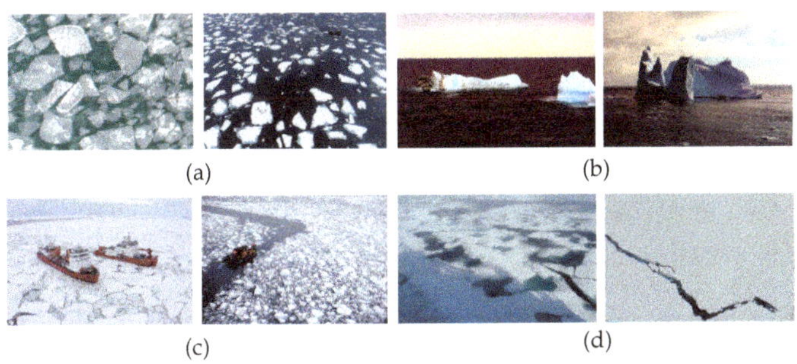

Figure 2. Local-scale polar multi-target dataset. (**a**) Sea ice; (**b**) icebergs; (**c**) icebreakers; (**d**) melt pond and inter-ice waterway.

2.4. Results

The model training and testing configurations are shown in Table 2. The detailed training parameters are shown in Table 3.

Table 2. Configurations and versions.

Configuration	Version
Operating System	Window10
Central processing unit CPU	Intel Xeon W-2255
Graphics GPU	NVIDIA Quadro P620
Deep Learning Platform	Pytorch
Pytorch version	1.10.2
CUDA version	11.3
CUDNN version	8.2.1
Python version	3.9

Table 3. Training parameters.

Parameters	Values
num_calsses	4
learning_rate_base	0.002
batch_size	4
momentum	0.937
num_workers	4
epoch	1000
weight_decay	0.0005

The steps in the training were as follows. Firstly, the training process was mainly used to predict the results and calculate the loss value by the forward propagation algorithm. Secondly, the parameter gradient value was calculated by backward propagation, and the parameters were optimized and updated. Finally, the training was completed by iterating the gradient descent algorithm to the maximum number of iterations. The process was stopped when the model reached loss convergence; then, a model was generated for the subsequent training and target detection tasks.

Average precision (AP), F1 score, and mean average precision (mAP) were determined to evaluate the detection accuracy [22]. The precision (P) value can quantify the effectiveness of sample classification, and the recall (R) value can evaluate the capacity to detect positive samples. Considering only precision or only recall is not sufficient to evaluate a model; so, the F1 score was used to harmonize P and R. The calculation of mAP can be divided into two steps: the first step consists of the calculation of the AP (average precision) of each category, while the second step involves determining the sum of the average precision values of each category and then its average value to obtain mAP. These parameters were calculated according to Equations (1)–(5):

$$\text{Precision} = \frac{\text{TP}}{\text{TP} + \text{FP}} \quad (1)$$

$$\text{Recall} = \frac{\text{TP}}{\text{TP} + \text{FN}} \quad (2)$$

$$\text{F1} = 2\frac{P \cdot R}{P + R} \quad (3)$$

$$\text{AP} = \int_0^1 P \cdot R dR \quad (4)$$

$$\text{mAP} = \frac{\sum_{i=1}^{k} \text{AP}_i}{\text{k}} \quad (5)$$

where TP (true positives) is the number of correctly classified positive samples, FP (false positives) is the number of incorrectly classified positive samples, TN (true negatives) is the number of correctly classified negative samples, and FN (false negatives) is the number of incorrectly classified negative samples; k is the category number.

After training, the model was applied to the test set, and finally, an mAP value of 70.19% was obtained. The accuracy of the icebreaker category was the highest at 92%, followed by those of the iceberg category, which was 85%, and of the fy (first-year ice) category, which reached 77%. The accuracies of channel and pool were the lowest, 52% and 45%, respectively, due to the low number of images or labels for these two categories. The detection results for each category are shown in Figure 3. Some of the test results are shown in Figure 4. The SSD model works well for the detection of large targets at close range, but it is not effective in detecting small targets at a distance.

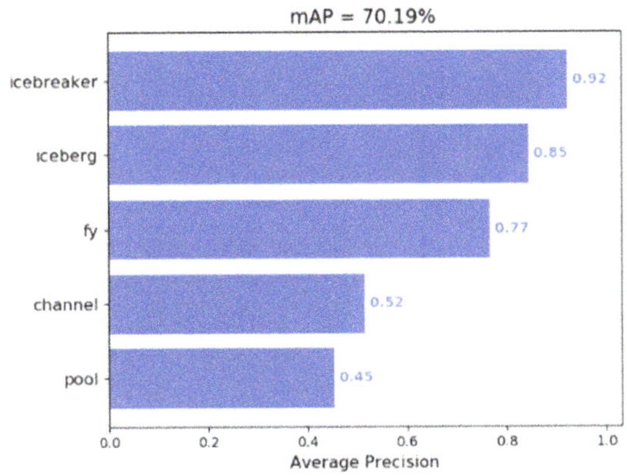

Figure 3. Detection results for each category.

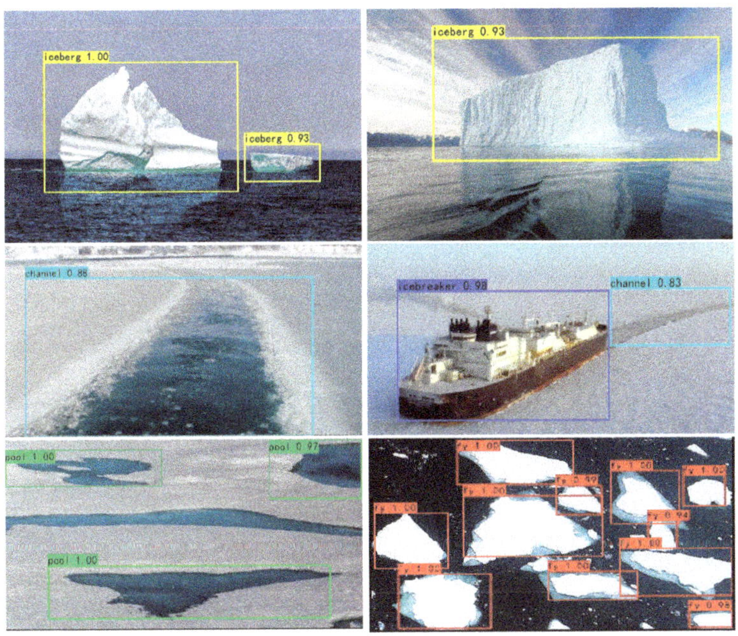

Figure 4. Example of the test results.

3. Sea Ice Detection by Remote Sensing

The detection on a local scale does not fully meet the requirements of navigating in polar regions, and using a single data source has certain limitations. The ship optical cameras cannot obtain large-scale and long-time series images and cannot monitor non-navigable areas. If sea ice in remote sensing images can be identified and located, and data fusion between local-scale and large-scale data can be performed, the advantages of different data sources can be fully utilized [23].

3.1. Introduction of the YOLOv5 Model

In remote sensing images, the ice masses appear very small and densely clustered, and the SSD model is not able to analyze them. After improving its accuracy and efficiency, the YOLOv5 model was applied to the detection of ice through remote sensing. The backbone, neck, and head are the three basic structural components of the YOLOv5 model, as shown in Figure 5.

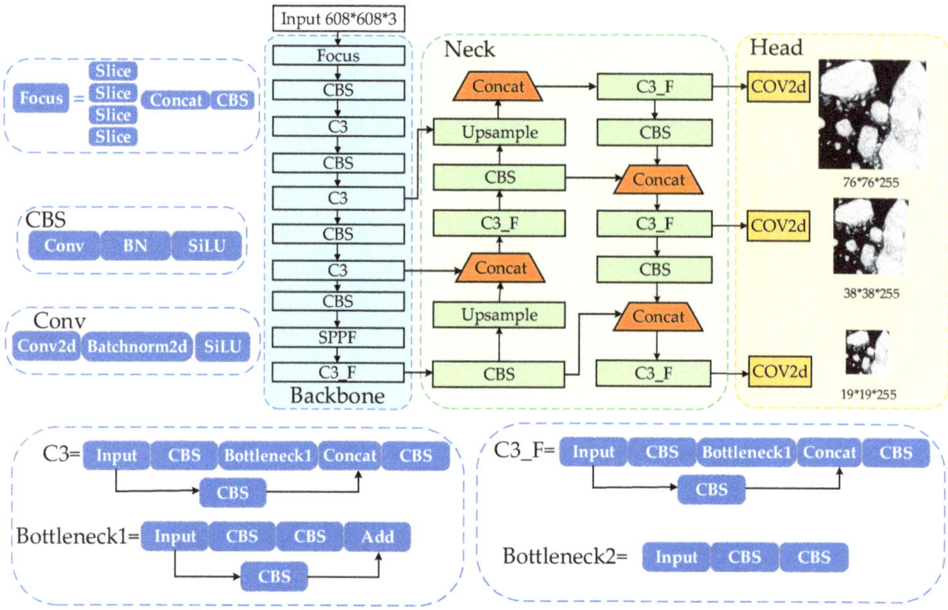

Figure 5. Structure of the YOLOv5 model.

The YOLOv5 backbone utilizes CSPDarknet as the backbone for extracting features from images, which is composed of cross stage partial networks. The focal module is responsible for efficiently downsampling the images. It is designed to transmit the images through the channel while maintaining primitive information. The backbone layer incorporates the utilization of the C3, C3_F, and Spatial Pyramid Pooling Fast (SPPF) modules. The C3 and C3_F modules can enhance the extraction of image features and augment the overall speed.

The neck module in YOLOv5 utilizes PANet to produce a feature pyramid network. These aggregated features are subsequently forwarded to the head module for prediction. The neck layer integrates the structures of the feature pyramid network (FPN) and the path aggregation network (PAN). Deep-feature images possess a higher degree of semantic information but a lower degree of location information, whereas shallow-feature images exhibit the reverse characteristics. The FPN model can transmit semantic information from a deep-feature image to a shallow-feature image. In contrast, PAN can transmit location information from a shallow-feature image to a deep-feature image. The integration of FPN and PAN enables the consolidation of parameters across various detection layers.

The YOLOv5 head is composed of layers that produce predictions from the anchor box. The head can be categorized into the loss function and non-maximum suppression (NMS). The binary cross entropy loss function is employed for the computation of classification loss and confidence loss, whereas the complete IoU (CIoU) loss function is utilized for the estimation of location loss. The CioU loss function incorporates three crucial parameters: the overlap area, the distance from the center, and the aspect ratio. NMS is employed to eliminate redundant detection while retaining the candidate box with the highest prediction probability as the ultimate prediction box.

3.2. *Improved YOLOv5 Model*

The YOLOv5 model was improved in three aspects. Firstly, the Squeeze-and-Excitation Networks (SE) attention module was added to the backbone of the original model. Secondly, the Fast Spatial Pyramid Pooling and Cross Stage Partial Network module (SPPCSPC-F) were used to augment the characterization capabilities. Finally, Funnel Activation (FReLU) was introduced to replace the Sigmoid-Weighted Linear Unit (SiLU) and improve the accuracy of ice detection.

3.2.1. Squeeze-and-Excitation Networks (SE) Attention Mechanism

Due to the large size of the remote sensing images and the small size of the ice targets, it is easy to lose some useful information. The Squeeze-and-Excitation Networks (SE) attention mechanism was added to the YOLOv5 backbone [24]. The SE module was inserted after the convolutional layers. The module consists of two operations: squeeze and excitation. It is integrated to adaptively adjust the importance of each channel by learning their weights. The structure of the SE attention mechanism is shown in Figure 6.

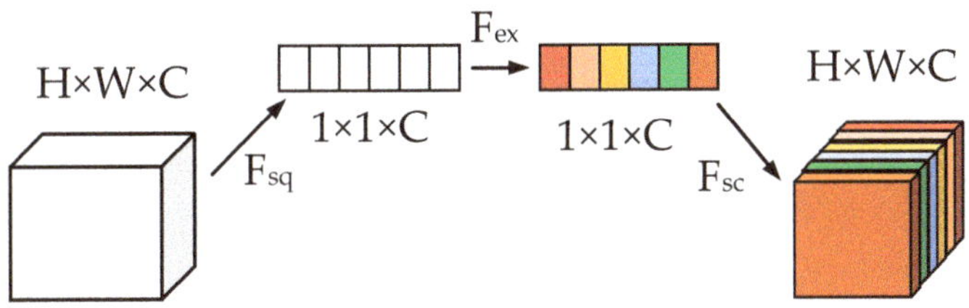

Figure 6. Structure of the SE attention mechanism.

In the squeeze phase (Fsq), global average pooling is applied to the input feature map, compressing it from three dimensions to one dimension. This one-dimensional tensor captures global information for each channel. In the excitation phase (Fex), a set of fully connected layers operates on the output of the squeeze phase. These layers model the importance of each channel and generate a channel attention vector. Finally, a rescale operation (Fsc) normalizes the weights and multiplies them onto each feature channel.

3.2.2. SPPCSPC-F

The Spatial Pyramid Pooling Fast (SPPF) is a module designed to enhance feature representation. SPPF is the improved version of Spatial Pyramid Pooling (SPP) and is faster than SPP under the same conditions. The structure of the SPPF module is shown in Figure 7.

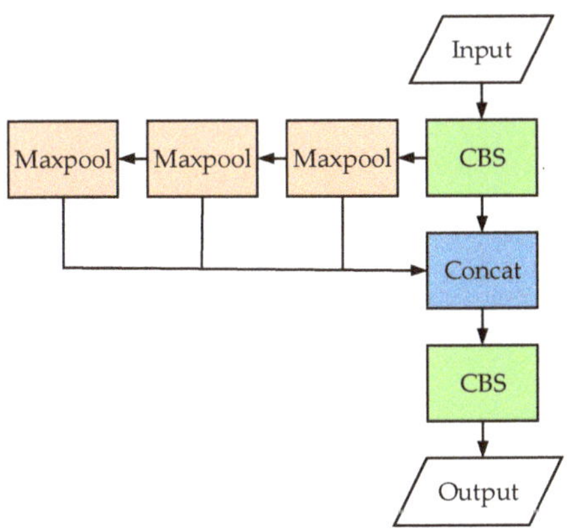

Figure 7. Structure of the SPPF module.

The input feature map passes through three 5 × 5 maximum pooling layers, and three different sizes of receptive fields are obtained. Although maximum pooling can expand the receptive field, it will reduce the resolution of the feature map and cause the loss of some useful information. SPPCSPC is a structural module that combines the concepts of SPP and Cross Stage Partial Network (CSP) [25]. In this paper, we present the SPPCSPC-F to replace the SPPF concerning the idea of SPPCSPC. The structure of SPPCSPC-F is shown in Figure 8.

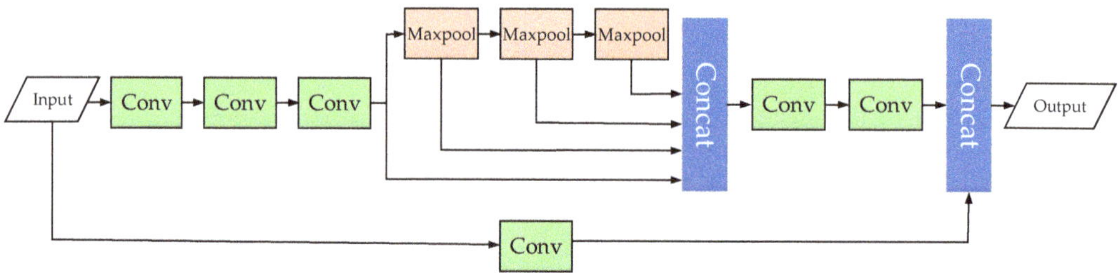

Figure 8. Structure of the SPPCSPC-F module.

The input feature map is passed through the SPPCSPC-F module, with one path performing convolutional operations to extract lower-level features, and the other path preserving the original features. Next, the module performs multi-scale pooling operations on the feature map to capture features with different receptive fields. Finally, the fused features are further processed by subsequent convolutional layers. The order of pooling is modified to increase the speed while keeping the feeling field constant.

3.2.3. FReLU Activation Function

In the YOLOv5, the Sigmoid-Weighted Linear Unit (SiLU) is used as the activation function. When the input values move away from zero, the derivative of the SiLU can approach zero, leading to gradient saturation. It is difficult for the network to converge or cause training instability. The FReLU was used to replace the SiLU. The FReLU activation function incorporates learnable parameters, enabling the network to adaptively adjust the shape of the activation function through learning [26]. This flexibility enhanced the model's learning capacity and improved its adaptation to the sea ice characteristics. Combining SE attention with FReLU enables YOLOv5 to extract high-quality features, concentrate on key objects, reduce overfitting, and improve generalization ability, especially for detecting small objects in polar regions. The FReLU is defined by Equations (6) and (7):

$$f(x_c, i, j) = \max(x_c, i, j, T(x_c, i, j)) \tag{6}$$

$$T(x_c, i, j) = x^w_{c,i,j} \cdot p^w_c \tag{7}$$

where $T(\cdot)$ denotes the funnel condition, $x^w_{c,i,j}$ denotes a $k_h \times k_w$ Parametric Pooling Window centered on $x^w_{c,i,j}$, p^w_c denotes the coefficient on this window which is shared in the same channel, and $(\cdot)$ denotes dot multiply. The FReLU activation function is shown in Figure 9.

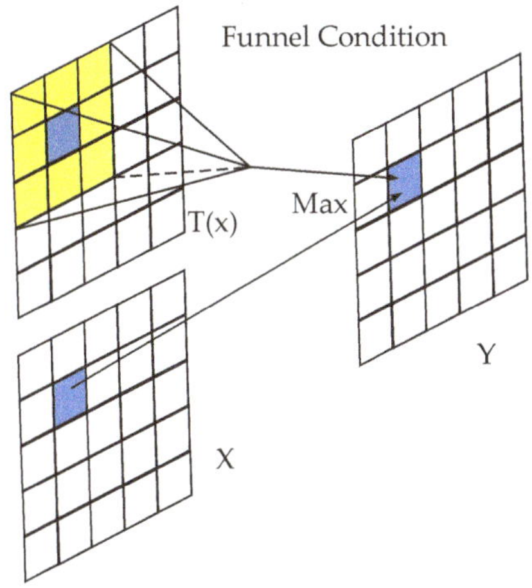

Figure 9. FReLU activation function.

3.3. *Construction of a Remote Sensing Sea Ice Dataset*

The remote sensing sea ice dataset was mainly derived from the Google Earth (http://earthengine.google.com/, accessed on 25 December 2022) and the Northwestern Polytechnical University (NWPU) datasets [27]. A total of 600 images, obtained after de-duplication, annotation, and review, constituted the dataset. The tag name was ice, and the number of tags was 15,948. It was randomly divided into a training set and a test set at the data ratio of 8:2. Some of the sample images in the dataset are shown in Figure 10.

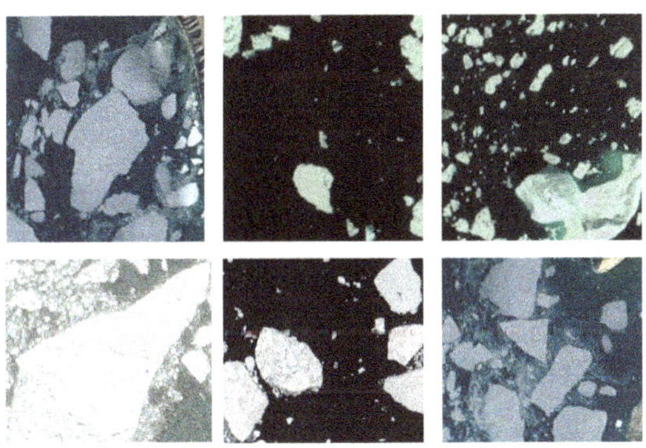

Figure 10. Remote sensing sea ice dataset.

Neural networks need a large amount of data and a high data quality to improve their performance and robustness. The YOLOv5 uses Mosaic, adaptive cutout, and other data processing methods for data enhancement [28].

The main idea of Mosaic is to randomly crop and scale several images and then randomly arrange and splice them to form a single image, to enrich the dataset and improve the training speed of the network. In the normalization operation, several images are calculated at one time, which can reduce the demand for computer memory. The data augmentation process is shown in Figure 11.

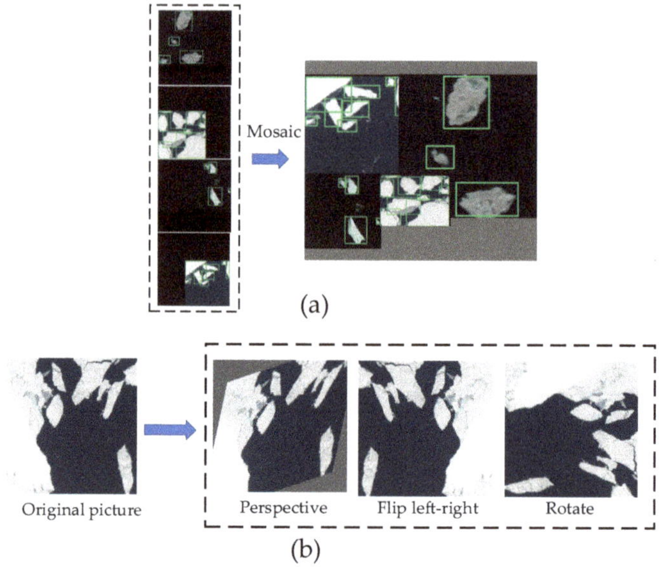

Figure 11. Data augmentation. (**a**) Mosaic; (**b**) perspective, flip left–right and rotation processing.

There are many challenges in the detection of remote sensing images, as some targets are relatively small in size and usually clustered together. If the images are directly sent into the network for detection, many small targets cannot be effectively identified.

To solve this problem, in the detection stage, a sliding window was used to cut a specified-size (such as a 416 × 416) image as the input. The cutout adjacent images had a 15% overlap. The slicing operation on the remote sensing image is shown in Figure 12. The purpose of the overlap is to ensure that every region is completely detected. Although this causes duplicate detection, overlapping sections can be filtered out by the NMS. Finally, the results of each cutout image were combined to obtain the detection results.

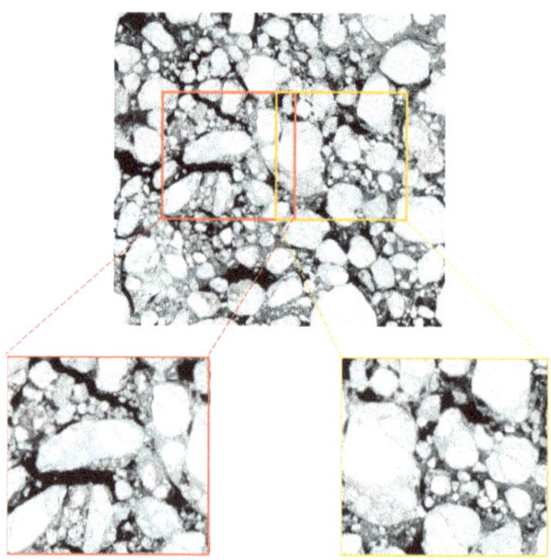

Figure 12. Slicing operation.

In order to verify the accuracy of the improved YOLOv5 model, we combined simulated sea ice images and real sea ice images into a hybrid dataset. The simulated images were constructed as follows. Firstly, we built a large flat ice field. Secondly, we fragmented the flat ice field to obtain a broken ice field. The Voronoi diagram is morphologically similar to an ice field with large pieces of broken ice and consists of a set of continuous polygons formed by the perpendicular bisectors of lines connecting two neighboring points. We used the RayFire plug-in of 3ds Max to fragment the flat ice field according to the Voronoi diagram, as shown in Figure 13a. Finally, the size of the broken ice field was reduced by 80% to enlarge the gaps between the ice blocks, as shown in Figure 13b.

Figure 13. Sea ice modeling process. (**a**) Broken sea ice field; (**b**) operation of enlarging the gaps between the ice blocks.

3.4. Results

3.4.1. Ablation Study

The ablation study was conducted to facilitate the comparison of the different improvement methods. They were trained with the same configuration used in the local-scale polar objection. The epoch was set as 300, the initial learning rate was 0.001, the momentum parameter was 0.9, the weight decay parameter was 0.0005, and the NMS threshold was 0.5. The evaluation was carried out after every 30 training epochs. The results are shown in Table 4.

Table 4. Ablation study results.

Method	P	R	F1	mAP
YOLOv5	0.719	0.684	0.701	0.719
YOLOv5+SE	0.731	0.701	0.716	0.738
YOLOv5+CSPCF	0.737	0.688	0.712	0.743
YOLOv5+FReLU	0.723	0.706	0.714	0.747
YOLOv5+SE+SPPCSPC-F+FReLU	0.753	0.703	0.727	0.754

In Table 5, it can be observed that the mAP of the original YOLOv5 model was 0.719, the lowest among those of the evaluated models. The implementation of SE resulted in an increase in the mAP to 0.738, i.e., by 1.9%. The inclusion of SPPCSPC-F resulted in a 2.4% increase in the mAP, which reached the value of 0.743. However, the R value was relatively low, i.e., 0.688. The inclusion of FReLU resulted in a 2.8% increase in the map, to the value of 0.747. When adding SE, SPPCSPC-F, and FReLU, the mAP was improved by 3.5%, reaching the highest value among those of all the examined models.

Similarly, the P, R, and F1-scores of the original YOLOv5 model were 0.719, 0.684, and 0.701. However, for the proposed method, the P, R, and F1-scores were 0.753, 0.703, and 0.727, that is, they increased by 3.4%, 1.9%, and 1.8%, respectively. Therefore, the improved YOLOv5 model revealed superior accuracy and enhanced performance in the domain of remote sensing sea ice detection.

Table 5. The performance of different models in sea ice detection.

Method	P	R	F1	mAP
YOLOv3	0.858	0.407	0.552	0.604
YOLOv4-tiny	0.757	0.548	0.636	0.648
Faster-RCNN	0.641	0.632	0.636	0.655
YOLOv5	0.719	0.684	0.701	0.719
Ours	0.753	0.703	0.727	0.754

3.4.2. Contrast Study

In order to further validate the advantage benefits and efficacy of the improved YOLOv5 model, incorporating the three mentioned modules, a comparative experiment was conducted. We compared the improved model with other conventional models, such as Faster-RCNN, YOLOv3, and YOLOv4-tiny; the values of loss and mAP are shown in Figure 14.

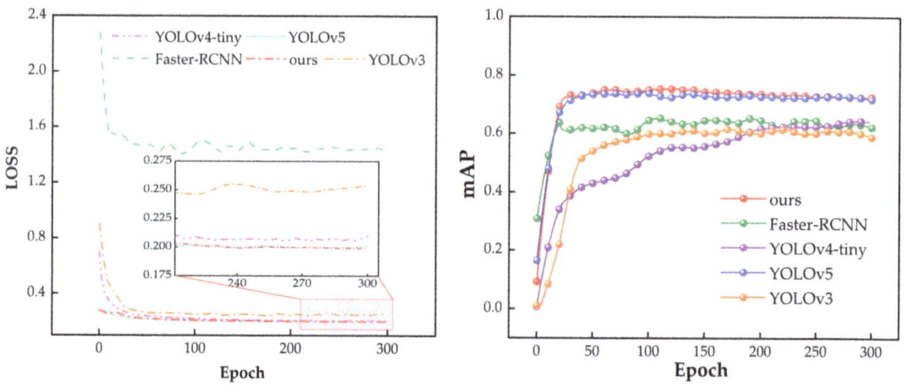

Figure 14. The values of loss and mAP of the different models tested.

In the first 40 epochs, the loss of each model fell quickly, indicating that the training did not achieve a stable state. When the training is stable, the loss in the curve is flat rather than sharp. The loss of our model was lower than that of the others when training reached a steady stage. The mAP rose sharply in the first 80 epochs. All models tended to become more stable after 250 training epochs, and the mAP of our model was the highest.

The values of the evaluation indicators are shown in Table 5. Compared with those of YOLOv3, YOLOv4-tiny, Faster-RCNN, and original YOLOv5, the mAP of YOLOv3 was the lowest, at 60.4%, whereas the mAP of our model was the highest, at 75.4%. YOLOv3 and YOLOv4-tiny showed a higher P value but a lower R value, which indicated that these two models largely miss their ice targets when detecting sea ice. Based on the above results, the improved YOLOv5 can better perform in sea ice detection.

The improved YOLOv5 was used to test a remote sensing image with a resolution of 3660 × 3660. Since some sea ice targets were too dense, the confidence degree was hidden in the results. The detection results of the original YOLOv5 are shown in Figure 15a. Figure 15b shows zoomed-in local images using the original YOLOv5, in which the number of detected sea ice masses was 14 and 55. Figure 15c shows zoomed-in local views of the image detected by the improved YOLOv5, in which the number of detected sea ice masses was 53 and 88. When using the improved YOLOv5, the number of detected ice targets increased by 39 and 33 units, and most of them were small.

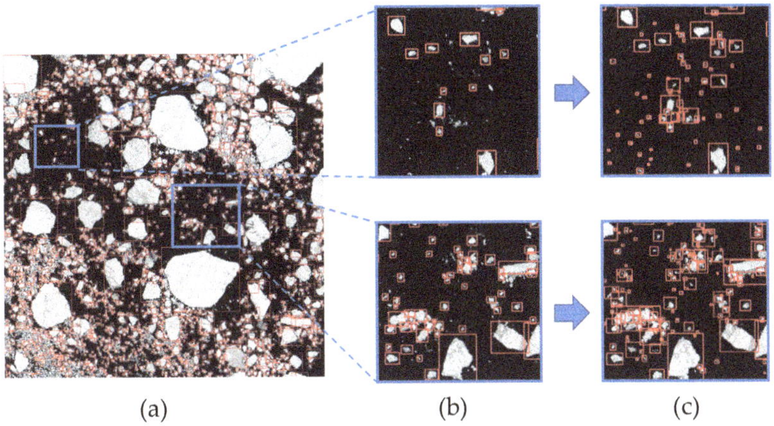

Figure 15. Comparison of the detection results. (**a**) Detection results for the original YOLOv5; (**b**) localized zoomed-in views using the original YOLOv5; (**c**) localized zoomed-in view using the improved YOLOv5.

The results with the confidence degree are shown in Figure 16. Both a real image and a simulated sea ice image are presented. The results demonstrated that the improved YOLOv5 model was able to detect ice targets in simulated sea ice images with strong generalization ability and robustness.

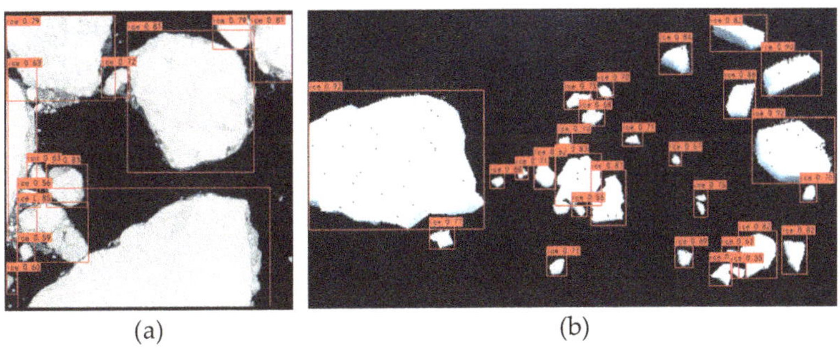

Figure 16. Test detection results using the hybrid dataset. (**a**) Detection result for a real ice image; (**b**) detection result for a simulated ice image.

Local scale detection covers from tens to hundreds of meters. Correspondingly, remote sensing scale detection covers from tens to hundreds of kilometers. If a ship navigates in the polar regions using only local-scale data, the planned path may be optimal at the local scale but not on the whole, as it could be unnecessarily long. If only remote sensing data are used, the planned path may be the best on a large scale, but it may miss some obstacles that will jeopardize the safety of ship navigation on a local scale. In this paper, local-scale and remote sensing data were combined to take advantage of their respective strengths. Our results indicated that the use of this combination for the detection of obstacles can improve the safety and efficiency of polar navigation.

4. Discussion

The instability of polar condition makes navigation difficult. Sea ices which float on the surface are difficult to detect and are prone to collision with the hull or the propeller. In this paper, polar datasets at different scales were constructed. The SSD model was used for multi-target detection at the local scale. For remote sensing images, hybrid datasets were constructed and a slicing operation was performed, the YOLOv5 model was improved and tailored to detect sea ices. Ablation and comparison experiments were conducted to verify the proposed model.

For the source of data, most studies mainly adopt remote sensing or optical image as a single data source. For example, Li et al. [29] who developed a novel method to extract sea ice cover using Sentinel-1 data based on the support vector machine (SVM). Xu et al. [30] proposed a Recurrent Attention Convolutional Neural Network (RA-CNN) to classify different ships. In this paper, the fusion of remote sensing and optical images is used to take advantage of the complementary strengths.

For ice detection, some studies did not change their model according to the characteristic of ices. Moreover, many studies used only real datasets to verify the accuracy of their model. For example, Frederik et al. [31] proposed a deep learning model based on YOLOv3 for distinguishing icebergs and ships. Markus et al. [32] detected the ice on rotor blades. In this paper, the YOLOv5 mode was improved to ensure that small ices can be detected. The hybrid dataset was constructed to verify the proposed model and the results showed that the model had a good generalization ability.

Although this study successfully detected multi-scale polar objects, it still has some limitations. The lower detection accuracy of some categories on the local scale was due to the small amount of data. The datasets used can be expanded to increase the accuracy [33]. This study focused on rectangular detection boxes; if more detailed sea ice information is needed, in the future, the ice images can be processed with instance segmentation [34].

5. Conclusions

In order to avoid the limitations caused by the use of a single dataset, we constructed multi-scale datasets by combining data from different sources. The SSD model was used to detect local-scale targets, and the improved YOLOv5 model was used to detect remote sensing sea ice targets. The following conclusions can be drawn:

1. The SSD model can be used for the detection of polar targets on a local scale. The dataset it uses includes sea ice, icebergs, icebreakers, ice melt ponds, and inter-ice waterways; the mAP can reach 70.19%; icebergs and icebreakers were detected with the highest average accuracy of 84% and 81%.
2. An improved YOLOv5 model was obtained through Squeeze-and-Excitation Networks (SE), Funnel Activation (FReLU), Fast Spatial Pyramid Pooling, and Cross Stage Partial Network (SPPCSPC-F). The utilization of SE and SPPCSPC-F allowed the characteristics of objects to be strengthened, thereby augmenting the overall detection efficacy and precision of the model. The FReLU activation function was used to enhance the learning capacity and enable better adaptation to sea ice characteristics. A slicing operation was performed on remote sensing images to detect small ice masses. Simulated ice images were included to verify the precision of the proposed model.
3. In comparison to other conventional models such as Faster-RCNN, YOLOv3, and YOLOv4-tiny, the proposed model demonstrated higher accuracy, with an mAP of up to 75.4%, which verified its generalization ability and robustness. The proposed method is tailored to detect remote sensing sea ice, compared to the original model, the mAP value increased 3.5%.
4. For future research, large and diverse polar datasets need to be established. These datasets should contain polar images from different seasons, weather conditions, and periods, so that the model can better adapt to changes in the polar environment. Additionally, improved detection can provide support to avoid polar ship collision with ice masses and improve navigation path planning. It can also provide some help for the calculation of the ice pressure load for ships [35,36].

Author Contributions: Methodology, L.Z. and D.Z.; analysis, L.Z. and D.Z.; investigation, S.D.; resources, F.L. and S.D.; data curation D.Z. and Q.W.; software, D.Z and S.H.; writing—original draft preparation, D.Z.; writing—review and editing, L.Z. and F.L; visualization, L.Z. and F.L.; supervision, L.Z. and S.D.; project administration, L.Z. and S.D.; funding acquisition, L.Z. All authors have read and agreed to the published version of the manuscript.

Funding: This study was funded by the National Key Research and Development Program (2022YFE010700), General Projects of National Natural Science Foundation of China (52171259) and High-Tech Ship Research Project of the Ministry of Industry and Information Technology ([2021]342), CSSC-SJTU joint prospect funding (ZCJDQZ202307A01), and the Science and Technology Commission of Shanghai Municipality Project (22dz1204403).

Data Availability Statement: All analyzed data in this study are included in the manuscript.

Acknowledgments: The authors would like to thank the Jiangsu University of Science and Technology (JUST).

Conflicts of Interest: The authors declare no conflict of interest.

References

1. Chuah, L.F.; Mokhtar, K.; Ruslan, S.M.M.; Abu Bakar, A.; Abdullah, M.A.; Osman, N.H.; Bokhari, A.; Mubashir, M.; Show, P.L. Implementation of the energy efficiency existing ship index and carbon intensity indicator on domestic ship for marine environmental protection. *Environ. Res.* **2023**, *222*, 115348. [CrossRef] [PubMed]
2. Zuo, Q.; Qian, L.; Xu, X.; Yan, J.; Cheng, L.; Zhang, Z. Navigation strategy and economic research of the northeast passage in the Arctic. *Chin. J. Polar Res.* **2015**, *27*, 203.
3. Lu, Y.; Gu, Z.; Liu, S.; Chuang, Z.; Li, Z.; Li, C. Scenario-based optimization design of icebreaking bow for polar navigation. *Ocean Eng.* **2021**, *244*, 110365. [CrossRef]
4. Yu, M.; Lu, P.; Li, Z.; Li, Z.; Wang, Q.; Cao, X.; Chen, X. Sea ice conditions and navigability through the Northeast Passage in the past 40 years based on remote-sensing data. *Int. J. Digit. Earth* **2020**, *14*, 555–574. [CrossRef]
5. Oloruntobi, O.; Mokhtar, K.; Gohari, A.; Asif, S.; Chuah, L.F. Sustainable transition towards greener and cleaner seaborne shipping industry: Challenges and opportunities. *Clean. Eng. Technol.* **2023**, *13*, 100628. [CrossRef]
6. Mahadi, C.M.H.C.; Mokhtar, K.; Chuah, L.F.; Chan, S.R.; Suhrab, M.I.R.; Mubashir, M.; Asif, S.; Show, P.L. An organisational search and rescue performance assessment for a cleaner environment. *Clean. Eng. Technol.* **2023**, *14*, 100641. [CrossRef]
7. Ogishima, A.; Saiki, K. Development of a micro-ice production apparatus and NIR spectral measurements of frosted minerals for future lunar ice exploration missions. *Icarus* **2021**, *357*, 114273. [CrossRef]
8. Anderson, S. Remote Sensing of the Polar Ice Zones with HF Radar. *Remote Sens.* **2021**, *13*, 4398. [CrossRef]
9. Weissling, B.; Ackley, S.; Wagner, P.; Xie, H. EISCAM—Digital image acquisition and processing for sea ice parameters from ships. *Cold Reg. Sci. Technol.* **2009**, *57*, 49–60. [CrossRef]
10. Worby, A.; Comiso, J. Studies of the Antarctic sea ice edge and ice extent from satellite and ship observations. *Remote. Sens. Environ.* **2004**, *92*, 98–111. [CrossRef]
11. Lu, P.; Leppäranta, M.; Cheng, B.; Li, Z.; Istomina, L.; Heygster, G. The color of melt ponds on Arctic sea ice. *Cryosphere* **2018**, *12*, 1331–1345. [CrossRef]
12. Cai, J.; Ding, S.; Zhang, Q.; Liu, R.; Zeng, D.; Zhou, L. Broken ice circumferential crack estimation via image techniques. *Ocean Eng.* **2022**, *259*, 111735. [CrossRef]
13. Shi, L.; Liu, S.; Shi, Y.; Ao, X.; Zou, B.; Wang, Q. Sea Ice Concentration Products over Polar Regions with Chinese FY3C/MWRI Data. *Remote. Sens.* **2021**, *13*, 2174. [CrossRef]
14. Belchansky, G.I.; Douglas, D.C.; Alpatsky, I.V.; Platonov, N.G. Spatial and temporal multiyear sea ice distributions in the Arctic: A neural network analysis of SSM/I data, 1988–2001. *Geophys. Res. Atmos.* **2004**, *109*, C10017. [CrossRef]
15. Karvonen, J. Baltic Sea Ice Concentration Estimation Based on C-Band Dual-Polarized SAR Data. *IEEE Trans. Geosci. Remote Sens.* **2013**, *52*, 5558–5566. [CrossRef]
16. Ressel, R.; Frost, A.; Lehner, S. A Neural Network-Based Classification for Sea Ice Types on X-Band SAR Images. *IEEE J. Sel. Top. Appl. Earth Obs. Remote Sens.* **2015**, *8*, 3672–3680. [CrossRef]
17. Mei, H.; Lu, P.; Wang, Q.; Cao, X.; Li, Z. Study of the spatiotemporal variations of summer sea ice thickness in the pacific arctic sector based on shipside images. *Chin. J. Polar Res.* **2021**, *33*, 37–48. [CrossRef]
18. Ren, S.; He, K.; Girshick, R.; Sun, J. Faster R-CNN: Towards real-time object detection with region proposal networks. *IEEE Trans. Pattern Anal. Mach. Intell.* **2017**, *39*, 1137–1149. [CrossRef]
19. Liu, W.; Anguelov, D.; Erhan, D.; Szegedy, C.; Reed, S.; Fu, C.Y.; Berg, A.C. SSD: Single shot multibox detector. In Proceedings of the European Conference on Computer Vision, Amsterdam, The Netherlands, 11–14 October 2016. [CrossRef]
20. Morozov, E.G.; Krechik, V.A.; Frey, D.I.; Zamshin, V.V. Currents in the Western Part of the Weddell Sea and Drift of Large Iceberg A68A. *Oceanology* **2021**, *61*, 589–601. [CrossRef]
21. Morozov, E.; Zuev, O.; Zamshin, V.; Krechik, V.; Ostroumova, S.; Frey, D. Observations of icebergs in Antarctic cruises of the R/V "Akademik Mstislav Keldysh". *Russ. J. Earth Sci.* **2022**, *22*, ES2001. [CrossRef]

22. Redmon, J.; Farhadi, A. YOLO9000: Better, Faster, Stronger. In Proceedings of the IEEE Conference on Computer Vision & Pattern Recognition, Honolulu, HI, USA, 21–26 July 2017; pp. 6517–6525.
23. Li, W.; Liu, L.; Zhang, J. Fusion of SAR and Optical Image for Sea Ice Extraction. *J. Ocean Univ. China* **2021**, *20*, 1440–1450. [CrossRef]
24. Hu, J.; Shen, L.; Sun, G. Squeeze-and-Excitation Networks. In Proceedings of the 2018 IEEE/CVF Conference on Computer Vision and Pattern Recognition (CVPR), Salt Lake City, UT, USA, 18–22 June 2018. [CrossRef]
25. Wang, C.-Y.; Bochkovskiy, A.; Liao, H.-Y.M. YOLOv7: Trainable Bag-of-Freebies Sets New State-of-the-Art for Real-Time Object Detectors. In Proceedings of the IEEE/CVF Conference on Computer Vision and Pattern Recognition, Vancouver, BC, Canada, 18–22 June 2023; pp. 7464–7475. [CrossRef]
26. Qiu, S.; Xu, X.; Cai, B. FReLU: Flexible Rectified Linear Units for Improving Convolutional Neural Networks. In Proceedings of the 2018 24th International Conference on Pattern Recognition, Beijing, China, 20–24 August 2018; pp. 1223–1228. [CrossRef]
27. Cheng, G.; Han, J. A survey on object detection in optical remote sensing images. *ISPRS J. Photogramm. Remote Sens.* **2016**, *117*, 11–28. [CrossRef]
28. Zhao, B.; Wu, Y.; Guan, X.; Gao, L.; Zhang, B. An Improved Aggregated-Mosaic Method for the Sparse Object Detection of Remote Sensing Imagery. *Remote Sens.* **2021**, *13*, 2602. [CrossRef]
29. Li, X.-M.; Sun, Y.; Zhang, Q. Extraction of Sea Ice Cover by Sentinel-1 SAR Based on Support Vector Machine With Unsupervised Generation of Training Data. *IEEE Trans. Geosci. Remote Sens.* **2020**, *59*, 3040–3053. [CrossRef]
30. Xu, Z.; Sun, J.; Huo, Y. Ship images detection and classification based on convolutional neural network with multiple feature regions. *IET Signal Process.* **2022**, *16*, 707–721. [CrossRef]
31. Hass, F.S.; Arsanjani, J.J. Deep Learning for Detecting and Classifying Ocean Objects: Application of YoloV3 for Iceberg–Ship Discrimination. ISPRS Int. *J. Geo-Information* **2020**, *9*, 758. [CrossRef]
32. Kreutz, M.; Alla, A.A.; Eisenstadt, A.; Freitag, M.; Thoben, K.-D. Ice Detection on Rotor Blades of Wind Turbines using RGB Images and Convolutional Neural Networks. *Procedia CIRP* **2020**, *93*, 1292–1297. [CrossRef]
33. Wu, S.; Wang, J.; Liu, L.; Chen, D.; Lu, H.; Xu, C.; Hao, R.; Li, Z.; Wang, Q. Enhanced YOLOv5 Object Detection Algorithm for Accurate Detection of Adult Rhynchophorus ferrugineus. *Insects* **2023**, *14*, 698. [CrossRef]
34. Zhou, L.; Cai, J.; Ding, S. The Identification of Ice Floes and Calculation of Sea Ice Concentration Based on a Deep Learning Method. *Remote. Sens.* **2023**, *15*, 2663. [CrossRef]
35. Zhou, L.; Diao, F.; Sun, X.; Ding, S.; Zhu, A.; Song, M.; Han, Y. Numerical simulation of ice-breaking loads on ships in ice areas based on the circumferential cracking method. In Proceedings of the 19th China Marine (Shore) Engineering Symposium (Previous), Ningbo, China, 10 July 2019; pp. 197–202.
36. Xie, C.; Zhou, L.; Ding, S.; Liu, R.; Zheng, S. Experimental and numerical investigation on self-propulsion performance of polar merchant ship in brash ice channel. *Ocean Eng.* **2023**, *269*, 113424. [CrossRef]

water

MDPI

Article

Reconstruction of Snow Cover in Kaidu River Basin via Snow Grain Size Gap-Filling Based on Machine Learning

Linglong Zhu [1,2], Guangyi Ma [2], Yonghong Zhang [1,2,*], Jiangeng Wang [3] and Xi Kan [1]

1 School of Internet of Things Engineering, Wuxi University, Wuxi 214105, China
2 Jiangsu Collaborative Innovation Center of Atmospheric Environment and Equipment Technology (CICAEET), Nanjing University of Information Science & Technology, Nanjing 210044, China
3 Key Laboratory for Aerosol-Cloud-Precipitation of China Meteorological Administration, Nanjing University of Information Science & Technology, Nanjing 210044, China
* Correspondence: zyh@nuist.edu.cn

Abstract: Fine spatiotemporal resolution snow monitoring at the watershed scale is crucial for the management of snow water resources. This research proposes a cloud removal algorithm via snow grain size (SGS) gap-filling based on a space–time extra tree, which aims to address the issue of cloud occlusion that limits the coverage and time resolution of long-time series snow products. To fully characterize the geomorphic characteristics and snow duration time of the Kaidu River Basin (KRB), we designed dimensional data that incorporate spatiotemporal information. Combining other geographic and snow phenological information as input for estimating SGS. A spatiotemporal extreme tree model was constructed and trained to simulate the nonlinear mapping relationship between multidimensional inputs and SGS. The estimation results of SGS can characterize the snow cover under clouds. This study found that when the cloud cover is less than 70%, the model's estimation of SGS meets expectations, and snow cover reconstruction achieves good results. In specific cloud removal cases, compared to traditional spatiotemporal filtering and multi-sensor fusion, the proposed method has better detail characterization ability and exhibits better performance in snow cover reconstruction and cloud removal in complex mountainous environments. Overall, from 2000 to 2020, 66.75% of snow products successfully removed cloud coverage. This resulted in a decrease in the annual average cloud coverage rate from 52.46% to 34.41% when compared with the MOD10A1 snow product. Additionally, there was an increase in snow coverage rate from 21.52% to 33.84%. This improvement in cloud removal greatly enhanced the time resolution of snow cover data without compromising the accuracy of snow identification.

Keywords: reconstruction; snow cover; cloud removal; snow grain size; machine learning

Citation: Zhu, L.; Ma, G.; Zhang, Y.; Wang, J.; Kan, X. Reconstruction of Snow Cover in Kaidu River Basin via Snow Grain Size Gap-Filling Based on Machine Learning. *Water* **2023**, *15*, 3726. https://doi.org/10.3390/w15213726

Academic Editor: Juraj Parajka

Received: 26 August 2023
Revised: 6 October 2023
Accepted: 24 October 2023
Published: 25 October 2023

1. Introduction

Snow cover monitoring with fine spatial–temporal resolution has important guiding significance for watershed-scale snow water resource management and sustainable utilization, natural disaster assessment, and early warning in pastoral areas. Spaceborne optical and microwave sensors are important platforms for snow monitoring. However, an optical remote sensing image is sensitive to cloud cover, and it is not possible to obtain information of snow cover under clouds. Exploring the cloud removal algorithm has great significance for restoring the snow condition under the cloud [1,2].

A large number of studies have shown that snow cover monitoring using optical remote sensing performs with high accuracy. The principle is that snow shows high reflectivity in visible and infrared bands but low reflectivity in the shortwave infrared (SWIR), which is different from other land covers [3]. The normalized difference snow index (NDSI) distinguishes snow pixels by measuring the relative magnitude of the reflectance difference between the visible band (GREEN) and SWIR. The moderate resolution imaging spectroradiometer (MODIS) mounted on Terra and Aqua satellites has provided worldwide stable daily snow cover data for nearly 20 years with its excellent spatial–temporal resolution and good stability [4–6]. However, the similar spectral reflection characteristics of snow and cloud in the visible and near-infrared bands, especially the similar spectral response of cirrus cloud and snow in the whole infrared spectrum, results in the misjudgment of snow and cloud [7,8]. In addition, there are still a large number of cloud pixels in the daily snow cover data of MODIS, which affects the spatial scope of snow monitoring, the accuracy of snow cover mapping, and the temporal resolution of snow, and limits the further application of optical remote sensing snowpack products. Therefore, many scholars have carried out a lot of research on cloud removal of snow remote sensing to improve the temporal resolution of snow cover [9,10]. At present, four major methods of cloud removal from snow cover using satellite remote sensing are summarized, the first is temporal filtering-based cloud removal, the second is spatial filtering-based cloud removal, the third is the cloud removal algorithm based on multi-sensor fusion, and the fourth is the cloud removal using snowline elevation.

Gafuov and Bárdossy proposed a cloud removal algorithm from MODIS snow cover products based on temporal filtering [11,12], which assumes that the snow will not melt quickly in a short time, but the clouds will move quickly. By synthesizing the snow products from Terra and Aqua, the moving clouds are filtered to maximize the snow cover extent. Cloud removal based on temporal filtering can be deduced and calculated without other satellite or ground auxiliary data. For areas lacking multi-source satellite data or relevant geographical parameters, the real snow cover extent can also be calculated.

The aforementioned studies have proved that selecting the appropriate time window and synthetic days to conduct temporal filtering cloud removal can obtain snow cover recognition results with high accuracy, but the step of "Filtering cloud removal during snow accumulation and melting" will cause many false or missed judgments of snow pixels that are fragmented in time series, thus reducing the accuracy of snow cover recognition. Furthermore, the appropriate time window and synthetic days are uncertain and different for different regions and periods. Therefore, large errors will occur when this method is applied to regions with a wide spatial range, strong snow heterogeneity, and long-time series snow data sets. There will be large errors, and the applicability of the algorithm will be greatly reduced. The core of the spatial filtering cloud removal method is to select cloud-free pixels in the spatial neighborhood to estimate the ground coverage under the cloud. In the spatial filtering cloud removal strategy proposed by Gafurov and Bárdossy, there are the "Near four-pixel method", "Near eight-pixel method", and other methods [6]. The cloud removal algorithm based on spatial filtering can be deduced and calculated without other satellite data. Simultaneously, in practical application, although the amount of cloud removal of this algorithm is less, it also maintains the lowest error.

The cloud removal method based on temporal filtering and spatial filtering mainly uses the temporal and spatial variation of snow cover of the same optical remote sensor to extract the ground information under the cloud. In contrast, another cloud removal method based on multi-source data fusion uses complementary information between different data sources, such as optical remote sensing observation, microwave remote sensing observation, and station observation [13–16]. But the distribution and number of meteorological stations limit the prediction ability of this method for snow reconstruction. However, the above research can only qualitatively infer the distribution of snow cover under clouds, lacking quantitative characterization of snow cover parameters under clouds.

In recent years, researchers have been striving to use spatiotemporal information for one-step cloud removal algorithm research. Xia et al. [17] introduced variational interpolation to construct a three-dimensional implicit function containing five consecutive day data, which can easily obtain the shape of the snow cover boundary. The cloud removal method proposed by Poggio and Gimona [18] is a combination of a generalized additive model (GAM) and a geostatistical spatiotemporal model. The multidimensional spatiotemporal GAM models binary variables, and geostatistical kriging methods are used to explain spatial details. This method utilizes auxiliary data such as surface temperature, land cover, and soil type to effectively simulate the spatiotemporal correlation of snow cover and can achieve satisfactory reconstruction accuracy even under high cloud cover. The adaptive spatiotemporal weighting method [19] estimates the snow cover of cloud pixels by combining adaptive weights based on the probability of snow cover in space and time, which can completely remove the cloud layer. Huang et al. [20] established a hidden Markov random field framework to remove cloud pixels from MODIS binary snow cover data. This method effectively utilizes spatiotemporal information and achieves an overall accuracy of 88.0% for the restored snow cover range under the cloud. Additionally, the conditional probability interpolation method [13] can effectively calculate the conditional probability of snow pixels being covered by clouds and meteorological data to remove clouds, but this method has limited capacity for removing clouds in areas with few in situ observations. Furthermore, Chen [21] proposes a conditional probability interpolation method based on a space–time cube (STCPI), which takes the conditional probability as the weight of the space–time neighborhood pixels to calculate the snow probability of the cloud pixels, and then the snow condition of the cloud pixels can be recovered by the snow probability. However, existing one-step cloud removal algorithms that utilize spatiotemporal information have significant time computational costs and require multiple auxiliary data, which to some extent limits the application of the algorithm. In addition, the progress of machine learning and deep learning technology has also led to new developments in remote sensing snow cover mapping. Luan et al. proposed an m-day dynamic training strategy, which divides a long-term snow cover mapping task into multiple short-term tasks with consecutive m days and reduces the problems caused by changes in snow cover over time. This strategy is applied to random forest models for binary snow cover (BSC) mapping and fractional snow cover (FSC) mapping, achieving higher accuracy than other training strategies [22]. A new algorithm based on a machine learning method was designed to improve FSC retrieval from brightness temperature, considering other auxiliary information, including soil property, land cover, geography information, and the overall accuracy of the above method reached 0.88 [23]. Guo [24] trained the DeepLap v3+ model using a transfer learning strategy to overcome the computational time and resource consumption of deep learning models. The feasibility and effectiveness of automatically extracting snow cover were demonstrated on high-resolution remote sensing images. Hu trained random forest models using combinations of multispectral bands and normalized difference indices and generated sub-meter and meter-level snow maps based on very-high-resolution images [25]. Liu introduced a highly accurate snow map acquired by unmanned aerial vehicles as a reference to machine learning models, which significantly improved the MODIS fractional snow cover mapping accuracy [26]. Yang et al. are committed to designing a cloud snow recognition model based on a lightweight feature map attention network (Lw-fmaNet) to ensure the performance and accuracy of the cloud snow recognition model [27].

We propose a new strategy for reconstructing snow cover under clouds to achieve the following objectives: (a) exploring the correlation between snow particle size and geographic and meteorological information at the watershed scale, and (b) improving the temporal resolution of snow cover at the watershed scale by filling gaps in SGS and achieving snow mapping for the entire watershed. The organization of the entire article is as follows. After introducing the study area and data preprocessing in Section 2, the methodology for cloud removal and snow cover reconstruction is expounded in Section 3. The results of accuracy verification and mapping will be presented in Section 4. Ultimately, a summary of the research will be described in Section 5.

2. Study Area and Data Preprocessing

2.1. Study Area

The Kaidu River originates from the southern slope of the Central Tianshan Mountain in the northwestern region of China (Figure 1c), traverses through the Small Urdus Basin to the Bayanbulak hydrological station, then turns towards the southeast and passes through Husitaixili and the Big Urdus Basin and along a canyon, and finally flows through Yanqi basin into Bosten Lake. The KRB through which it flows is an important sub-basin of the Tarim River Basin and covers an area of approximately 1.9 × 10^4 km^2, with an elevation ranging from 1348 to 4709 m above sea level (m a.s.l). The KRB is located in the hinterland of Eurasia and features a typical warm temperate desert climate. The daily precipitation, temperature, and snow depth data were obtained from the Bayanbulak meteorological station at an elevation of 2458.9 m a.s.l. and were provided by the National Meteorological Information Center of the China Meteorological Administration (http://data.cma.cn/en (accessed on 26 August 2023)). The mean monthly precipitation, temperature, and snow depth for the normal period of 2000–2019 for Bayanbulak are presented in Figure 1a. In this period, the mean annual precipitation was 308.84 mm. The precipitation was mainly concentrated in summer (June, July, and August). More than two-thirds of the precipitation occurred in this season. Because the mean monthly T_{min} was above 0 °C, precipitation was liquid precipitation. However, the sum of the mean monthly precipitation was only 32.1 mm from October to March. Because the mean monthly T_{max} was below 0 °C, solid precipitation was mainly concentrated in this period. The snow depth of ground observations was measured manually with a wooden ruler at 8 o'clock every day when more than half of the ground in the field of view around the meteorological station was covered by snow. The measurements were made three times, and the distance between the three measurements was more than 10 m. The measured value was accurate to 0.1 cm. The final snow depth at the station was determined as the average of the three measurements, and an average snow depth of less than 1 cm was recorded as 0. Figure 1b shows the Bayanbulak meteorological station time series of daily snow depth from January 2000 to December 2019. The greatest values of snow depth (35 cm) were found in the winter of 2011. A monthly histogram of the mean snow depth from January 2000 to December 2019 is shown in the inset figure. There was snow in every month except July. The mean monthly snow depth increased from 0.06 to 10.62 cm during August to February and decreased from 7.1 to 0.05 cm during March to June [28,29].

2.2. Satellite Remote Sensing Data

SGS is one of the important parameters to describe the microphysical properties of snow [30,31]. SGS indicates the energy balance state of the snow and is a major input parameter for the snowmelt runoff model and climate model [32]. In addition, the variation of SGS helps to discriminate snow melt area, pure ice area, and snow mound, which is important for the retrieval of snow depth and snow density at the watershed scale. Satellite remote sensing data in the visible (0.4~0.7 μm), near-infrared (0.7~1.4 μm), and short-wave infrared (1.4~2.5 μm) have been widely used for SGS estimation [33–36].

In this study, the SGS was selected as an important parameter to characterize snow, and SGS under a cloud was estimated by a nonlinear relationship between the distribution of SGS and geographic data. Eventually, the snow information reconstruction under the cloud is implemented. The SGS data used in this study were estimated based on the snow grain size and pollution (SGSP) amount algorithm with careful topographic correction in our previous study [32] for the period of 2000 to 2020 with a spatial resolution of 500 m. The SGS data is calculated and exported based on the Google Earth Engine (GEE) platform after deploying the above algorithm. The specific properties are shown in Table 1.

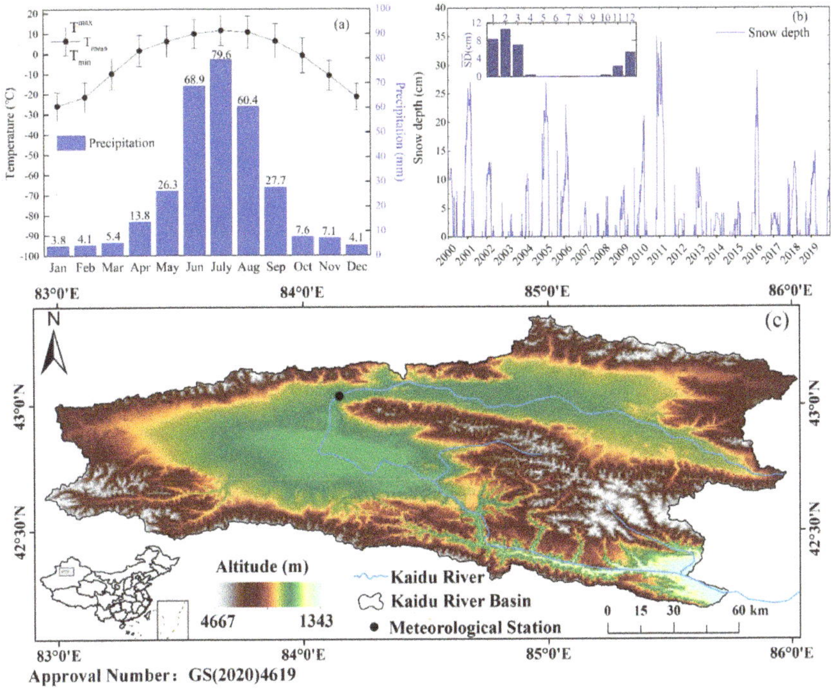

Figure 1. (**a**) The monthly mean precipitation and temperature at the Bayanbulak meteorological station from January 2000 to December 2019; (**b**) Bayanbulak meteorological station time series of daily snow depth from January 2000 to December 2019, with a monthly histogram of mean snow depth from January 2000 to December 2019 in the inset figure; (**c**) geographical location and topographic relief of the KRB. The different background colors indicate different elevations. The location of the Bayanbulak meteorological station and the river system are also shown.

Table 1. Attributes of SGS.

Properties	Variable Description
Numerical Range	0–1000 μm (0: Snow-free, Others: Snow)
Data Format	GeoTiff
Space Scope	82.33° E~88.12° E; 42.03° N~44.37° N
Spatial Resolution	0.005° (500 m)
Time Range	27 February 2000~7 May 2020
Time Resolution	Daily

The SGS data used in this study have two channels, of which channel 1 is the SGS data retrieval by SGSP and channel 2 is the image classification data. Combining these two channels, the details of the channels are shown in Table 2.

Table 2. Interpretation of SGS data channel.

Channel 1	Channel 2	Image Classification
0	0	Cloud
0	1	Snow-free
1 1000 μm	0	Snow-cover
1	1	None

2.3. Geographic Data

Geographic data includes topographic features such as altitude, slope, aspect, and land cover used to characterize the underlying surface of the snowpack. Topography will affect the accumulation and melting rate of snow, which has a significant impact on the redistribution of snow and the variation of SGS. The digital elevation model (DEM) used in this study originates from the 004 version of the Shuttle Radar Topography Mission (SRTM), with a spatial resolution of 90 m. It is drawn based on the Interferometric Radar of the United States Geological Survey (USGS) and can be downloaded from http://srtm.csi.cgiar.org/index.asp (accessed on 26 August 2023). In addition, topographic features such as slope and aspect are calculated from DEM through ArcGIS.

The difference in land cover will also change the size of the snow grain [37]. MODIS land cover (MCD12Q1) version 6 products are updated year by year for global land cover type. The product has a spatial resolution of 500 m and a period from 2001 to the present. According to the MCD12Q1 reclassification strategy for Western China [38], the land cover in the KRB is reclassified into five types: water, forest, grassland, farmland, and bare land. As shown in Figure 2c, the major land covers in the KRB are grassland and bare land, of which grassland accounts for 84.63% and the area of bare land accounts for 15.36%.

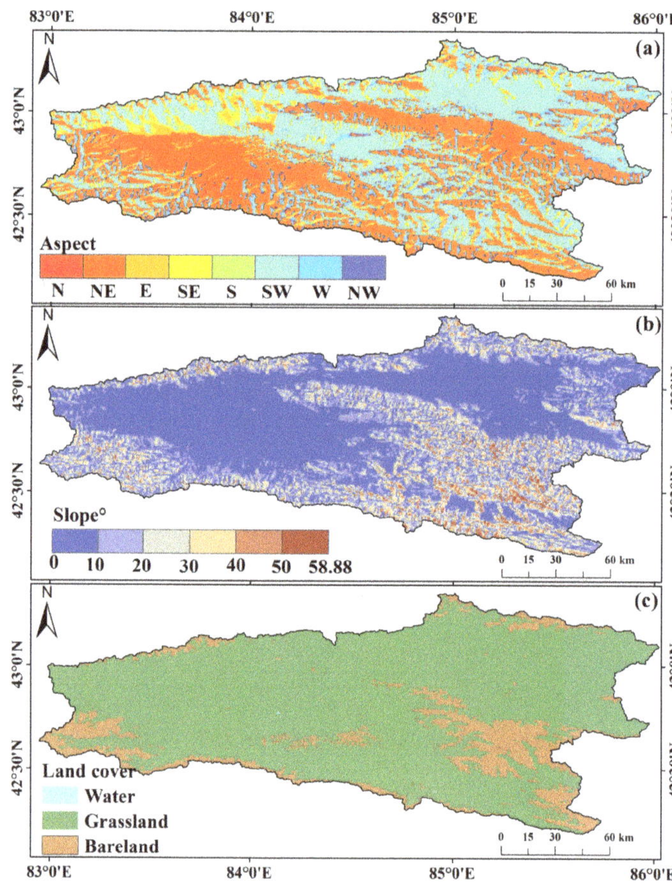

Figure 2. Slope (**a**), aspect (**b**), and land cover (**c**) in the KRB.

2.4. Ground Observation

The daily snow depth data from 2000 to 2020 observed at Bayanbulak station (Figure 1) of the China Meteorological Administration was used as the "ground truth" values for assessing the accuracy of the gridded snow cover reconstruction data [39,40]. The in situ snow depth measurements are the most accurate record of snow cover, when the snow depth is greater than or equal to 1 cm, the site is snowy, and vice versa, the site is snow-free. Thus, they are widely used for evaluating not only satellite-based snow depth products [38] but also snow cover products [41].

3. Methodology

3.1. Construction of Space–Time Extra Trees Model

Among the current mainstream machine learning algorithms, the random forest (RF) has outstanding performance in both regression and classification. A RF can effectively handle thousands of input samples with high-dimensional features without dimensionality reduction. It is also able to evaluate the importance of each input feature on the objective function. In the algorithm execution, an unbiased estimate of the internally generated error is obtained and a high tolerance for certain missing data is obtained. However, extra trees (ETs) make full use of all samples compared to the Bagging strategy applied with the RF, only the features are randomly selected due to the random splitting, leading to a better regression result than the RF [42]. Also, a RF is used to get the best splitting attributes within a random subset, but ET is used to get the splitting values completely randomly in the global data.

An ET is an integrated machine-learning method developed from an RF [43]. The algorithm is denoted by $\{T(V, X, D)\}$, where T is the final classifier model, D is the sample set, and V is the number of base classifiers. Each classifier produces a prediction based on the input samples. The execution steps of the ET are shown below [44]:

Step 1: Sample selection: Given the original data sample set D, the number of samples S, and the number of features W. In the ET classification model, each base classifier is trained using the full set of samples.

Step 2: Feature selection: The Base classifier is generated from the Classification and Regression Tree (CART) decision tree. At each node splitting, m features are randomly selected from W features, the optimal attribute is selected for each node for node splitting, and the splitting process is not pruned. Step 2 is performed iteratively on the subset of data generated by splitting until a decision tree is generated.

Step 3: Construction of additional trees: Create additional trees and repeat Steps 1 and 2 for V times to generate V decision trees and ET.

Step 4: Regression of result: Test results are generated from test samples based on designed ET, and the prediction results of all basic classifiers are counted. The final result is determined according to the average value of all decision tree outputs.

ET is the random bifurcation of rows and columns of data, which will lead to the generalization ability of ET being stronger than that of RF. At the same time, each regression tree in the ET makes full use of all the training samples and randomly selects the bifurcation attributes on the node bifurcation, which enhances the randomness of the node splitting of the base classifier.

In this study, an SGS filling model based on ET is constructed to implement the reconstruction of snow cover under clouds at the watershed scale, as shown in Figure 3. Geographic elements such as altitude, slope, aspect, and land cover are resampled to 500 m resolution using nearest-neighbor sampling technology as input, and the SGS under a clear sky is used as a sample label. The nonlinear mapping relationship between multi-source data is constructed based on ET.

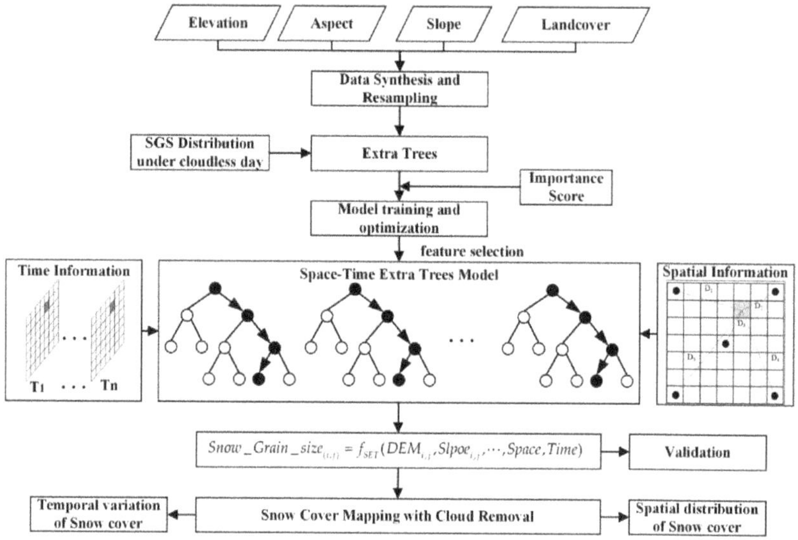

Figure 3. Flow chart of space–time extra tree model.

In the process of model training, importance scores given by the ET can help select the input factors with high importance, optimize the feature selection process of the model, reduce the amount of parameter calculation, and overcome the overfitting caused by parameter redundancy. Simultaneously, considering the spatial distribution information characterizing the SGS and the variation with time, the input data of two-dimensional characterized temporal information and two-dimensional characterized spatial information are designed when training the ET model, and the data about the spatial and temporal information are elaborated as shown in Section 3.2.

3.2. Design of Temporal and Spatial Dimensional Information

In many previous studies, especially for parameter retrieval at large spatial scales, latitude and longitude were introduced into the model as spatial information parameters to characterize the position of a grid in the whole region. However, it is difficult to accurately quantify the spatial information using latitude and longitude because of the proximity and fine resolution of the grids in the watershed-scale parameter retrieval. In this study, the KRB features a topographic landscape with mountains surrounding the basin, and the Elpin Mountains divide the basin into the Big Urdus Basin and Small Urdus Basins on the east and west sides, forming a typical geomorphic feature of mountains surrounding and blocking the basin. Wei et al. calculated the Haversine distance from each rater point to the upper left corner of the rectangular study area as D_1, and analogously, the distance to the upper right corner as D_2, the distance to the lower left corner as D_3, the distance to the lower right corner as D_4, and the distance to the center of the matrix noted as D_5, to improve the representation of the model for spatial information [45]. However, the Haversine distance is more suitable for the expression of large-scale spatial information. The study area of the KRB is characterized by high elevation around and low elevation in the middle. We divided the KRB into four quadrants and selected the highest elevation positions of the mountains in four directions in the basin, which are located at (83.7943° E, 43.12466° N), (85.43821° E, 43.23695° N), (83.02624° E, 42.68449° N), and (85.33491° E, 42.54525° N). The Euclidean distance from each grid to the four highest elevation positions was calculated and denoted as D_{1-1}, D_{1-2}, D_{1-3}, and D_{1-4}, respectively. $D_1 = D_{1-1} + D_{1-2} + D_{1-3} + D_{1-4}$ characterizes the weighted sum of each grid to the highest elevation position in the watershed, as shown by the green line in Figure 4.

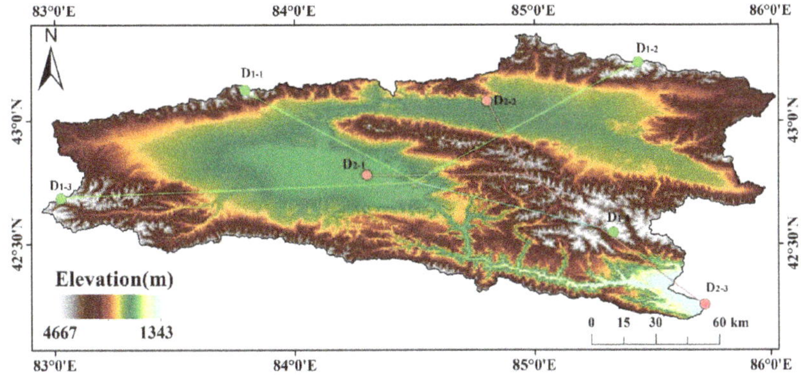

Figure 4. Construction of the spatial distance factor; the weighted sum of the four green segments is D_1, and the weighted sum of the three pink segments is D_2.

Following a similar idea, the lowest elevation positions in Big Urdus Basin, Small Urdus Basin, and Yanqi Basin, located at (84.30185° E, 42.77881° N), (84.80490° E, 43.07975° N), and (85.72118° E, 42.2533° N), respectively, were selected, and the Euclidean distances from each grid to the lowest positions were calculated and denoted as D_{2-1}, D_{2-2}, and D_{2-3}. $D_2 = D_{2-1} + D_{2-2} + D_{2-3}$ characterizes the weighted sum of each grid to the lowest elevation position in the watershed, as shown by the pink line in Figure 4. The aforementioned two data D_1 and D_2 are constructed to improve the model's representation of spatial information.

Theoretically, the hydrological year is based on the Earth's hydrological cycle, which begins at the point of return of runoff, which is usually the beginning of the flood season and the end of the dry season. There is some variability in the start and end times in different regions and between years. In the research of snowpack phenology, the hydrological year can also be defined by using the day of onset of snowpack accumulation and the day of final melt as a boundary [46]. In this study, a hydrological year from 1 Sept to 31 Aug of the following year was delineated, taking into account the snowpack characteristics and the characteristics of the study area. The hydrological year is divided into four seasons: spring (March to May), summer (June to August), autumn (September to November), and winter (December to February). The snow cover days (SCDs) of the hydrological year in which the single-view data are located are used as one-dimensional, temporal information, and the SCDs characterize the number of days a grid is covered by snow in a hydrological year. Areas with high SCDs generally have lower temperatures, more snowfall, and more abrupt variability in SGS.

The principle of the second-dimensional temporal information construction is as follows: the grain size of new snow is tiny when in contact with the ground, and the distribution of its grain size shows obvious distribution characteristics that change with altitude and slope. Therefore, the second-dimensional data characterizes the number of consecutive days with snow cover (Snow Duration Index, SDI) at the current moment of a grid, with daily resolution. Compared with the SCDs, the SDI can better reflect the snow status at the current moment. As shown in Figure 5, it represents the state presented by a grid on the time series, where white indicates a snow-free grid and blue indicates a snow grid. Based on the duration of the presence of snow on the time series, the SDI constructed from the 1st grid is $[0, 1, 0, 1, 2, 3, 0, \ldots, 0, 1, 2]$, the SDI constructed from the 2nd grid is $[1, 2, 3, 0, 0, 1, 0, \ldots, 1, 2, 0]$, the 1, 2, 3, and 4 in arrays indicate that the snow has existed for 1, 2, 3, and 4 days, respectively, at the current moment.

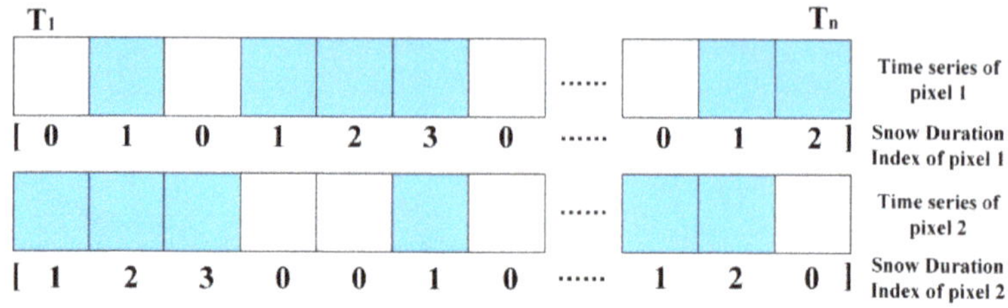

Figure 5. Construction of snow duration index. White indicates snow-free at this time point, blue indicates snow-covered at this time point, and the constructed index values are shown in square brackets.

3.3. Applicability Evaluation and Factor Optimization of the Model

Daily SGS data of the KRB were generated in batches based on the asymptotic radiative transfer model in the GEE for a long series. However, for the SGS data with too much cloud, the mapping relationship between geographic, spatial–temporal information, and SGS cannot be fitted better due to the limited effective training data. Therefore, the training set and test set are divided according to the ratio of 8:2, and the experiments are conducted with different data missing rates (i.e., cloud percentage in the watershed). It can be seen that in Figure 6 the test error increases abruptly when the cloud percentage is greater than 75%. The data missing rate increases to a certain threshold, resulting in too few effective training samples and causing underfitting in model training. Thus, SGS data with cloud coverage below 70% are selected for snow reconstruction in this study.

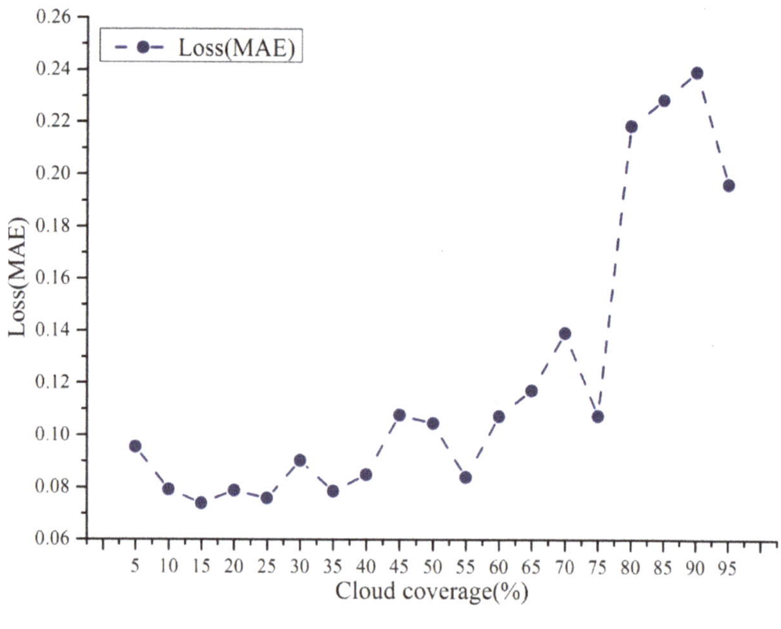

Figure 6. Loss variation of the model training under cloud coverage change (i.e., Data Loss Rate).

Under the premise of determining the missing data rate applied to the model, the corresponding altitude, slope, aspect, land cover, spatial dimension data D_1 and D_2, and temporal dimension data of SCDs and SDI are extracted according to the latitude and longitude, and are used to construct an SGS filling model. The points with an SGS of zero, (i.e., no snow) are also added to balance the snow cover and snow-free surfaces in the model.

In the model training phase, the importance scores are used to evaluate the contributions of the input factors to the results and to optimize the input factors. Since the model is trained for daily images and fills in the missing SGS information to achieve snow reconstruction, the difference in snow status and cloud percentage due to the environmental changes will make the importance scores of input factors ranked differently on each day. In this study, the ranking of the mean importance scores obtained from model training for consecutive years was calculated, as shown in Figure 7. It can be seen that altitude, as the most significant topographic element, is the most important factor influencing the retrieval results, with a score of 0.192. The importance scores of the SDI and SCDs are 0.14126 and 0.13605, respectively, indicating that the present time of the grid at the current moment and the distribution of snow accumulation throughout the hydrological year show a great role in the filling of SGS. While the importance scores of aspects, D_1 and land cover were all closer. The lowest importance score is the slope, which is only 0.077, one order of magnitude behind other input factors. Therefore, the importance scores of the hydrological year scale are combined, and altitude, SDI, SCD, D_2, aspect, D_1, and land cover are finally selected as the inputs of the model for retrieval of SGS under the cloud layer, to realize snow reconstruction. Altitude, aspect, and land cover are geographic information inputs that come into direct contact with snow cover, thereby affecting the SGS through surface temperature conduction, gravity accumulation caused by terrain, and the amount of solar radiation received from different aspects. D_1 and D_2 further characterize the spatiotemporal properties of snow particles within the watershed, which helps to improve the accuracy of SGS estimation at the watershed scale. SDI and SCD characterize the phenology of snow cover on short and annual time scales, respectively, especially SDI, which is closely related to the evolution and size of the snow grain.

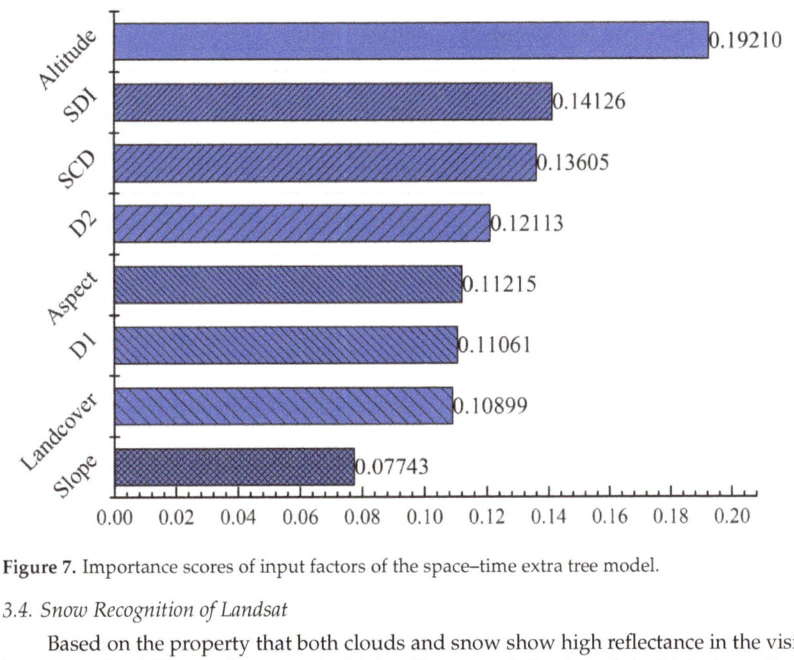

Figure 7. Importance scores of input factors of the space–time extra tree model.

3.4. Snow Recognition of Landsat

Based on the property that both clouds and snow show high reflectance in the visible band, and the difference between the high reflectance of clouds and the high absorption of snow in the short-wave infrared band, the SNOWMAP algorithm [47] was used to identify the snow cover in the Landsat images. In this study, the Normalized Difference Snow Index

(NDSI) is calculated in the GEE platform for the green band (band 3) and the short infrared band (band 6) of Landsat-OLI images that have completed radiometric calibration and atmospheric correction, and the threshold of the NDSI for snow identification is set greater or equal to 0.4. The calculation method of the NDSI is shown in Equation (1):

$$\text{NDSI} = \frac{\text{band 3} - \text{band 6}}{\text{band 3} + \text{band 6}} \tag{1}$$

Given the low reflectance of the water body in both visible and short-wave infrared bands, the threshold of band 5 is set as greater than 0.11 as a way to eliminate the interference of the water body. The combined criterion of NDSI $\geq$ 0.4 and band 5 $>$ 0.11 can achieve snow identification at 30 m resolution. In the snow binary map, 1 denotes a snow element and 0 denotes a snow-free element.

3.5. Metrics for Evaluating the Accuracy of Snow after Cloud Removal

The accuracy evaluation of snow reconstruction is divided into two parts. The first is the accuracy evaluation of the SGS estimated based on the machine learning model, and the second is the accuracy evaluation of the snow reconstruction results. The root mean squared error (RMSE) and mean absolute error (MAE) between the predicted SGS and the measured SGS are evaluated employing ten-fold cross-validation, and the above indexes are calculated as shown in Equations (2) and (3):

$$\text{RMSE} = \sqrt{\frac{1}{m}\sum_{i=1}^{m}\left(y_i - \hat{y}_i\right)^2} \tag{2}$$

$$\text{MAE} = \frac{1}{m}\sum_{i=1}^{m}\left|y_i - \hat{y}_i\right| \tag{3}$$

In the accuracy evaluation system of snow reconstruction, the measured snow depth at the meteorological station is taken as the ground truth, and the ground is judged to be snowy when the snow depth is greater than 1 cm, and vice versa. Based on the aforementioned criteria, the results of snow reconstruction have the following four cases: True Positive (TP), True Negative (TN), False Positive (FP), and False Negative (FN), and the detailed definitions of TP, TN, FP, and FN are described in Table 3. TP refers to both snow reconstruction data and ground observation being judged as snow, and TN refers to both being snow-free. FP means that the snow reconstruction data shows snow, while ground observation is snow-free, which generally occurs when the reconstruction misidentifies cirrus clouds as snow, and FN means that the ground observation shows snow, while the snow reconstruction is snow-free, which belongs to snow omission.

Table 3. Four categories of snow reconstruction compared to ground-based observations.

Ground True / Snow Reconstruction	Snow	Snow-Free
Snow	TP	FP
Snow-free	FN	TN

Based on the four categories of snow reconstruction data compared with ground observations, a series of performance metrics were introduced to evaluate the accuracy of snow reconstruction by the algorithm in this study and to compare the accuracy with existing snow cover products from MODIS. Based on the ground-based meteorological station observations in the study area, the values of TP, TN, FP, and FN are counted for the period of hydrological years from 2000 to 2020, and the four perspectives of overall accuracy (OA), precision, recall and the combined performance index of F1-score are used to evaluate the performance of snow reconstruction data. The specific formulas for calculating the above four types of indicators are shown in Equations (4)–(7):

$$\text{Overall Accuracy(OA)} = \frac{\text{TP} + \text{TN}}{\text{TP} + \text{TN} + \text{FP} + \text{FN}} \quad (4)$$

$$\text{Precision} = \frac{\text{TP}}{\text{TP} + \text{FP}} \quad (5)$$

$$\text{Recall} = \frac{\text{TP}}{\text{TP} + \text{FN}} \quad (6)$$

$$\text{F1} - \text{Score} = 2 \times \frac{\text{Precision} \times \text{Recall}}{\text{Precision} + \text{Recall}} \quad (7)$$

Using the measured snow depth from ground-based meteorological stations as the ground truth to assess the effect of snow reconstruction will result in the problem of underrepresentation due to the sparse stations, taking the KRB as an example, with 20,507 km^2, while there is only one meteorological station in the basin, Bayanbulak (51,542), with an elevation of 2458 m. As a result, there is a lack of ground truth to evaluate the effect of snow reconstruction in other areas of the basin, especially in high mountain regions. Therefore, after evaluating the accuracy by ground truth, 30 m resolution snow cover derived from Landsat was also used to assess the 500 m snow reconstruction data. The Kappa coefficient is used to calculate the image agreement between snow cover derived from Landsat and snow reconstruction data derived from MODIS, and the formula is shown in Equations (8)–(10):

$$K = \frac{P_o - P_c}{1 - P_c} \quad (8)$$

$$P_o = s/n \quad (9)$$

$$P_c = (a_1 \times b_1 + a_0 \times b_0)/(n \times n) \quad (10)$$

where P_o is the actual consistency rate and P_c is the theoretical consistency rate. In Equations (8) and (9), the total pixels of the remote sensing image is n, the number of snow pixels in the Landsat image representing the real situation of the ground is a_1, and the number of snow-free pixels is a_0. In the corresponding snow reconstruction data, the number of snow pixels is b_1, and the number of snow-free pixels is b_0, s represents the same number of the corresponding pixels in the two images.

According to the literature [48], the Kappa test can be used to represent different levels of consistency using five sets of classifications. Table 4 illustrates the level of consistency of the two images corresponding to Kappa in different intervals.

Table 4. Comparison of image consistency levels based on Kappa coefficients.

The Value Interval for Kappa	The Consistency Level of the Image
$0 < \text{Kappa} \leq 0.20$	Very low consistency
$0.20 < \text{Kappa} \leq 0.40$	General consistency
$0.40 < \text{Kappa} \leq 0.60$	Medium consistency
$0.60 < \text{Kappa} \leq 0.80$	High consistency
$0.80 < \text{Kappa} \leq 1$	Almost complete consistency

4. Results

4.1. Mapping of Snow Cover Reconstruction

The error of SGS estimation will also indirectly affect the result of snow cover reconstruction under the cloud. Therefore, some known SGS data were randomly masked in the model training stage, and the size of the masked snow grain data will be estimated and compared with the original truth. Table 5 lists the RMSE and MAE of SGS based on the space–time ET model and other mainstream machine learning methods. It can be seen that the space–time ET model proposed in this study performs the best result, with the lowest RMSE and MAE of 52.751 μm and 40.109 μm, respectively.

Table 5. Accuracy comparison of SGS filled by different methods.

Methods	RMSE (μm)	MAE (μm)
Classification and Regression Tree, CART	68.048	49.934
K-Nearest Neighbor, KNN	57.108	42.516
Random Forest, RF	55.822	41.692
Ridge Regression, RR	57.054	45.048
Support Vector Regression, SVR	56.710	42.880
Denoising Autoencoder Artificial Neural Network, DAANN	54.186	40.852
Space–Time Extra Randomized Trees	52.751	40.109

Based on SGS filling, the minimum value of SGS in the image of that day is taken as the threshold for snow discrimination. If the estimation result is greater than the threshold, it is judged as snow, otherwise, it is judged as snow-free. Two environments of "more snow and less bare land" and "less snow and more bare land" are selected to compare the accuracy of snow cover reconstruction. First, the reconstruction of snow under clouds in the KRB on 6 March 2014, is taken as an example. It can be seen from Figure 8c that the cloud accounts for about 45.62% of the KRB, especially in the mountain area, a large number of pixels are blocked by clouds, and the area of snow accounts for about 47.8%. The snow product is directly used for the estimation of snow areas, which will lead to large errors. After filling the SGS, the complete reconstruction of the snow cover under clouds in the watershed scale is realized, as shown in Figure 8d, and the proportion of snow area has reached 81.5%, increasing the proportion by 33.79%. The kappa coefficient is calculated to evaluate the consistency between the reconstructed snow cover and the Landsat snow cover map. The black part in Figure 8a,b is the area that has not been scanned by Landsat and does not participate in the calculation of consistency. The Kappa coefficient between Landsat's snow cover map and the reconstructed snow cover map is 32.84%, which is generally consistent but compared with the extremely low consistency of 17.79%, it has been improved by 15.05%. In addition, the scores of OA and F1-score after snow reconstruction increased by 16.88% and 14.5%, respectively, compared with those before cloud removal. The specific accuracy data are listed in Table 6. The snow reconstruction on 21 April 2016, is shown in Figure 8f–j as the second case, with 60.73% cloud coverage. The Big Urdus Basin and Elpin Mountains are covered by generous continuous clouds, and the traditional spatiotemporal filtering makes it difficult to achieve snow reconstruction under a cloud with high accuracy. Based on the SGS filling, the spatial distribution of snow in the Elpin Mountains and the southwest of the basin is restored, and the proportion of snow cover has increased from 7.42% to 8.77%, as shown in Figure 8i, which has good consistency with the Landsat 30 m resolution snow map in Figure 8g. The kappa coefficient of snow reconstruction results and Landsat snow map increased from 6.83% before cloud removal to 86.74%, with almost complete consistency, and the OA index and F1-score also increased by 50.19% and 38.39%, respectively. In addition, this study also selected the snow cover of 2 November 2017 as Case 3. The cloud coverage rate in the KRB region is only 13.08%, which effectively achieves the reconstruction of snow cover in the watershed under low cloud conditions. From Figure 8n, it can be seen that the mountains in the eastern part of the KRB have sustained cloud cover. After filling with SGS, it is estimated that the reconstruction of snow cover on the eastern mountain of the watershed has been achieved relatively well. Compared with the snow cover map of Landsat in Figure 8l, it can better match the spatial details. The kappa coefficients of both, increased from 69.48% to 85.54% after reconstruction. This study also added widely recognized snow cover extent (SCE) data [49] for spatial comparison with the reconstructed snow cover, as shown in Figure 8e,j,o. Taking the Landsat snow cover map in Figure 8b,g,l as the true value, it can be seen that the snow cover reconstruction results based on SGS filling have a more detailed representation in complex mountainous terrain compared to Hao's SCE and can more accurately depict the distribution characteristics of snow cover in mountainous environments.

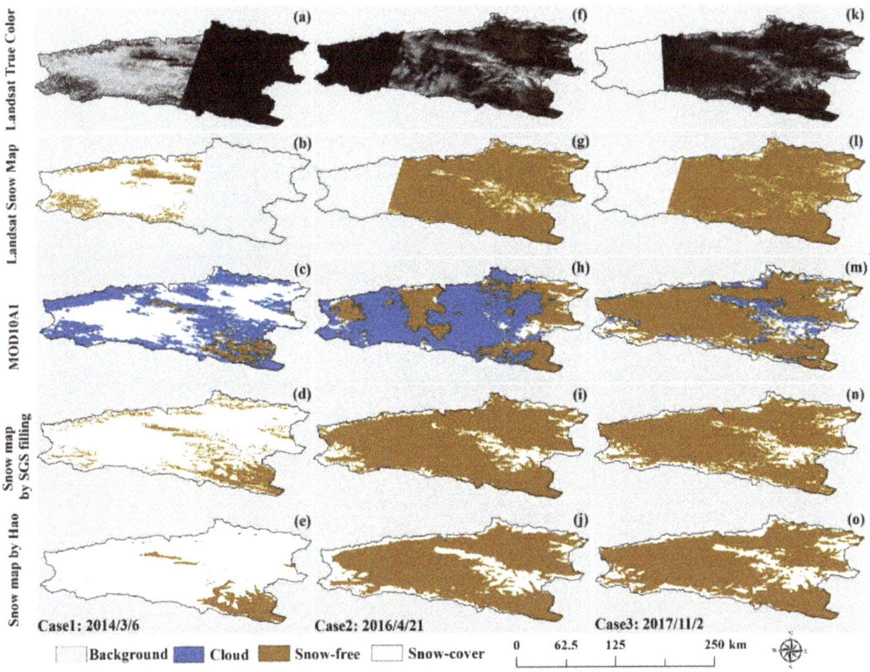

Figure 8. Comparison of cloud removal effect based on SGS filling, (**a**,**f**,**k**): Landsat true color map; (**b**,**g**,**l**): Landsat snow cover map; (**c**,**h**,**m**): MODIS snow cover product; (**d**,**i**,**n**): snow cover by SGS filling; and (**e**,**j**,**o**): snow map by Hao.

Table 6. Accuracy comparison between Case 1, Case 2, and Case 3 before as well as after snow cover reconstruction.

Accuracy (%)	Case 1: 6 March 2014		Case 2: 21 April 2016		Case 3: 2 November 2017	
	Before Reconstruction	After Reconstruction	Before Reconstruction	After Reconstruction	Before Reconstruction	After Reconstruction
OA	64.15	81.03	38.26	88.45	77.46	85.23
Precision	65.28	82.86	4.38	35.74	82.48	88.53
Recall	85.79	95.28	34.24	65.16	81.24	86.68
F1-Score	74.14	88.64	7.77	46.16	81.86	87.60
Kappa	17.79	32.84	6.83	86.74	81.46	86.93

Based on the above three cases, whether in the environment of "more snow and less bare land" or "less snow and more bare land", and under different cloud coverage rates, the snow cover reconstruction based on the SGS filling has achieved good performance.

4.2. Accuracy of Snow Reconstruction and Cloud Coverage Variations

Cloud removal based on SGS filling can realize complete snow reconstruction under clouds with cloud coverage of less than 70% and maintain high accuracy. The method greatly improves the cloud removal rate on the hydrological annual scale. Figure 9 illustrates the comparison between the snow cover days of MODIS products and reconstructed snow cover days on the hydrological annual scale from 2000 to 2020. The annual snow cover days of MODIS products are shown in the first and third columns, and the reconstructed snow cover days are in the second and fourth columns. It can be seen from the figure that the mountains around the basin and the Elpin Mountains in the middle are the areas with the most significant increase in snow cover days. As the region with the most abundant snow resources in the basin, the good performance of cloud removal in mountains is helpful to the accurate calculation of snow resources. According to the comprehensive analysis in Table 7, the number of days in the KRB where the daily cloud coverage is less than 70% ranges from 215 to 256 during the period of 2000~2020, and the mean days that can be applied to the SGS filling is 243.65, accounting for 66.75% of the whole year. Bayanbulak (51,542), the only meteorological station in the KRB, has an average annual snow cover of 115.64 days from 2000 to 2020, while the average annual snow cover days calculated based on MODIS Snow products are only 51.3 days, which is only 44.36% of the true observation data. The method in this study only reconstructs the snow cover under a cloud for an average of 66.75% of the data every hydrological year and increases the average annual snow cover days of Bayanbulak based on remote sensing observation to 84.7 days, accounting for 73.24% of the true data. In addition, the difference between the reconstructed SCD after cloud removal and the SCD before cloud removal is shown in Figure 10. The SCD difference on the mountains around the KRB and the southeast side of the watershed is the largest, indicating that snow mapping in complex mountainous environments is prone to interference from cloud cover. However, the method used in this study effectively achieves snow removal and reconstruction in complex mountainous environments.

Table 7. Comparison of SCD before and after snow reconstruction at the hydrological year scale.

Hydrological Year	Trainable Days	Un-Trainable Days	SCD of MODIS	SCD of Reconstructed Snow	SCD of the Station
2000–2001	216	149	46	104	159
2001–2002	244	121	22	83	111
2002–2003	234	131	3	3	16
2003–2004	252	114	44	72	78
2004–2005	229	136	59	103	144
2005–2006	246	119	58	108	148
2006–2007	243	122	29	40	49
2007–2008	248	118	31	39	29
2008–2009	227	138	58	75	110
2009–2010	225	140	67	90	127
2010–2011	256	109	78	142	173
2011–2012	238	127	69	77	89
2012–2013	236	129	39	74	124
2013–2014	250	115	69	99	137
2014–2015	225	140	63	87	116
2015–2016	215	150	35	94	134
2016–2017	231	134	64	96	141
2017–2018	229	136	54	92	146
2018–2019	228	137	53	94	136
2019–2020	221	144	85	122	146
Mean days	234.65	130.45	51.30	84.7	115.65

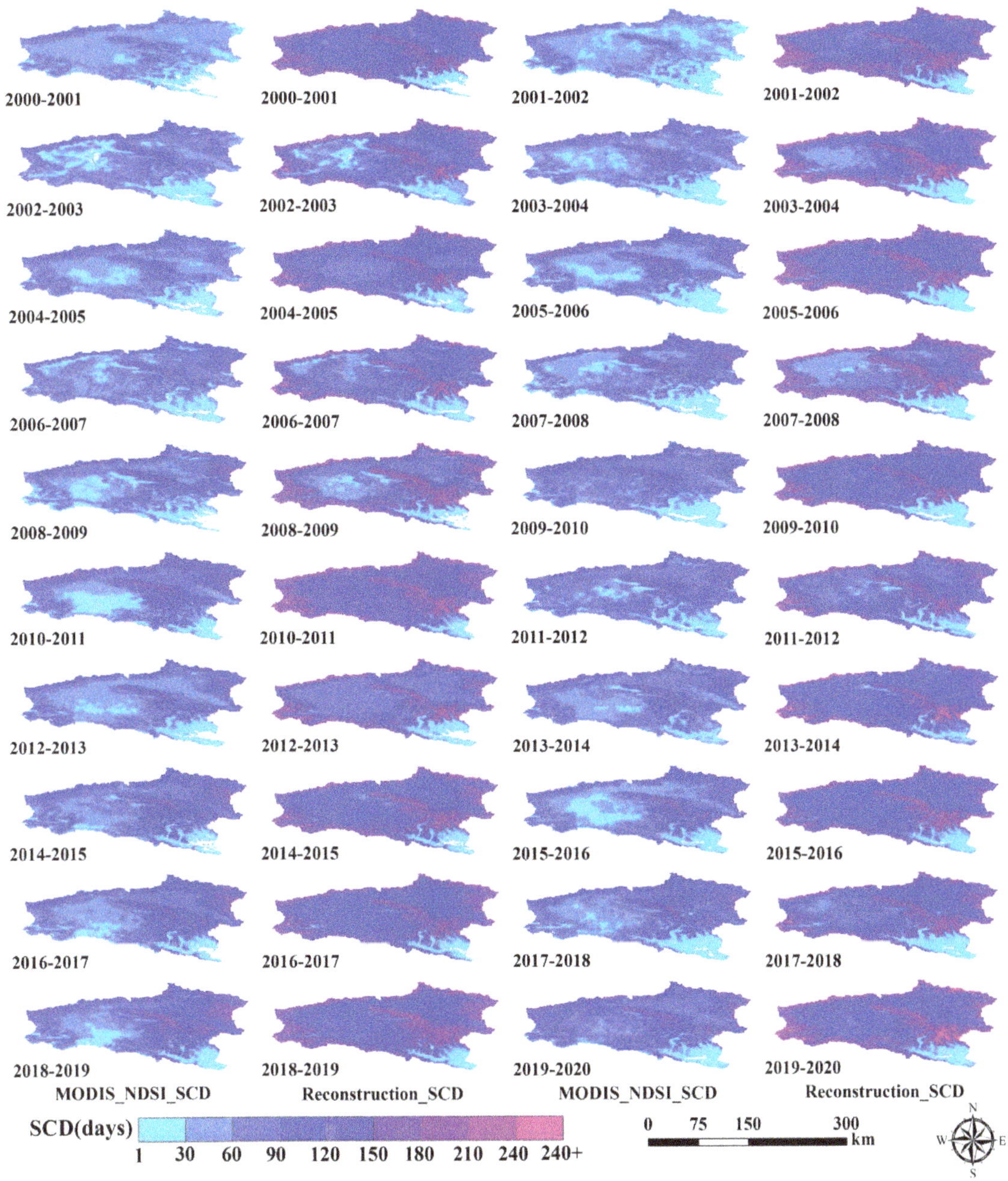

Figure 9. Comparison of snow cover days between MODIS snow products and snow cover data based on SGS filling.

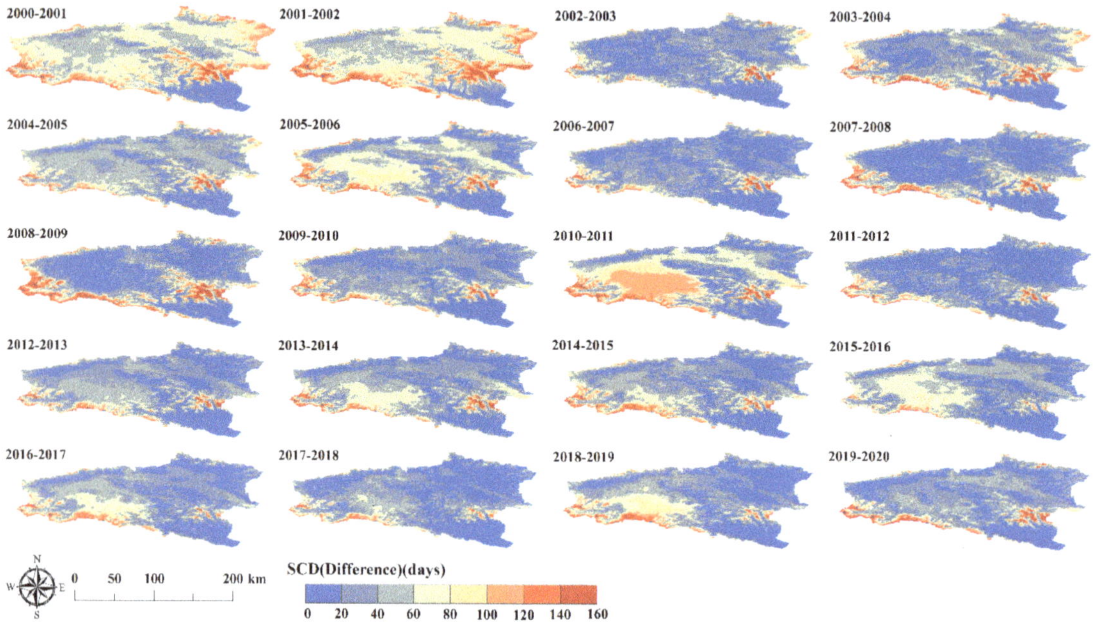

Figure 10. Spatial distribution of differences in snow cover days among different hydrological years.

Based on the analysis of the increase in snow cover days in the whole basin and meteorological stations on the hydrological annual scale, the accuracy of reconstructed snow cover data from 2000 to 2020 is further analyzed based on the station data. As shown in Table 8, compared with the MODIS snow products, the reconstructed snow cover data only decreased by 0.73% in the OA index, and the precision, recall, and F1-scores increased by 1.37%, 2.35%, and 1.84%, respectively. Combining 66.75% reconstructed snow cover data and 33.25% unreconstructed snow cover data, the average annual cloud coverage decreased from 52.46% to 34.41%, while the average annual proportion of snow cover and snow-free surface increased to 33.84% and 31.75%, respectively.

Table 8. Comparison of performance of snow data before and after snow cover reconstruction.

Snow Cover Data	Annual Average Accuracy (%)				Average Annual Coverage (%)		
	OA	Precision	Recall	F1	Snow	Cloud	Snow-Free
MODIS Snow	93.69	82.54	86.67	84.55	21.52	52.46	26.02
Reconstructed snow	92.96	83.91	89.02	86.39	33.84	34.41	31.75

This study focuses on the weekly scale to analyze the cloud removal and snow cover reconstruction. As shown in Figure 11, it can be seen that the weekly average cloud coverage ranges from 35% to 65%, reaching the highest around the 30th week (i.e., around March). After the SGS filling, the weekly average cloud coverage decreased by about 20%. During the period of mid-November to mid-March, the average snow cover increased by more than 20%. The rest of the time, the increased proportion of snow cover is limited due to less snowfall in the KRB. In general, the weekly average cloud coverage decreases by 20%, while the removed clouds are mainly filled with snow in winter and spring, and mainly filled with a snow-free surface in spring and autumn, which is also consistent with the temporal distribution of snow in the basin.

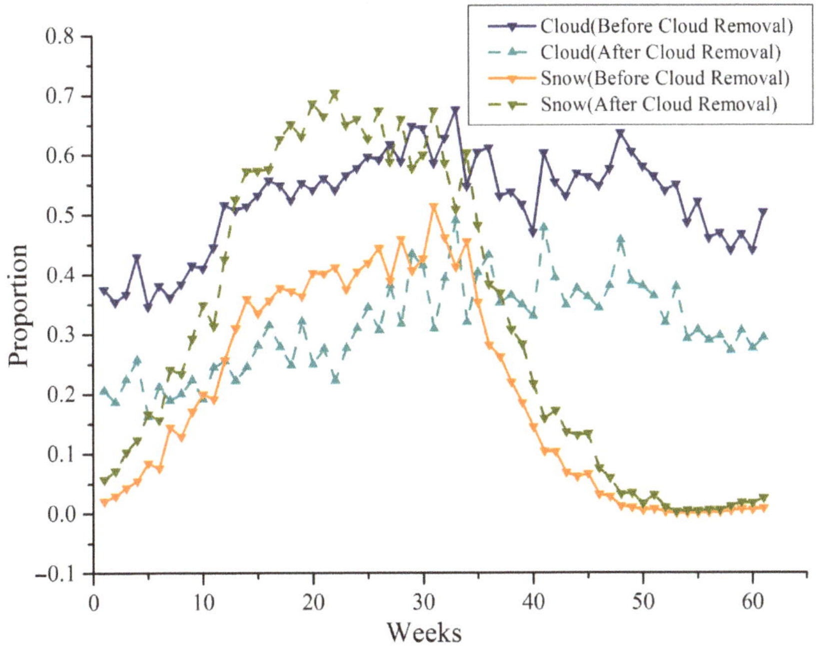

Figure 11. Variation in the proportion of weekly average cloud coverage and snow coverage before and after cloud removal in the KRB.

4.3. Validation of Individual Snow Cover Mapping

Since there is only one ground observation station in the study area, observations based on one station cannot provide representative evaluations of the advantages and limitations of the proposed method. The original snow cover data with the least cloud cover were used for validation purposes [11]. Several days with the least cloud-covered snow data were filled by clouds of other dense cloud-covered snow data. In this way, we could generate "observed" snow cover products where the performance of the proposed method can be validated.

For validation purposes, 8 October 2015 (validation day 1), 20 April 2016 (validation day 2), 25 April 2018 (validation day 3), and 1 June 2018 (validation day 4) were selected because of the least cloud coverage observed on these days. For validation day 1, the original snow cover data from this day with 0.25% cloud coverage was filled by the cloud cover values of 9 October 2015 snow cover data. The cloud cover fraction of the study area on 9 October 2015 was 54.62%. For validation day 2, the satellite observed less cloud cover with only 1.05% over the KRB. The original cloud cover pixels from 21 April 2016 with 60% cloud fraction were assigned to this day for validation purposes. For validation day 3, the original snow cover data from this day with 3.44% cloud coverage was filled by the cloud cover values of 24 April 2018 snow cover data. The cloud cover fraction of the study area on 24 April 2018 was 45.89%. For validation day 4, the satellite observed less cloud cover with only 0.05% over the KRB. The original cloud cover pixels from 2 June 2018 with 41.58% cloud fraction were assigned to this day for validation purposes. The generated snow cover maps with assigned cloud cover pixels were used as an input for a proposed method and the results were compared with the original snow cover data without cloud filling. The detailed validation results are shown in Figure 12 and Table 9.

Table 9. Validation of individual snow cover mapping.

Accuracy (%)	Validation Day 1		Validation Day 2		Validation Day 3		Validation Day 4	
	Our's	Hao's	Our's	Hao's	Our's	Hao's	Our's	Hao's
OA	92.89	94.43	86.99	92.58	91.44	89.45	95.77	94.96
Precision	70.12	75.23	87.55	78.54	80.34	69.02	73.13	63.72
Recall	90.40	87.65	54.69	91.52	90.84	96.62	76.21	65.80
F1-Score	78.98	80.97	67.32	84.53	85.27	80.52	74.64	64.75
Kappa	74.78	77.72	59.74	79.69	79.26	73.56	72.33	62.03

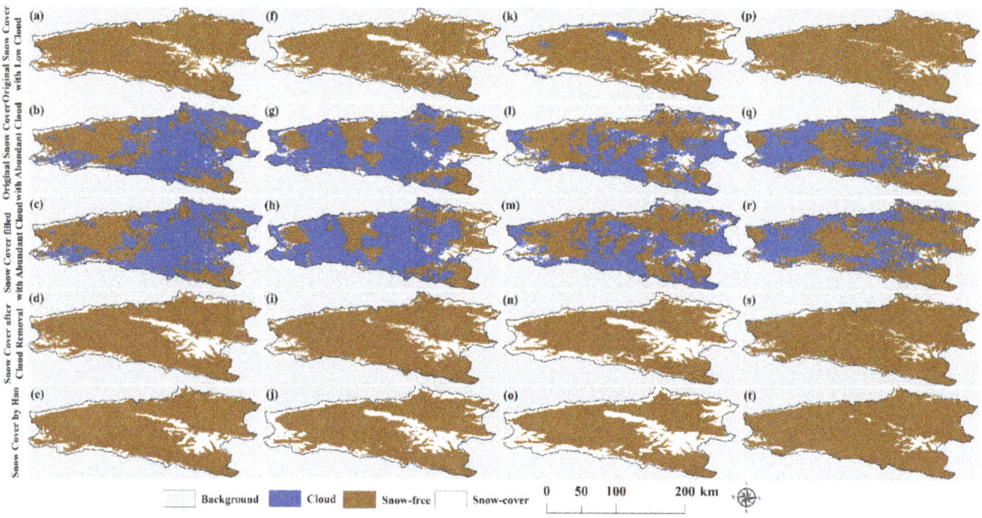

Figure 12. Validation results: (**a**) original snow cover for 8 October 2015, (**b**) original snow cover for 9 October 2015, (**c**) snow cover data from 8 October 2015 filled with cloud values of 9 October 2015 snow cover data, and (**d**) snow cover data from (**c**) after cloud removal; (**e**) snow cover data from (**c**) by Hao's method; (**f**) original snow cover for 20 April 2016, (**g**) original snow cover for 21 April 2016, (**h**) snow cover data from 20 April 2016 filled with cloud values of 21 April 2016 snow cover data, (**i**) snow cover data from (**h**) after cloud removal, and (**j**) snow cover data from (**h**) by Hao's method; (**k**) original snow cover for 25 April 2018, (**l**) original snow cover for 24 April 2018, (**m**) snow cover data from 25 April 2018 filled with cloud values of 24 April 2018 snow cover data, (**n**) snow cover data from (**m**) data after cloud removal, and (**o**) snow cover data from (**h**) by Hao's method; and (**p**) original snow cover for 1 June 2018, (**q**) original snow cover for 2 June 2018, (**r**) snow cover data from 1 June 2018 filled with cloud values of 2 June 2018 snow cover data, (**s**) snow cover data from (**r**) data after cloud removal, and (**t**) snow cover data from (**h**) by Hao's method.

As can be seen from the table and figure, the validation results reflect that the proposed cloud removal algorithm achieves or even surpasses Hao's method. Taking validation day 3 as an example, the proposed method outperforms Hao's method in all indicators except Recall. More importantly, the detailed features of the snow cover data reconstructed by the proposed method for cloud removal are more significant, demonstrating the changes in the trend of snow cover along mountain ranges. However, traditional spatiotemporal filtering and multi-sensor fusion methods have achieved good results in cloud removal and snow cover reconstruction, but they cannot effectively represent fine snow pixels. The proposed method is based on the spatiotemporal distribution of SGS within the entire watershed system, and then it reconstructs the distribution of snow under the cloud. Therefore, the reconstruction results consider the distribution characteristics of snow within the watershed. However, the methods of spatiotemporal filtering and multi-sensor fusion only focus on cloud pixels and their surrounding pixels, lacking large field observations and trade-offs.

Overall, the proposed method is a new exploratory study of the snow cover reconstruction of cloud removal. Its advantage lies in the accuracy index of snow reconstruction reaching or even surpassing traditional cloud removal schemes. However, its limitation is that it is only applicable when the cloud coverage is less than 70%, as excessive cloud cover can lead to insufficient training samples and affect the cloud removal effect.

5. Discussion

The objective of this study was to remove cloud-covered areas from the original MODIS snow cover products to obtain snow cover information for data-limited regions such as the KRB where no abundant snow cover data is available locally. As stated in Section 4.2, combining 66.75% reconstructed snow cover data and 33.25% unreconstructed snow cover data, the average annual cloud coverage decreased from 52.46% to 34.41%, while the average annual proportion of snow cover and snow-free surface increased to 33.84% and 31.75%, respectively. As a comparison, the spatiotemporal filtering method in a previous study [49] can remove 21.47% of cloud coverage from the KRB, increasing the annual snow cover rate from 20.34% to 41.81%. The SCE data, which has undergone spatiotemporal filtering and cloud removal, has an average annual accuracy of approximately 93% based on site validation in the KRB, slightly higher than the 92% OA of the proposed method. Multi-sensor fusion can further remove 2.59% of cloud coverage and achieve cloud-free mapping of the KRB, with an average annual snow cover rate of 44.40%. Under the condition of complete cloud removal, the overall accuracy based on site verification decreased to 89%, indicating that the uncertainty of multi-sensor cloud removal in complex mountain environments is relatively high. From this point of view, the proposed method in this study maintains higher snow recognition accuracy after cloud removal. The limitation of this study is that it cannot achieve cloud removal of all data on an annual scale.

Except for differences in specific accuracy indicators, the proposed method provides more detailed and consistent mountain trends in restoring snow cover under clouds, as shown in the comparison between Figures 12i and 12j. This may be because this method is based on continuous numerical retrieval, which is then used to obtain the value of SGS and determine the snow cover. Moreover, the method in the article mainly relies on cloud pixels themselves or small-scale neighborhood information, lacking overall feature extraction for watershed scale. Based on the above expression, the snow reconstruction-based method on SGS gap filling can be extended to the reconstruction of other missing data (continuous values), such as the reconstruction of ground temperature under clouds and the filling of soil moisture under clouds. For the reconstruction of snow cover under clouds, the proposed method can also be extended to use parameters such as snow density and snow wetness to reconstruct snow cover under clouds [18].

6. Conclusions

The "spatiotemporal" dimensional data that can fully characterize the geomorphic characteristics of the KRB and temporal characteristics of snow was designed and constructed as input data. At the same time, based on the physical characteristics of the variation of SGS with altitude, slope, aspect, and land cover in the watershed scale, the daily SGS filling algorithm of the space–time ET model is constructed and trained, so that the snow cover reconstruction under clouds in the KRB is realized. The algorithm applies to the daily data with a missing rate of less than 70% (i.e., the cloud coverage is less than 70%). A total of 66.75% of snow products have realized the snow cover reconstruction under the cloud based on this method from 2000 to 2020. Compared with MOD10A1 snow cover products, the average annual cloud coverage rate decreased from 52.46% to 34.41%, while the snow coverage rate increased from 21.52% to 33.84%. The data source of this study is SGS data derived from MODIS L1B data. The cloud removal algorithm based on SGS gap-filling can be applied to daily data with cloud coverage less than 70% and achieve one-step cloud removal. Daily data with cloud coverage greater than 70% has poor performance when applying this algorithm. Therefore, this part of the data is not within the scope of this study. In summary, on a year-round scale, the cloud removal algorithm reduced the average annual cloud cover rate of the KRB from 52.46% to 34.41%, and 68.25% of the removed cloud pixels were classified as snow cover, while the remaining 31.75% were classified as snow-free surface.

The main contribution of this study is to carry out SGS filling based on the physical characteristics of SGS distribution at the watershed scale for the first time. Different from the traditional spatiotemporal filtering cloud removal algorithm, the proposed method focuses on considering the spatial and temporal distribution characteristics of snow cover across the entire watershed scale and reconstructed snow cover that fits the geographic and meteorological characteristics of the watershed. Therefore, it better realizes the reconstruction of snow information under continuous large-scale clouds, improves the time resolution of snow, and realizes the deep integration of physical mechanisms and machine learning in the field of snow remote sensing. This article is an exploratory study designed and conducted based on the strong correlation between the spatiotemporal distribution of SGS within a small-scale watershed (KRB) and the terrain and meteorological elements of the watershed. On a larger spatiotemporal scale (such as the entire northern Xinjiang), the strong spatiotemporal heterogeneity of snow grain size can lead to the failure of the method in this study. Meanwhile, the algorithm in this study relies on training with a large amount of effective data. When the cloud coverage rate is higher than 70%, the training effect of the model will deteriorate due to the reduction in data volume, which is also a limitation of this study. In the future, we will attempt to conduct relevant research on a broader basin scale and explore the differences in our research methods across different basin scales to obtain more applicable strategies and methods.

Author Contributions: Conceptualization, L.Z. and J.W.; methodology, L.Z. and J.W.; software, L.Z.; validation, L.Z., G.M. and X.K.; formal analysis, L.Z.; investigation, L.Z. and J.W.; resources, L.Z. and J.W.; data curation, L.Z. and G.M.; writing—original draft preparation, L.Z.; writing—review and editing, L.Z.; visualization, L.Z.; supervision, J.W. and Y.Z.; project administration, Y.Z.; funding acquisition, Y.Z. All authors have read and agreed to the published version of the manuscript.

Funding: This research was funded by the National Natural Science Foundation of China (No. 42305158, No. 42105143), the National Key Research and Development Program of China (No. 2021YFE0116900), the Natural Science Foundation of the Jiangsu Higher Education Institutions of China (No. 23KJB170025), and the Wuxi University Research Start-up Fund for Introduced Talents (No. 2022r035).

Data Availability Statement: The data presented in this study are available upon request from the corresponding author.

Acknowledgments: We thank the China Meteorological Administration (http://cdc.cma.gov.cn (accessed on 26 August 2023)) for providing meteorological data and snow depth observation, the United States Geological Survey (USGS) (http://www.edc.usgs.gov (accessed on 26 August 2023)) for providing digital elevation model data, and NASA (https://ladsweb.modaps.eosdis.nasa.gov (accessed on 26 August 2023)) for providing MODIS reflectance and land cover data.

Conflicts of Interest: The authors declare no conflict of interest.

References

1. Zhu, Z.; Wang, S.X.; Woodcock, C.E. Improvement and expansion of the Fmask algorithm: Cloud, cloud shadow, and snow detection for Landsats 4-7, 8, and Sentinel 2 images. *Remote Sens. Environ.* **2015**, *159*, 269–277. [CrossRef]
2. Yuan, C.; Li, Q.; Nie, W.; Ye, C. A depth information-based method to enhance rainfall-induced landslide deformation area identification. *Measurement* **2023**, *219*, 113288. [CrossRef]

3. Hall, D.K.; Riggs, G.A.; Salomonson, V.V. Development of methods for mapping global snow cover using moderate resolution imaging spectroradiometer data. *Remote Sens. Environ.* **1995**, *54*, 127–140. [CrossRef]
4. Li, C.; Su, F.; Yang, D.; Tong, K.; Meng, F.; Kan, B. Spatiotemporal variation of snow cover over the Tibetan Plateau based on MODIS snow product, 2001–2014. *Int. J. Climatol.* **2018**, *38*, 708–728. [CrossRef]
5. Zhang, H.; Zhang, F.; Zhang, G.; Che, T.; Yan, W.; Ye, M.; Ma, N. Ground-based evaluation of MODIS snow cover product V6 across China: Implications for the selection of NDSI threshold. *Sci. Total Environ.* **2019**, *651*, 2712–2726. [CrossRef]
6. Wu, M.; Ba, Z.; Liang, J. A procedure for 3D simulation of seismic wave propagation considering source-path-site effects: Theory, verification and application. *Earthq. Eng. Struct. Dyn.* **2022**, *51*, 2925–2955. [CrossRef]
7. Bian, J.; Li, A.; Liu, Q.; Huang, C. Cloud and snow discrimination for CCD images of HJ-1A/B constellation based on spectral signature and spatio-temporal context. *Remote Sens.* **2016**, *8*, 31. [CrossRef]
8. Liu, H.; Li, J.; Meng, X.; Zhou, B.; Fang, G.; Spencer, B.F. Discrimination Between Dry and Water Ices by Full Polarimetric Radar: Implications for China's First Martian Exploration. *IEEE Trans. Geosci. Remote Sens.* **2022**, *61*, 5100111. [CrossRef]
9. Li, X.; Jing, Y.; Shen, H.; Zhang, L. The recent developments in cloud removal approaches of MODIS snow cover product. *Hydrol. Earth Syst. Sci.* **2019**, *23*, 2401–2416. [CrossRef]
10. Zhuo, Z.; Du, L.; Lu, X.; Chen, J.; Cao, Z. Smoothed Lv Distribution Based Three-Dimensional Imaging for Spinning Space Debris. *IEEE Trans. Geosci. Remote Sens.* **2022**, *60*, 5113813. [CrossRef]
11. Gafurov, A.; Bárdossy, A. Cloud removal methodology from MODIS snow cover product. *Hydrol. Earth Syst. Sci.* **2009**, *13*, 1361–1373. [CrossRef]
12. Zhu, W.; Chen, J.; Sun, Q.; Li, Z.; Tan, W.; Wei, Y. Reconstructing of High-Spatial-Resolution Three-Dimensional Electron Density by Ingesting SAR-Derived VTEC Into IRI Model. *IEEE Geosci. Remote Sens. Lett.* **2022**, *19*, 4508305. [CrossRef]
13. Dong, C.; Menzel, L. Producing cloud-free MODIS snow cover products with conditional probability interpolation and meteorological data. *Remote Sens. Environ.* **2016**, *186*, 439–451. [CrossRef]
14. Shen, H.; Meng, X.; Zhang, L. An integrated framework for the spatio–temporal–spectral fusion of remote sensing images. *IEEE Trans. Geosci. Remote Sens.* **2016**, *54*, 7135–7148. [CrossRef]
15. Wang, W.; Huang, X.; Deng, J.; Xie, H.; Liang, T. Spatio-temporal change of snow cover and its response to climate over the Tibetan Plateau based on an improved daily cloud-free snow cover product. *Remote Sens.* **2014**, *7*, 169–194. [CrossRef]
16. Yin, L.; Wang, L.; Keim, B.D.; Konsoer, K.; Yin, Z.; Liu, M.; Zheng, W. Spatial and wavelet analysis of precipitation and river discharge during operation of the Three Gorges Dam, China. *Ecol. Indic.* **2023**, *154*, 110837. [CrossRef]
17. Xia, Q.; Gao, X.G.; Chu, W.; Sorooshian, S. Estimation of daily cloud-free, snow-covered areas from MODIS based on variational interpolation. *Water Resour. Res.* **2012**, *48*, 9523. [CrossRef]
18. Poggio, L.; Gimona, A. Sequence-based mapping approach to spatio-temporal snow patterns from MODIS time-series applied to Scotland. *Int. J. Appl. Earth Obs. Geoinf.* **2015**, *34*, 122–135. [CrossRef]
19. Li, X.H.; Fu, W.X.; Shen, H.F.; Huang, C.L.; Zhang, L.P. Monitoring snow cover variability (2000–2014) in the Hengduan Mountains based on cloud-removed MODIS products with an adaptive spatio-temporal weighted method. *J. Hydrol.* **2017**, *551*, 314–327. [CrossRef]
20. Huang, Y.; Liu, H.; Yu, B.; Wu, J.; Kang, E.L.; Xu, M.; Wang, S.; Klein, A.; Chen, Y. Improving MODIS snow products with a HMRF-based spatio-temporal modeling technique in the Upper Rio Grande Basin. *Remote Sens. Environ.* **2018**, *204*, 568–582. [CrossRef]
21. Chen, S.Y.; Wang, X.Y.; Guo, H.; Xie, P.Y.; Wang, J.; Hao, X.H. A Condi-tional Probability Interpolation Method Based on a Space-Time Cube for MODIS Snow Cover Products Gap Filling. *Remote Sens.* **2020**, *12*, 3577. [CrossRef]
22. Luan, W.; Zhang, X.; Xiao, P.; Wang, H.; Chen, S. Binary and Fractional MODIS Snow Cover Mapping Boosted by Machine Learning and Big Landsat Data. *IEEE Trans. Geosci. Remote Sens.* **2022**, *60*, 4305714. [CrossRef]
23. Xiao, X.; He, T.; Liang, S.; Zhao, T. Improving Fractional Snow Cover Retrieval from Passive Micwave Data Using a Radiative Transfer Model and Machine Learning Method. *IEEE Trans. Geosci. Remote Sens.* **2021**, *60*, 4304215.
24. Guo, X.; Chen, Y.; Liu, X.; Zhao, Y. Extraction of snow cover from high-resolution remote sensing imagery using deep learing on a smll dataset. *Remote Sens. Lett.* **2020**, *11*, 66–75. [CrossRef]
25. Hu, J.; Shean, D. Improving Mountain Snow and Land Cover Mapping Using Very-High-Resolution (VHR) Optical Satellite Images and Random Forest Machine Learning Models. *Remote Sens.* **2022**, *14*, 4227. [CrossRef]
26. Liu, C.; Huang, X.; Li, X.; Liang, T. MODIS fractional snow cover mapping using machine learning technology in a mountainous area. *Remote Sens.* **2020**, *12*, 962. [CrossRef]
27. Yang, C.; Zhang, Y.; Xia, M.; Lin, H.; Liu, J.; Li, Y. Satellite Image for Cloud and Snow Recognition Based on Lightweight Feature Map Attention Network. *ISPRS Int. J. Geo-Inf.* **2022**, *11*, 390. [CrossRef]
28. Chen, Z.; Chen, Y.; Li, B. Quantifying the effects of climate variability and human activities on runoff for Kaidu River Basin in arid region of northwest China. *Theor. Appl. Climatol.* **2013**, *111*, 537–545. [CrossRef]
29. Haiyan, C.; Yaning, C.; Weihong, L.; Xinming, H.; Yupeng, L.; Qifei, Z. Identifying evaporation fractionation and streamflow components based on stable isotopes in the Kaidu River Basin with mountain–oasis system in north-west China. *Hydrol. Process.* **2018**, *32*, 2423–2434. [CrossRef]
30. Domine, F.; Albert, M.; Huthwelker, T.; Jacobi, H.W.; Kokhanovsky, A.A.; Lehning, M.; Picard, G.; Simpson, W.R. Snow physics as relevant to snow photochemistry. *Atmos. Chem. Phys.* **2008**, *8*, 171–208. [CrossRef]

31. Guo, Z.M.; Wang, N.L. Progress in the research on snow grain size retrieved from remote Sensing. *J. Glaciol. Geocryol.* **2011**, *33*, 539–545.
32. Wang, J.; Feng, X.; Xiao, P.; Ye, N.; Zhang, X.; Cheng, Y. Snow grain-size estimation over mountainous areas from modis imagery. *IEEE Geosci. Remote Sens. Lett.* **2017**, *15*, 97–101. [CrossRef]
33. Fan, C.; Cheng, C.; Qi, P.; Liu, X.; Wang, X. Retrieval of Snow Grain Size and Albedo Using Two Radiative Transfer Models. *Acta Opt. Sin.* **2020**, *40*, 15–24.
34. Wei, Y.; Hao, X.; Wang, J.; Li, H.; Zhao, H.; Gao, Y.; Shao, D.; Zhong, X.; Li, H. Retrieval and analysis of spatiotemporal variation of snow black carbon and snow grain size in Northern Xinjiang based on MODIS data. *J. Glaciol. Geocryol.* **2019**, *41*, 1192–1204.
35. Hao, X.H.; Wang, J.; Wang, J.; Zhang, P.; Huang, C.L. The measurement and retrieval of the spectral reflectance of different snow grain size on Northern Xinjiang, China. *Spectrosc. Spectr. Anal.* **2013**, *33*, 190–195.
36. Wang, J.; Li, W.P. Snow Grain and Snow Fraction Retrieval Algorithm based on Asymptotic Radiative Transfer Model. *Remote Sens. Technol. Appl.* **2017**, *32*, 64–70.
37. Zhang, Y.; Chen, S.; Wang, J.; Zhu, L.; Chen, S. Cloud removal method for snow cover products based on denoising autoencoder artificial neural network. *J. Nanjing Univ. Inf. Sci. Technol.* **2023**, *15*, 169–179. (In Chinese) [CrossRef]
38. Wang, Y.; Huang, X.; Wang, J.; Zhou, M.; Liang, T. AMSR2 snow depth downscaling algorithm based on a multifactor approach over the Tibetan Plateau, China. *Remote Sens. Environ.* **2019**, *231*, 111268. [CrossRef]
39. Yan, D.; Ma, N.; Zhang, Y. Development of a fine-resolution snow depth product based on the snow cover probability for the Tibetan Plateau: Validation and spatial–temporal analyses. *J. Hydrol.* **2022**, *604*, 127027. [CrossRef]
40. Zhu, L.; Zhang, Y.; Wang, J.; Tian, W.; Liu, Q.; Ma, G.; Kan, X.; Chu, Y. Downscaling snow depth mapping by fusion of microwave and optical remote-sensing data based on deep learning. *Remote Sens.* **2021**, *13*, 584. [CrossRef]
41. Kan, X.; Zhang, Y.; Zhu, L.; Xiao, L.; Wang, J.; Tian, W.; Tan, H. Snow cover mapping for mountainous areas by fusion of MODIS L1B and geographic data based on stacked denoising auto-encoders. *Comput. Mater. Contin.* **2018**, *57*, 49–68. [CrossRef]
42. Geurts, P.; Ernst, D.; Wehenkel, L. Extremely randomized trees. *Mach. Learn.* **2006**, *63*, 3–42. [CrossRef]
43. Huang, C.; Chen, B.; Ma, C.; Wang, T. WRF-CMAQ-MOS Studies Based on Extremely Randomized Trees. *Acta Meteorol. Sin* **2018**, *76*, 779–789.
44. Fu, W.; Zhang, Z.; Huang, W. Research on Fault Early Warning Algorithm of Reheater in Thermal Power Plant Based on Extreme Random Tree. *J. Shanghai Univ. Electr. Power* **2020**, *36*, 445–450.
45. Wei, J.; Li, Z.; Lyapustin, A.; Sun, L.; Peng, Y.; Xue, W.; Sun, T.; Cribb, M. Reconstructing 1-km-resolution high-quality $PM_{2.5}$ data records from 2000 to 2018 in China: Spatiotemporal variations and policy implications. *Remote Sens. Environ.* **2021**, *252*, 112136. [CrossRef]
46. Zhu, L.; Ma, G.; Zhang, Y.; Wang, J.; Tian, W.; Kan, X. Accelerated decline of snow cover in China from 1979 to 2018 observed from space. *Sci. Total Environ.* **2022**, *814*, 152491. [CrossRef]
47. Dozier, J. Spectral signature of alpine snow cover from the Landsat Thematic Mapper. *Remote Sens. Environ.* **1989**, *28*, 9–22. [CrossRef]
48. Klein, A.G.; Barnett, A.C. Validation of daily MODIS snow cover maps of the Upper Rio Grande River Basin for the 2000–2001 snow year. *Remote Sens. Environ.* **2003**, *86*, 162–176. [CrossRef]
49. Hao, X.; Huang, G.; Zheng, Z.; Sun, X.; Ji, W.; Zhao, H.; Wang, J.; Li, H.; Wang, X. Development and validation of a new MODIS snow-cover-extent product over China. *Hydrol. Earth Syst. Sci.* **2022**, *26*, 1937–1952. [CrossRef]

water

Article

Research on the Evolution of Snow Crystal Necks and the Effect on Hardness during Snowpack Metamorphism

Jie Wei, Peng Lu *, Shengbo Hu, Qiuming Zhao, Shunqi Yuan, Puzhen Huo and Qingkai Wang

State Key Laboratory of Coastal and Offshore Engineering, Dalian University of Technology, Dalian 116024, China; jiewei@mail.dlut.edu.cn (J.W.); hushengboandy@163.com (S.H.); qiumingzhao@mail.dlut.edu.cn (Q.Z.); yuansq_rex101@mail.dlut.edu.cn (S.Y.); hpz_ice@mail.dlut.edu.cn (P.H.); wangqingkai@dlut.edu.cn (Q.W.)
* Correspondence: lupeng@dlut.edu.cn

Abstract: To study the snow microstructure at various metamorphism times and extract the snow neck area, a constant density (200 kg/m^3) snow metamorphism experiment was conducted. The findings show that the neck region is mostly influenced by temperature, sun radiation, snow density and specific humidity, with wind speed having little effect. Additionally, we developed a multiple linear regression equation for the neck area under atmospheric forcing: "$S = 288T + 2E + 189\rho + 12{,}194V - 20{,}443RH - 42{,}729$". This equation accounts for solar radiation (E), temperature (T), snow density (ρ), specific humidity (RH) and wind speed (V). Notably, the above five factors can account for 84% of the factors affecting the neck area, making it a crucial factor. The relationship between snow hardness and neck area is correlated at 71%, and in later stages of metamorphism, the correlation may increase to 91%. Based on the neck area, the following hardness value prediction is made: "$H = 0.002764S + 67.922837$". This study documents the growth variations in the neck region of the metamorphic snow cover and elucidates the process by which outside factors impact the microstructure and macroscopic physical characteristics of the snow cover.

Keywords: snow crystals; microstructure; neck growth; metamorphism; hardness

check for updates

Citation: Wei, J.; Lu, P.; Hu, S.; Zhao, Q.; Yuan, S.; Huo, P.; Wang, Q. Research on the Evolution of Snow Crystal Necks and the Effect on Hardness during Snowpack Metamorphism. *Water* **2024**, *16*, 48. https://doi.org/10.3390/w16010048

Academic Editor: Richard Smardon

Received: 19 November 2023
Revised: 15 December 2023
Accepted: 20 December 2023
Published: 22 December 2023

1. Introduction

Snow is a significant part of the cryosphere and serves as a proxy for climate change [1]. As a result of global warming's effects on the cryosphere, snow's physical characteristics have changed. The thermodynamic properties and mechanical strength of snow are mostly determined by the interior crystal structure of the snow [2]. Complex ice skeletons are formed by the different shapes and binding interactions of snow crystals, and understanding the microstructure of snow can help with the advancement of snow engineering [3]. The thermal characteristics and mechanical indices of snow are crucial parameters to be considered in cold-region science and engineering. The process of how snow evolves will undoubtedly be significantly impacted by changing climate conditions as global warming progresses [4,5], requiring a careful examination of snow mechanics and snow hazard issues [6–10]. In the future, a crucial strategy for dealing with snow science and snow risks will be to link the microstructure and apparent physical properties of snow.

A morphological viewpoint has been used by earlier researchers to study the physical features of snowpacks, given the significance of these qualities and the impact of metamorphism on them. The primary factor impacting the macroscopic physical properties of snow during snow metamorphism is the modification of the microstructure of snow crystals, notably the necks. A neck is one of the many microstructural characteristics of snow. The narrow area between two ice crystal grains is referred to as the neck, and it starts where the surface curvature changes from being outwardly convex to inwardly concave [11]. Therefore, the neck area is the region in the connecting point of the two necks of two adjacent ice crystals. Hardness is one of the mechanical factors of snow accumulation, and the neck, being the weakest region of the snow crystal, greatly influences its mechanical properties. According to the earlier definition of hardness as the resistance to plastic deformation, subsequent research by Mellor [12] and Lee [13] suggested that the hardness of the snowpack can be thought of as a particular type of compression in the presence of side limits, with a strong correlation between the hardness value's magnitude and the snowpack's density and the state of its lateral confinement. Lateral confinement refers to the limitation of lateral displacement of snow by using restraint components in experiments. Its deformation characteristics are determined by the stresses that are transferred between the grains and the movement of the grains under natural conditions. The snow's continuous densification over time causes the snow hardness value to continuously increase over a certain period of time, with a tendency to increase with the snow depth [14]. Like other physical characteristics, snow hardness is largely influenced by the metamorphism between snow crystals. The bonding or necking between adjacent grains during the metamorphic process results in ice deposition at the contact points, and with densification and particle coarsening [15], the snow crystal necks become larger, and the snow becomes denser. Scholars have researched the microstructure of snowpacks, particularly the necks of snow crystals. Libbrecht et al. [16] conducted experimental studies on the growth rate of ice crystals on different surfaces, determined the morphology of snow crystals at different temperatures and supersaturation, and obtained the growth patterns of snow crystal shapes and neck bonds. Gubler et al. [17] studied chain-shaped snow crystals and used ice chains to describe the load-bearing capacity of snow by calculating the length of grain bonds, thus improving the model for analyzing snow intensity using snow crystal shape and neck bond growth. Edens and Brown et al. [18] examined the reasons for alterations in the neck area caused by significant deformation as well as the impact of new bond creation on the neck area. Brown et al. [19] extended the neck growth model to porous metals and proposed an intrinsic law based on the deformation of grains and grain bonding to explain the changes in the microstructure of snow under pressure. Snow deformation behavior was explained by Kry [20] using the idea of chaining, and Hansen and Brown [21] created a statistical model utilizing bond radii to describe the structure of snow particles. Less research has been performed on the impact of snow microstructural changes on snow hardness during metamorphism; however, past studies have concentrated on the development of an ontological model for snow. An important background for ice and snow engineering problems is the connection between the microstructure of snow and its apparent mechanical hardness.

In contrast to previous studies, this study continuously observed snow samples in the field to determine how their microstructure and macroscopic physical properties, such as hardness, changed as the snow underwent natural metamorphosis. The results, which were obtained via the stereology method to analyze the impact of metamorphism on snow from the viewpoint of microstructure, contribute to a deeper comprehension of snow mechanics and provide a point of reference for describing changes in snow crystal morphology and overall strength [22].

2. Materials and Methods

2.1. Overview of the Research Area and Research Equipment

The experiment was carried out at the Songhua River Dadingzi Mountain Hydroelectric Hub Comprehensive Test Site in Harbin City, Heilongjiang Province, China, between 24 December 2022 and 12 February 2023. The site coordinates are 46°39′ N and 127°25′ E (location of the red star marker in Figure 1). For approximately 130 days of the year, the Harbin section of the Songhua River is frozen over [23], and many snowfall events and extended periods of cold weather have led to the growth of local snow and ice sports, including skiing, cold-water fishing, winter fishing, and winter rallies. The thermal and physico-mechanical characteristics of snow are of guiding significance for the economic development of the area as well as the prevention and control of natural disasters because snow is a significant tourism resource in the region. A temperate monsoon environment with notable seasonal changes prevails in the research region. In the cold, dry winter, it is dominated by temperate continental air masses, and in the hot, humid summer, it is dominated by temperate oceanic air masses or modified tropical oceanic air masses. The average temperature over the course of the year is 4.3 °C, with the average temperature in January being approximately −18.3 °C and the average wind speed being approximately 2.8 m/s. The average temperature and wind speed in February are approximately −14 °C and 3.0 m/s, respectively. The snowfall season is centralized from November to January. During this period, the greatest snow depth is approximately 41 cm, with an average precipitation (snowfall) of 23.6 mm [24].

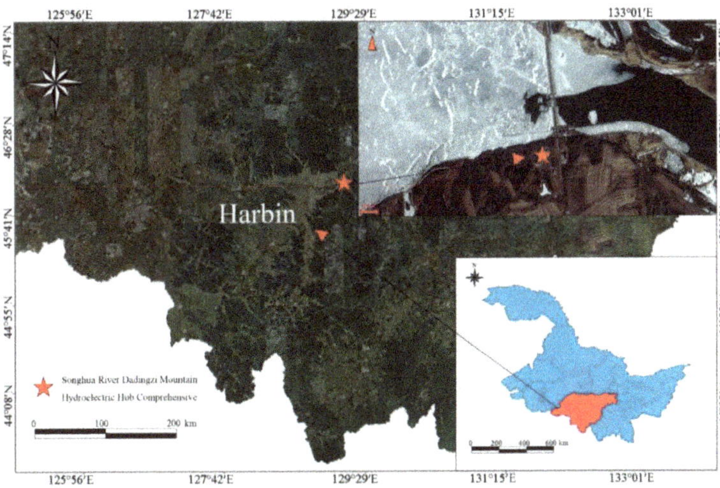

Figure 1. Location of the research area.

Snow temperature field observations were performed using temperature sensors for platinum resistance temperature sensors (accuracy of ±0.1 °C), while snow depth was measured using an ultrasonic ranging sensor (SR50A type (Campbell, Logan, UT, USA)) (accuracy of ±1 cm), both of which used a data acquisition instrument (Campbell CR3000 type(Campbell, Logan, UT, USA)) to collect data. The temperature chain layout is shown in Figure 2a. A JT H6 industrial microscope(Jingtuo Youcheng Co., Ltd., Shenzhen, China) was used to observe the snow crystal microstructure, as illustrated in Figure 2b. The magnifying range of the microscope ranged from 0.7 to 4, and calibration was necessary to establish the scale before viewing. A dynamic resistive strain gauge (YD-28A type(East China Electronic Instrument Factory, Shanghai, China)) was used to connect an electronically controlled penetrometer [25] to the acquisition equipment (VK701H+(Weijingyi Electronics Co., Ltd, Shenzhen, China)), as illustrated in Figure 2c, to measure the hardness of the metamorphic snow. Figure 2d illustrates how the ring knife approach was used to sample the whole snowpack to determine the snowpack density.

2.2. Experimental Design

Drawing upon previous research and with the aim of facilitating a closer examination of the microstructure of snow crystals, snow was collected from the same snowfall, and its density was established at 200 kg/m^3. The prepared snow samples were compacted layer by layer in a detachable wooden test box, which had a cross-sectional area of 25 × 25 cm and a height of 60 cm. To ensure the maximum possible initial uniformity of internal density in the snow samples, each layer was 10 cm thick. The initial internal density of the snow specimens was preserved as uniformly as possible through layered compression.

Snow samples that received solar radiation were placed in an open, unobstructed area for natural metamorphism, whereas snow samples without solar radiation were placed on the shaded side of the building and covered with a shading canvas to ensure they were not influenced by solar radiation.

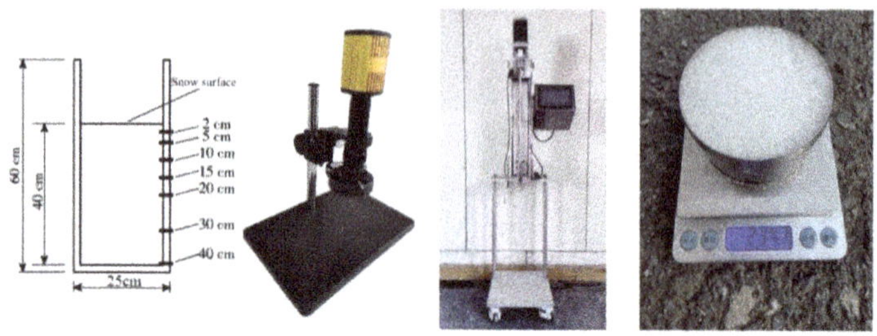

(a) (b) (c) (d)

Figure 2. Parts of the test equipment. (**a**) Snow test chamber and sidewall temperature sensor arrangement (the uppermost snow surface as the origin); (**b**) industrial microscope; (**c**) electronically controlled penetrometer; (**d**) ring knife.

Temperature sensors were positioned in holes drilled in the wall of the test chamber exposed to solar radiation to monitor temperature changes in the vertical direction of the snow layers. Additionally, a temperature sensor was installed for continuous monitoring of the air temperature, while a sensor for determining the ambient temperature was placed in a radiation shield to prevent any temperature increase due to sunlight exposure. Snow temperature and air temperature were sampled once per minute. The upper boundary of the specimen was in direct contact with the atmosphere and solar radiation, whereas the lower boundary was separated by a wooden board and a thick layer of prepacked snow on top of the soil. A significant amount of snow was accumulated at the bottom and around the periphery of the test chamber to reduce the impact of temperature variations on the underlay and to ensure uniformity of the test results.

To observe the microstructure of the snow specimen, it was necessary to open the sidewall of the wooden box and use an ice scraper to gently remove the snow particles in layers to avoid destroying the original structure between the snow crystals. The extracted snow crystals were laid flat on a black acrylic plate after cooling and adjusting the magnification and focal length of the microscope to obtain the best image. The lower end of the electronically controlled penetrometer penetration rod was connected to a conical cone with a tip angle of 60° and a base diameter of 30 mm, and the penetration rate was set at 15 mm/s. The penetration point was the center of the specimen, which was penetrated vertically to the bottom of the test box to stop the test, and the test data collected by the tension transducer were recorded. At the beginning of the metamorphic effect, the snow crystals changed more rapidly. The test interval was two days, and four days later, the test interval was changed to five days to ensure that the number of snow samples was sufficient. The test process and test snow sample storage are shown in Figure 3.

For each test, photographs of the snow crystals were taken in accordance with the four-layer stratification that occurred during compaction. When processing the photographs, preference was given to clearer photographs. An industrial microscope magnification scale was used to measure the actual area, circumference, and equivalent diameter of the neck region in the photographs. To ensure the accuracy of neck region selection, overlapping and mixed snow crystals were avoided, and necks with high resolution and obvious features were chosen. The method of selecting the neck area relied on manual identification and extraction. First, the area where two ice crystals were connected was found, and the four inflection points where the edge profile of the ice crystal changed from convex to concave were identified, i.e., the four points where the curvature of the edge profile of snow crystals with positive and negative signs changed (e.g., the four yellow dots in Figure 4b).

Afterwards, the two inflection points between different ice crystals were connected by smooth curves along the outer contour of the ice crystals (e.g., two red curves in Figure 4b), and the two inflection points of the same ice crystal were connected by straight lines (e.g., two blue straight lines in Figure 4b). The closed area formed by the two curves and two straight lines was the neck area needed for this test, and finally, the area, perimeter, and equivalent diameter of the box selected were calculated using image processing software. During calibration, the pixels occupied by the scale bar were marked as the corresponding actual length, and then the actual length and area of the selected area could be determined. Although errors in manual recognition are unavoidable, the change in neck area caused by incorrectly identifying the inflection point was much smaller than the total neck area, and the technique of calculating the average value by counting many different neck areas was also effective in minimizing the error.

Figure 3. Observation of microstructure of metamorphic snow and penetration hardness test process and storage of snow samples. (a) New snow compaction; (b) compacted snow sample; (c) observation of the microstructure of snow crystals; (d) metamorphic snow penetration test; (e) measurement of temperature and snow depth of deteriorated snow samples from the solar radiation group; (f) storage of deteriorated snow samples from solar radiation groups.

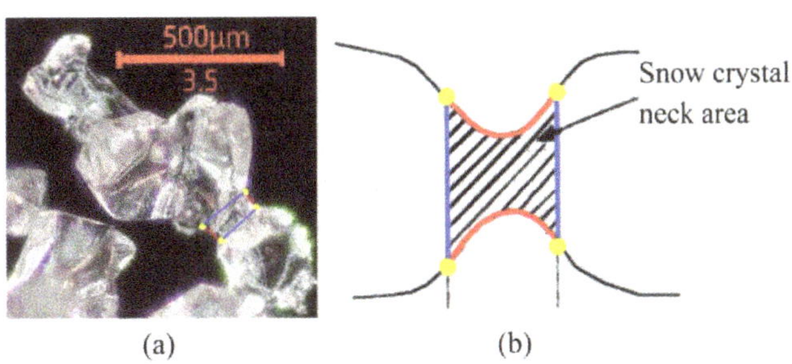

Figure 4. (a) Actual measurement of the neck area (magnification effect at 3.5 magnification). (b) Schematic diagram of the neck area.

3. Results and Discussion

3.1. Snow Metamorphosis

Figure 5 depicts the evolution of the air temperature, solar radiation group snow box profile temperature, snow density, hardness, and snow crystal neck area observed in the field over time.

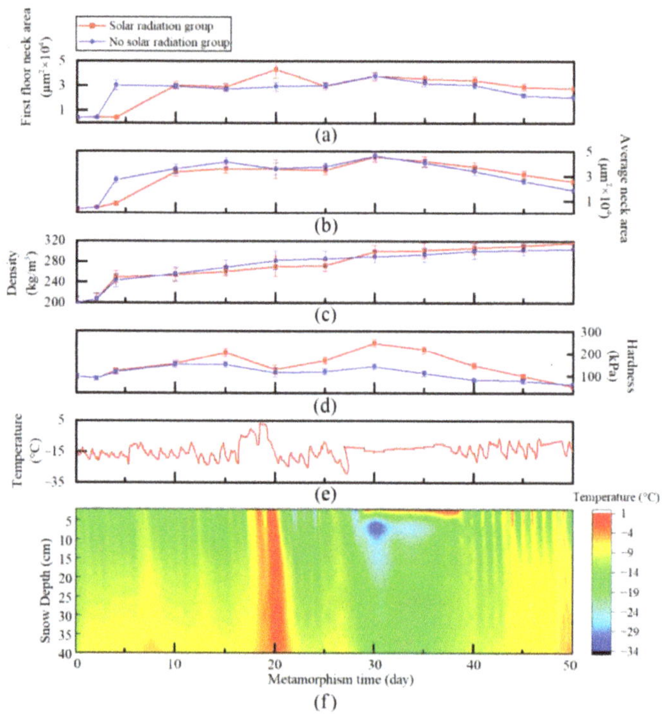

Figure 5. Changes in physical quantities over time during the experiment (**a**) neck area of the first layer of snow; (**b**) average neck area of snow; (**c**) density of snow; (**d**) hardness of snow; (**e**) temperature; (**f**) snow profile temperature of solar radiation group samples.

The test lasted for three months, from December to February of the following year, with average temperatures of −15.76, −14.48, and −12.86 °C, respectively. On the eighteenth day of the test, the temperature increased significantly, moving to above zero. The average snow crystal neck area fell over the following three days, and the profile temperatures inside the snow box climbed to various degrees, but the neck area of the top layer of snow grew. The diurnal temperature fluctuated significantly from the forty-first day to the forty-fifth day of metamorphosis, and the temperature of the snow in the snow box displayed alternating warm and cold oscillations.

The neck area of the snow crystals rose prior to thirty days of metamorphism, and the hardness values also gradually increased, as can be observed by combining plots a, b, and d in Figure 5. The first layer's neck region was more susceptible to the effects of temperature, particularly when warming caused a decrease in the neck area of the lower snow crystals. The snow crystal neck area decreased, and the hardness value also decreased at the later stage of metamorphism or after thirty days. As shown in Figure 5c, the snowpack density rose over time; however, its rate of change was concentrated in the metamorphic period, since the collapse rate in the latter phase slowed.

The error bar records the error situation during repeated experiments. For the penetration test of snow hardness determination, limited by the number of wooden boxes and the long metamorphism time, three wooden boxes were used for hardness testing each time. A single wooden box underwent three to four different penetrations, and its hardness value was the average of the final results, with relatively small dispersion. When measuring the neck size of snow crystals, the selected snow crystals came from multiple locations on the sidewalls of the snow layer. For snow layers at the same depth, we selected three to five locations parallel to the profile to extract snow crystals. We sprinkled them separately on the acrylic board and selected clear positions to take photos, extracting about ten images at a time. After summarizing the snow crystal photos of the same layer, 40–60 neck areas were selected by framing. Except for a few days when the temperature surged, the error in the neck area remained within a small range.

3.2. Analysis of Changes in Snow Crystal Neck Area

3.2.1. Effects of Solar Radiation

Figure 5a,b show the differences between the first layer and mean neck area over time curves for the two control groups, and it is clear that only the first layer, or the neck area of the 10 cm snow depth on the upper surface, of the snow specimens stored in wooden boxes was strongly affected by solar radiation (Figure 5a), with a difference of 13,913 μm^2 on the twentieth day. With a difference of 434 μm^2 on the twentieth day, the difference between the sun radiation group and the control group for the mean value of the neck area of each layer was not significant (Figure 5b).

The capacity of solar radiation to penetrate the snow layer makes it a crucial physical factor in changing the temperature of snow. Even if 90% of the incident radiation is reflected by snow, solar radiation will modulate the snow temperature and alter the snow temperature gradient either directly or indirectly [26]. The most significant and direct factor determining the warming of the snow specimen is the warming brought on by the absorption of solar radiation in the snow layer, which is more effective than heat conduction within the snow layer. This is reflected in the obvious temperature change in the snow in the upper 20 cm of the snow layer, and the closer to the snow surface it is, the more drastic the change in temperature is [27]. The borders of the wooden box that were utilized in this experiment always had a masking effect on the incoming sun radiation due to its height of 60 cm and initial snow thickness of 40 cm. Only the neck area of the snow crystals in the top 10 cm of the surface layer fluctuated considerably throughout the experiment, in contrast to the natural snowpack, which has an open surface without boundaries and solar radiation can affect a snow depth of up to 20 cm.

3.2.2. Effects of Temperature

Only the snow specimen profile temperature samples that received sun radiation were monitored for temperature because there were only a restricted number of collection devices in this experiment. The process of metamorphosis continued until the eighteenth day, when temperatures climbed above zero (Figure 5e). As a result of the temperature increase, the temperature of all layers of the snow specimen increased, and sublimation and condensation intensified. While the area of the surface snow necks rose, the sizes of the second, third, and fourth snow necks all shrank significantly (Figure 5a). The influence of heat flow in the underlying land determines temperature changes at the lower boundary of the snowpack, whereas temperature changes at the upper boundary of the snowpack are primarily influenced by solar radiation received at the snow surface and heat absorption and excretion during heat exchange on the snow–atmosphere contact surface [28]. Due to high-temperature sublimation, all snow layers other than the surface layer have a greater vapor flux and a contraction of the neck area. When the first snow crystals come into contact with the rising vapor flux from the lower snow surface, they precipitate, increasing the neck area of the first snow surface. The decrease in neck area caused by sublimation, however, is generally significantly greater for the entire snow sample than the rise in neck area caused by precipitation from adsorbed ice crystals (Figure 5b).

The curves of the snow crystal neck area relative to the air temperature for the solar radiation group are shown below, which combine the average neck area and air temperature during metamorphism in Figure 5b,e. The fitted curves are presented in Figure 6. Since the power metamorphism at the start of the fresh snow metamorphism is significantly larger than the temperature metamorphism, the fine snow whose metamorphism was carried out for up to ten days and beyond was selected for the statistics. Figure 6 shows that although the temperature and the neck are are adversely associated, the relationship is not particularly significant.

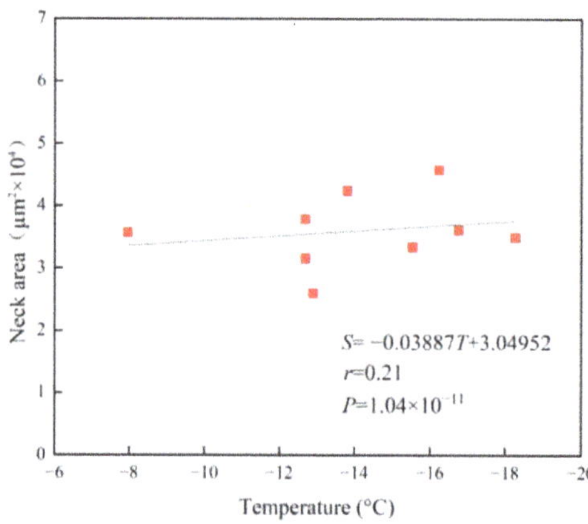

Figure 6. Relationship curve between temperature and neck area of the solar radiation group.

Due to the snow layer's strong insulation properties and low thermal conductivity, there is a temperature differential between the snow layer and the surrounding air. As a result, the snow crystal neck area's response to changes in air temperature lags slightly. Temperature is still one of the most significant factors influencing snow metamorphosis, despite the low correlation coefficient between the change in neck area and air temperature during snow metamorphosis. Lehning [29] considered grain growth and cross-layer mass transport as two distinct processes, with temperature serving as a crucial physical parameter that influences the transport of vapor fluxes within the snowpack, which has the potential to induce the transportation of water vapor, resulting in significant localized sublimation and precipitation within the snowpack. The growth of ice particles is a result of vapor diffusion among them, caused by the temperature gradient applied to the snow cover [30]. Furthermore, recrystallization leads to the formation of ice crystals, the strength of which is significantly influenced by temperature [31]. The impact of the snowpack's internal temperature on the neck region will be further examined in Section 3.2.4.

3.2.3. Effects of Snow Density

In the natural world, snow has a wide variety of densities, ranging from 300 to 550 kg/m^3. In this field environment, the recorded fresh snow density ranged from 70 to 150 kg/m^3 towards the conclusion of the snowstorm, and there were still significant voids after the snow was compacted with a density of 200 kg/m^3. As seen in Figure 5d, the rise in snow density in the region receiving solar radiation accounted for 45.2% of the overall density increase in the first ten days following the start of metamorphism, whereas the necking area quickly rose to 72.9% of the maximum necking area. In addition, the neck area of the snow in the region without sun radiation expanded to 77.7% of its maximum neck area, accounting for 51.9% of the overall density increase.

Following the statistics, Figure 7 displays the resulting curve of variation in the neck area and density of metamorphosed snow. In the thirty days between the start of the metamorphosis of the new snow and the start of the metamorphosis, the correlation coefficients of density and neck area were 0.90 for the solar radiation group and 0.96 for the no solar radiation group. The correlation coefficients for density and neck area fell after thirty days of metamorphosis, and throughout the duration of the test, they were 0.77 for the group that received solar radiation and 0.69 for the group that did not. The necks can already exist as separate crystal structures in the later phases of metamorphism because of their long-term growth, and the newly formed crystals interact with the original crystals to generate new necks. Thus, the overall neck area appears to decrease as the number of old necks decreases and the number of new necks increases; however, the density of the snow always rises once the snow collapses, resulting in the opposite pattern of change in the neck area and density in the late stage of snow metamorphism and a nonmonotonic trend throughout the process.

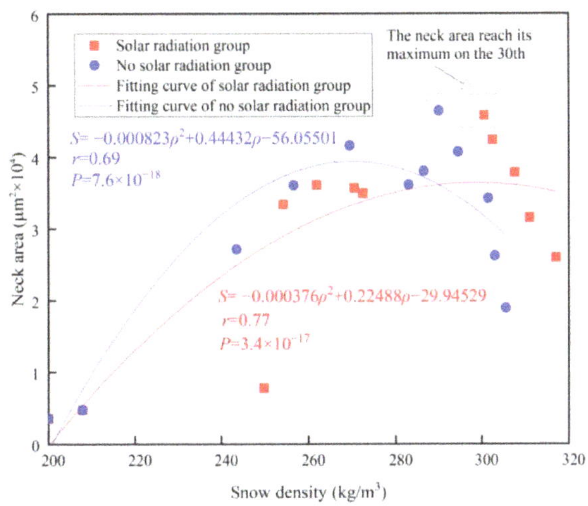

Figure 7. Relationship curve between snow density and the average neck area of each layer of snow crystals.

During the entire metamorphosis process, the experimental control groups exhibited the maximum neck area on the thirtieth day. Due to the exposure to light, the solar radiation group underwent a more intense metamorphosis, resulting in a more rapid elevation in density compared to that without solar radiation, with the peak in the neck area occurring at a later time. During the later stage of metamorphosis, the snow crystal morphology underwent a transformation from coarse snow to deep frost, with a notable decrease in the rate of density change compared to the initial stage.

3.2.4. Effects of Snow Depth

Based on stratifying the four layers of snow samples on the solar radiation surface and without the solar radiation surface (Figure 8), the area of the snow neck in each layer increases sequentially from top to bottom when there is no temperature fluctuation or when there is only a small amount of temperature fluctuation. The phenomenon described above in the relationship between temperature and snow neck area occurs when the temperature rises quickly for a brief period, causing the increase in the first layer's neck area to display a different pattern from the other three layers. The growth of the neck is a result of a fusion of pressure sintering, vapor diffusion, volume diffusion, and grain boundary diffusion. Over the course of seventeen to twenty-one days, the snow profile undergoes drastic temperature changes (Figure 5f), during which vapor diffusion takes over, leading to the generation of vapor from the lower snow crystals beyond the surface layer, which later condenses out of the surface snow crystals. If the external temperature does not experience significant changes, pressure sintering will be the dominant mechanism, where the underlying snow crystals will carry the load from the self-weight of the upper snow layers. As a result, the bottom snow crystal neck experiences increased compression, normal force, and shear force transfer, which causes the grain structure to form early and with a larger area of the neck to accommodate the above additional strain.

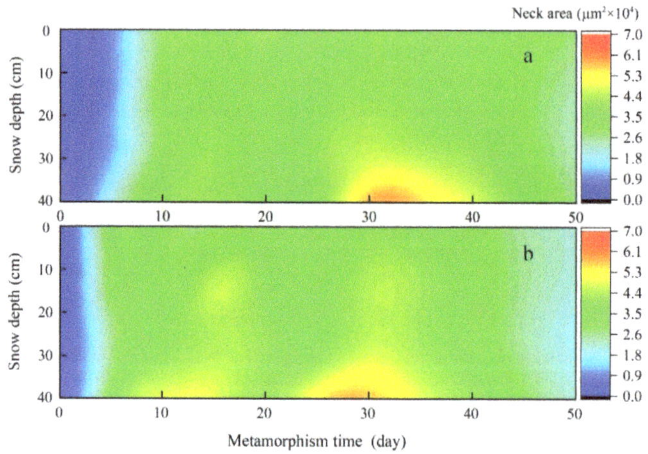

Figure 8. Changes in the neck area of snow crystals at different depths over time. (**a**) Changes in the neck area of snow crystals at different depths under solar radiation; (**b**) changes in neck area of snow crystals at different depths without solar radiation over time.

Figure 9a was created by plotting the temperature of the snow layer against the neck area of the snow crystals as they were measured in stages. The vertical snow pressure on the snow crystals in each layer was derived from the density and was used to determine the snow thickness using the snow layer settlement data collected by the ultrasonic range sensor. The four points from left to right in Figure 9b represent the average of the four layers of snow crystals in the snow box from the surface to the bottom at all metamorphic times. The points in Figure 9a cover the temperature and neck area of each layer during metamorphism, and since the snow layer temperature always varies with air temperature during metamorphism, the points are more discrete along the transverse axis and cannot be fitted after averaging the layers. The change in vertical load due to metamorphism is much smaller than the difference between snow layers, and so it can be calculated by averaging each snow layer. On the other hand, Figure 9b divides the changes in vertical load and neck area during metamorphosis of the four-layer snow specimens into four points after averaging and plots the error lines due to the changes in neck area and vertical load due to metamorphosis. The correlation in Figure 9a is better than the correlation of temperature in the neck area in Figure 6, and the snow profile temperature better reflects the effect on the neck area after accounting for the thermal conductivity of snow. However, the temperature is still less effective in describing the neck area, mainly because the effect of temperature on the neck of snow crystals is not an instantaneous process, and the description by the average value of the temperature does not fully reflect the metamorphic effect caused by vapor diffusion. Overall, the snow crystal neck area increases with snow depth, and the increase in vertical pressure contributes to the rise in the snow crystal neck area.

3.3. Relationship between Snow Crystal Neck Area and Total Crystal Area

Snow crystals undergo changes from needle-like, broad slab, and columnar morphologies to deep frost during metamorphosis, and the size of the snow crystals themselves constantly changes [32]. In order to explore the effect of snow crystal neck area on snow crystal size, Figure 10a statistically shows the change in the average size area of snow crystals (S_t) with metamorphosis time (D). It can be seen that, excluding the days at the beginning of metamorphosis and the days when the temperature increased from the 18th day, the snow crystal area showed an increasing trend. The surface area of snow crystals was initially larger because new snow retained the intact dendritic structure. With the onset of metamorphism, the dendritic structure was compressed and gradually disappeared, and the snow crystal area contracted. During the days of rapid temperature rise, as analyzed in Section 3.2.2, the vapor diffusion of the snow crystals increased and the area contracted. However, for the entire metamorphic period, the individual fine-grained snow crystals gradually coalesced to display an increasing trend in overall snow crystal area.

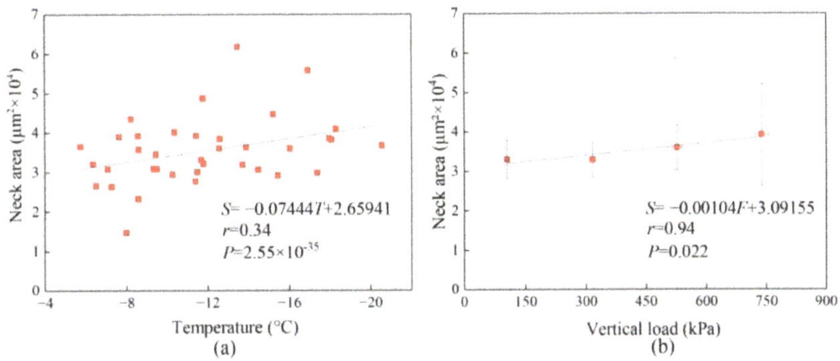

Figure 9. Relationship curve between neck area and solar radiation group snow (**a**) temperature and (**b**) vertical load.

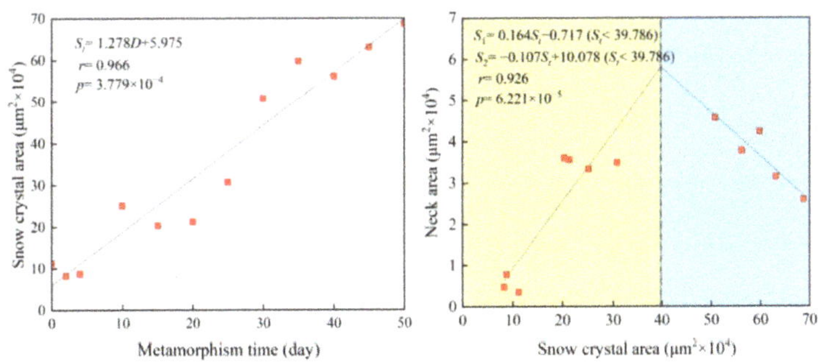

(**a**) (**b**)

Figure 10. Solar radiation group's (**a**) curve of mean snow crystal area with time and (**b**) curve of mean snow crystal area versus neck area (yellow and blue backgrounds represent different trends in the neck area and the snow crystal area).

As shown in Figure 10b, the neck area of snow crystals undergoes two stages with the variation of snow crystals. In the first stage (points covered by a yellow background in Figure 10, where the neck area of snow crystals can be described as the rising segment of S_1), the neck area increases with the increase in snow crystal area, and material exchange occurs between the two snow crystals through the neck. As the snow crystals increase in size, the neck, which serves as a material exchange channel, also gradually increases in area. In the second stage (points covered by a blue background in Figure 10, where the neck area of the snow crystal can be described as the S_2 descending segment), the neck area shows a downward trend as the snow crystal area increases. When the snow neck develops to a certain extent, its connection with the snow crystal changes, no longer presenting the shape described in Figure 4, but instead becoming an irregular shape close to the snow crystal.

More and more snow necks undergo the above evolution, leading to a decrease in neck area, while snow crystals continue to grow. At this time, the two show a negative correlation.

3.4. Relationship between Snow Crystal Neck Area and Hardness

The snow crystal neck region displayed an overall tendency to grow and then decrease as the number of metamorphism days increased. The snowpack hardness value peaked at approximately one month into the metamorphic process, and then it steadily declined as the neck region shrank, resulting from the formation of new crystals and necks at a later stage (Figure 5a,b,d). The following Pearson correlation analysis was performed using the snow hardness of the snow that received solar radiation from day 0 to day 50 of the metamorphic time, as well as the equivalent snow crystal grain diameter, neck circumference, neck area, and snow specimen density data, to further investigate the relationship between snow crystal microstructure and snow hardness (Table 1). With a correlation coefficient of 0.711 and passing the significance test at the 0.05 level, the results demonstrate a positive link between snow hardness and the neck area of snow crystal particles. To perform additional regression analysis on the influence of hardness, this article chooses the neck area of the snow samples.

Table 1. Correlation analysis of the hardness of metamorphic snow with the density and microscopic parameters of snow crystals.

	Snow Hardness (kPa)	Equivalent Neck Diameter (μm)	Neck Circumference (μm)	Snowpack Density (kg/m^3)	Neck Area (μm^2)
Snow hardness (kPa)	/	0.443	0.459	0.041	0.650 *
Equivalent neck Diameter (μm)	0.443	/	0.997 **	0.804 **	0.919 **
Neck circumference (μm)	0.459	0.997 **	/	0.783 **	0.926 **
Snowpack density (kg/m^3)	0.041	0.804 **	0.783 *	/	0.688 **
Neck area (μm^2)	0.650 *	0.919 **	0.926 **	0.688 **	/

Notes: * $p < 0.05$.; ** $p < 0.01$.

Based on a multiple linear regression analysis of the observed data, the correlation coefficient R between metamorphic snow hardness and the neck area of snow crystals was 0.711 (Figure 10), indicating a strong linear relationship between the two variables and positive overall regression effects. The regression equation's p value within 0.01, which was statistically significant and passed the significance test, showed that the neck area of the snow crystals in the snow specimens had a substantial impact on the snow hardness. Consequently, the regression equation for snow hardness in this examination scenario is presented below:

$$H = 0.002764S + 67.922837 \quad (1)$$

where H represents the predicted value of the snow hardness regression equation in kPa and S represents the snow crystal neck area in μm^2.

The snow crystal neck area and snow hardness have a good correlation, as shown in Figure 11. This finding demonstrates how many of the physical and mechanical characteristics of snow, including hardness and shear strength, are significantly influenced by the area around the snow crystal neck. The explanation for this is that ice particles may transfer loads to snow necks without experiencing significant deformation because they have a higher mass and are more rigid structurally than snow necks. The stresses within the snow crystals are exposed to the loading peak at the neck, similar to any cohesive granular material; in other words, the neck region bears the most stress during penetration tests. On the other hand, snow necks are more likely than ice grains to deform when they are put under additional strain and start to give and flow. However, snow triaxial measurements reveal that the stresses at the snow neck are 5 to 50 times greater than those at the grains [33], which is sufficient to demonstrate the influence of the neck area on the physical characteristics of the entire snow body.

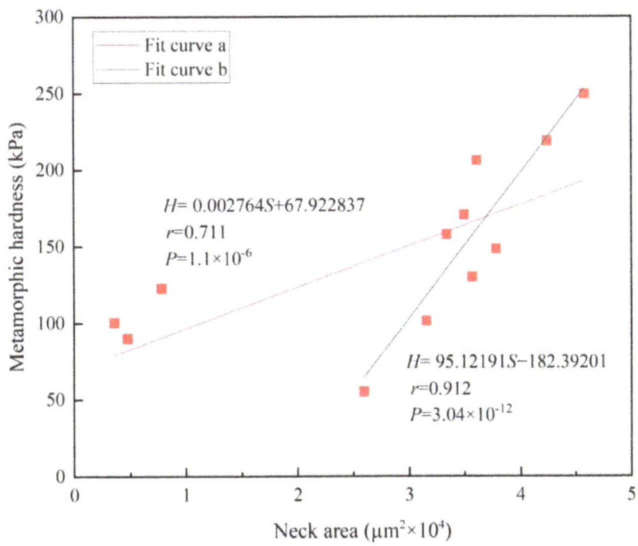

Figure 11. Curve of the relationship between the neck area and hardness of metamorphic snow in the solar radiation group (red line: fitting curve obtained based on all metamorphism times; black line: fitting curve obtained after ten days of metamorphism).

The three leftmost points in Figure 11 should be noted as they dramatically diverge from the general trend. These three measurements are the hardness and neck area taken on days 0, 2, and 5 at the start of the metamorphosis. The quick slumping and sintering modifications of the new snow intensify the heterogeneity of the metamorphosis of the layers and make the three measurements more distinct. If we account for only the metamorphism that occurs after ten days, as indicated by the black line, the correlation (r = 0.912) is noticeably larger when the same method is used to analyze the data. This result better illustrates how the neck of the snow crystals affects the snowpack's hardness.

4. Discussion

Combining field measurements and meteorological satellite data from the European Centre for Medium-Range Weather Forecasts, Table 2 lists the main meteorological conditions and the pertinent physical characteristics in the snow box of the solar radiation group during the test period. The mean ambient temperature readings, for instance, are based on the arithmetic mean of the temperature data collected in the field between the two tests. Two of the atmospheric forcing data for controlling snow cover metamorphism, solar radiation and wind speed, were sourced from the ERA5 Land dataset of the European Centre for Medium Range Weather Forecasts. The selected grids (46°36′ N–46°42′ N, 127°24′ E–127°30′ E) covered the experimental site (46°39′ N, 127°25′ E), with a time resolution of one hour and a spatial resolution of 0.1° × 0.1°. The data were collected by the Copernicus Climate Change Service (https://cds.climate.copernicus.eu/ (accessed on 1 April 2023)) and distributed at no cost through the Climate Data Store. The data from ERA5-Land have been applied to northeastern China on several occasions [34], and the data are in good agreement with the measured values [35]. Snow is a porous three-phase structure, and air humidity directly affects the surface of snow to change its moisture content and affect its metamorphism [36,37]. Considering the lack of humidity data in ERA5 Land, data from the nearest Bayan County meteorological station (46°3′ N, 127°23′ E) were selected, and the data were collected from the National Meteorological Information Center China Meteorological Data Network (http://www.nmic.cn/ (accessed on 1 April 2023)).

Table 2. Changes in the neck area of deteriorated snow cover and its influencing factors over time.

Days to Metamorphism (day)	Average Neck Area (μm^2)	Average Ambient Temperature (°C)	Cumulative Solar Radiation ($W \cdot h/m^2$)	Snow Density (kg/m^3)	Average Wind Speed (m/s)	Average Specific Humidity (g/kg)
0	3593.9	−13.5	1137.4	200.0	2.9	0.9
2	4750.5	−16.3	1668.5	207.9	1.5	0.6
4	7791.9	−17.1	2086.2	249.8	1.8	0.7
10	33,403.8	−15.5	7002.8	254.2	2.8	0.6
15	36,121.9	−16.7	5639.1	261.9	1.9	0.6
20	35,700.4	−8.0	5586.1	270.5	3.3	1.5
25	34,969.3	−18.3	6280.4	272.5	2.3	0.5
30	45,791.3	−16.2	6335.2	300.5	2.3	0.5
35	42,391.0	−13.8	7763.6	302.5	1.8	0.3
40	37,831.3	−12.7	8366.3	307.6	1.8	0.7
45	31,541.3	−12.7	8270.2	311.0	1.7	0.8
50	25,977.3	−12.9	8117.7	317.0	2.1	1.2

After a significance test, the regression equation for the snowpack neck area had a correlation coefficient of $r = 0.919$. Below is the multiple linear regression equation for the test conditions' snowpack neck area:

$$S = 287.883T + 2.188E + 188.983\rho + 12,194.499V - 20,443.081RH - 42,729.115 \quad (2)$$

where S represents the snow crystal neck area in μm^2; T represents the average ambient temperature in °C; E represents the cumulative value of solar radiation in $W \cdot h/m^2$; ρ represents the snow density in kg/m^3; V represents the average wind speed in m/s and RH represents the average specific humidity in g/kg.

Despite the high correlation coefficient of Equation (2), it is not directly evident how the five variables are correlated with the degree of influence on the neck area. Because of this rationale, the data analysis below involved a principal component analysis and factor analysis for the five variables that impact neck area in this statistic. The results of the principal component analysis that were obtained are presented in Table 3.

Table 3. Principal component results: total variance in explanation.

Ingredient	Initial Eigenvalue			Extraction of the Sum of Squares and Load		
	Total	Percentage of Variance (%)	Cumulative Percentage (%)	Total	Variance	Cumulative Percentage (%)
1	2.201	44.024	44.024	2.201	44.024	44.024
2	1.997	39.945	83.970	1.997	39.945	83.970
3	0.524	10.487	94.457	/	/	/
4	0.213	4.259	98.715	/	/	/
5	0.064	1.285	100.000	/	/	/

Table 3 shows that the first two principal components account for 84% of the variance, demonstrating that the neck region may be evaluated with some degree of certainty as the two extracted principal components can account for 84% of the five variables indicated above. According to the aforementioned study, the two major components were identified as $Y1$ and $Y2$, and the linear combination of $Y1$ and $Y2$ was then obtained as follows:

$$Y1 = 0.613T + 0.286E + 0.273\rho - 0.404V + 0.551RH \tag{3}$$

$$Y2 = -0.057T + 0.611E + 0.629\rho - 0.382V - 0.284RH \tag{4}$$

The four atmospheric forcing factors mentioned above, along with snow density, responded to 83.97 percent of the variables that determine how much the snow crystals' necks vary, indicating that they may be the main factors influencing the neck area. Further factor analysis was carried out on Equations (3) and (4) to determine the normalized principal component score coefficients based on the coefficient analyses of the mean ambient temperature (T), cumulative solar radiation (E), snow density (ρ), wind speed (V) and specific humidity (RH), as shown in Figure 12. From Figure 12, it can be seen that the three factors with the highest proportion are density, solar radiation, and temperature. The specific humidity plays a supplementary role in explaining the changes in neck area, with wind speed contributing the lowest, at only 2.2%, and the neck area is not sensitive to its influence. Microstructure, temperature, and sedimentation all have intricate feedback relationships with one another [29]. Four different forms of atmospheric forcing changes affect the direction and rate of internal microstructural changes. The neck area represents microstructural changes that are intrinsic to changes in the internal physical properties of the snowpack. For instance, pressure sintering causes the growth of snow crystal necks and bonds [38], and the growth of the neck region boosts the thermal conductivity and heat transmission of snow while allowing the bonding strength between grains to grow and the settling rate to decrease. Since there is less of a temperature difference in the snow's vertical direction because of the enhanced heat transfer, there is less vapor flow, which prevents the growth of the snow neck. Additionally, when pressure sintering slows due to the decrease in settling, neck growth will also slow. Then, the aforementioned processes can be reversed or cycled in response to changes in the ambient temperature, with snow layer thermal anisotropy being a key factor in temperature gradient metamorphism [39]. The solar radiation directly heats the snow layer to affect the metamorphism of snow cover, and its contribution cannot be ignored. Humidity and wind speed also supplement the changes in the neck. Environmental humidity, as an important factor in the variation of moisture content in snow, plays a role in the growth of snow crystals. The wind speed directly affects the particle morphology and bonding mode on the top snow surface, leading to changes in transmittance and thermal conductivity, affecting the microstructure of surface snow and altering the energy input of the snow sample from the outside.

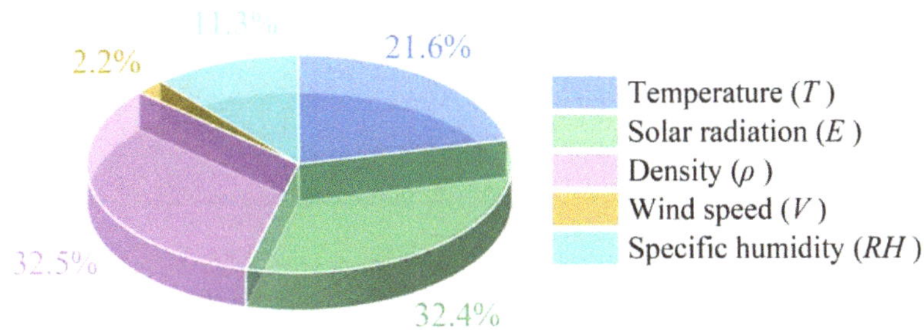

Figure 12. Normalized principal component score coefficient.

5. Conclusions

(1) The metamorphosis of the snow neck area is influenced by environmental factors including temperature, solar radiation, humidity and wind stress. The change in neck area is significantly affected by solar radiation and temperature, with the surface neck

region absorbing water vapor from lower layers, resulting in an increase in area during temperature surges. Conversely, the remaining snow necks undergo varying degrees of decrease due to vapor diffusion. The overall neck area displayed a declining pattern, which accounts for the fluffy appearance of the snow samples and the reduction in hardness numbers as the surrounding temperature rose. Humidity complemented the change in the neck area, while wind speed had the least effect.

(2) After ten days of metamorphosis, snow hardness had a strong positive association with the snow crystal neck area ($r = 0.912$). The weak link in the snow crystal linkage is the neck area, and one of the key determinants of the snow's mechanical properties is the area of the neck. The neck region better explains the anomaly that the metamorphosed snow's hardness initially rises and then falls, in contrast to the snow density, which constantly grows throughout the metamorphic process. Additionally, this demonstrates how modifications in the snowpack microstructure are inextricably linked to modifications in its macroscopic physical attributes.

(3) When atmospheric forcing is more steady, the neck area of each layer of the snowpack grows with snow depth. Pressure sintering, or the impact of the snowpack's makeup on the neck area, is the primary source of this phenomenon. The electronically controlled penetrometer hardness test further revealed that the hardness value increased as the depth of penetrated snow increased. After penetration, there was no discernible accumulation in the hole's vertical direction, indicating that the increase in hardness value was primarily brought on by an expansion of the snow neck's surface area in the depth direction as opposed to compression brought on by the vertical accumulation of snow during the penetration process.

In contrast to previous studies on snow microstructure, the present experiment was conducted to investigate the statistical changes in neck area during snow metamorphosis and the meteorological factors that affect it. Furthermore, the experiment aimed to elucidate how the snow neck area is affected by density, ambient temperature, wind speed, and solar radiation, as well as how the neck area affects snow hardness. In the future, further recording of more detailed environmental meteorological data and more microstructural parameters can be conducted. In addition, it is necessary to use improved methods to observe the snow accumulation during the melting period. At this stage, the nature of snow changes rapidly and is complex, which is of great significance for studying snow metamorphism.

Author Contributions: Conceptualization, P.L.; methodology, P.L. and J.W.; investigation, J.W., S.H., Q.Z. and Q.W.; data curation., S.H. and Q.Z.; formal analysis, J.W. and P.H.; visualization, J.W. and S.Y.; writing—original draft preparation, J.W.; writing—review and editing, J.W., P.L. and P.H.; project administration, P.L. All authors have read and agreed to the published version of the manuscript.

Funding: This research was funded by the National Key Research and Development Program of China (grant number 2022YFE0107000), the National Natural Science Foundation of China (grant number 42320104004) and the Major Scientific and Technological Projects of the Ministry of Water Resources of China (grant number SKS-2022017).

Data Availability Statement: Data are contained within the article.

Conflicts of Interest: The authors declare no conflict of interest.

References

1. Li, H.; Zhong, X.; Zheng, L.; Hao, X.; Wang, J.; Zhang, J. Classification of snow cover persistence across China. *Water* **2022**, *14*, 933. [CrossRef]
2. Calonne, N.; Flin, F.; Geindreau, C.; Lesaffre, B.; Rolland, D.R.S. Study of a temperature gradient metamorphism of snow from 3-D images: Time evolution of microstructures, physical properties and their associated anisotropy. *Cryosphere* **2014**, *8*, 2255–2274. [CrossRef]
3. Schneebeli, M. The importance of the microstructure of snow in nature and engineering. In Proceedings of the 1st International Conference on Design and Nature, Udine, Italy, 21 January 2002. [CrossRef]

4. Sarmiento, J.; Slater, R.; Barber, R.; Bopp, L.; Doney, S.; Hirst, A.; Kleypas, J.; Matear, R.; Mikolajewicz, U.; Monfray, P.; et al. Response of ocean ecosystems to climate warming. *Glob. Biogeochem. Cy.* **2004**, *18*, GB3003. [CrossRef]
5. Huo, P.; Lu, P.; Cheng, B.; Zhang, L.; Wang, Q.; Li, Z. Monitoring Ice Phenology in Lake Wetlands Based on Optical Satellite Data: A Case Study of Wuliangsu Lake. *Water* **2022**, *14*, 3307. [CrossRef]
6. Zhao, Q.; Li, Z.; Lu, P.; Wang, Q.; Wei, J.; Hu, S.; Yang, H. An Investigation of the Influence on Compacted Snow Hardness by Density, Temperature and Punch Head Velocity. *Water* **2023**, *15*, 2897. [CrossRef]
7. Xie, F.; Lu, P.; Li, Z.; Wang, Q.; Zhang, H.; Zhang, Y. A floating remote observation system (FROS) for full seasonal lake ice evolution studies. *Cold Reg. Sci. Technol.* **2022**, *199*, 103557. [CrossRef]
8. Han, H.; Yang, M.; Liu, X.; Li, Y.; Gao, G.; Wang, E. Study on the Constitutive Equation and Mechanical Properties of Natural Snow under Step Loading. *Water* **2022**, *15*, 3271. [CrossRef]
9. Ballesteros-Canovas, J.; Trappmann, D.; Madrigal-Gonzalez, J.; Eckert, N.; Stoffel, M. Climate warming enhances snow avalanche risk in the Western Himalayas. *Proc. Natl. Acad. Sci. USA* **2018**, *115*, 3410–3415. [CrossRef]
10. Medeu, A.; Blagovechshenskiy, V.; Gulyayeva, T.; Zhdanov, V.; Ranova, S. Interannual Variability of Snowiness and Avalanche Activity in the Ile Alatau ridge, northern Tien Shan. *Water* **2022**, *14*, 2936. [CrossRef]
11. Brown, R.; Edens, M. On the relationship between neck length and bond radius during compression of snow. *J. Glaciol.* **1991**, *37*, 203–208. [CrossRef]
12. Mellor, M. Engineering properties of snow. *J. Glaciol.* **1977**, *19*, 15–66. [CrossRef]
13. Lee, J.H.; Huang, D. Material point method modeling of porous semi-brittle materials. In Proceedings of the 9th World Congress on Computational Mechanics and 4th Asian Pacific Congress on Computational Mechanics, Sydney, Australia, 19–23 July 2010. [CrossRef]
14. Edens, M.; Brown, R. Measurement of microstructure of snow from surface sections. *Def. Sci. J.* **1995**, *45*, 107–116. [CrossRef]
15. Zhou, S.; Nakawo, M.; Hashimoto, S.; Sakai, A.; Narita, H.; Ishikawa, N. Densification and grain coarsening of melting snow. *J. Glaciol.* **2002**, *24*, 275–281. [CrossRef]
16. Libbrecht, K.G. Morphogenesis on ice: The physics of snow crystals. *Rep. Prog. Phys.* **2001**, *1*, 10–19. [CrossRef]
17. Gubler, H. Determination of the Mean Number of Bonds per snow grain And of the Dependence of the Tensile Strength of Snow on Stereological Parameters. *J. Glaciol.* **1978**, *20*, 329–341. [CrossRef]
18. Edens, M.Q.; Brown, R.L. Changes in microstructure of snow under large deformations. *J. Glaciol.* **1991**, *37*, 193–202. [CrossRef]
19. Brown, R.L. A volumetric constitutive law for snow based on a neck growth model. *J. Appl. Phys.* **1980**, *51*, 161–165. [CrossRef]
20. Kry, P.R. The relationship between the visco-elastic and structural properties of fine-grained snow. *J. Glaciol.* **1975**, *14*, 479–500. [CrossRef]
21. Hansen, A.C.; Brown, R.L. The granular structure of snow: An internal-state variable approach. *J. Glaciol.* **1986**, *32*, 434–438. [CrossRef]
22. Shapiro, L.H.; Johnson, J.B.; Sturm, M.; Blaisdell, G.L. Snow mechanics review of the state of knowledge and applications. In *US Army Cold Regions Research and Engineering Laboratory (CRREL) Report 97-3*; Cold Regions Research and Engineering Laboratory: Hanover, NH, USA, 1997. [CrossRef]
23. Yang, Q.; Song, K.; Hao, X.; Wen, Z.; Tan, Y.; Li, W. Investigation of spatial and temporal variability of river ice phenology and thickness across Songhua River Basin, northeast China. *Cryosphere* **2020**, *11*, 3581–3593. [CrossRef]
24. Wang, E.; Fu, X.; Han, H.; Liu, X.; Xiao, Y.; Leng, Y. Study on the mechanical properties of compacted snow under uniaxial compression and analysis of influencing factors. *Cold Reg. Sci. Technol.* **2021**, *182*, 103215. [CrossRef]
25. Feng, Z. Experimental Study on Snow Hardness and Its Testing Technology. Master's Thesis, Dalian University of Technology, Dalian, China, 2019. (In Chinese). [CrossRef]
26. Casado, M.; Landais, A.; Picard, G.; Arnaud, L.; Dreossi, G.; Stenni, B.; Prié, F. Water Isotopic Signature of Surface Snow Metamorphism in Antarctica. *Geophys. Res. Lett.* **2021**, *48*, e2021GL093382. [CrossRef]
27. Ma, H.; Liu, Z.; Sun, L.; Cai, G. A one-dimensional solution to the effect of solar radiation on the temperature field of snow cover: A case study of seasonal snow cover in the Western Tianshan Mountains of China. *Sci. Bull.* **1992**, *37*, 1205–1208. (In Chinese) [CrossRef]
28. Apaloo, J.; Brenning, A.; Bodin, X. Interactions between Seasonal Snow Cover, Ground Surface Temperature and Topography (Andes of Santiago, Chile, 33.5° S). *Permafrost Periglac.* **2013**, *23*, 277–291. [CrossRef]
29. Lehning, M.; Bartelt, P.; Brown, B.; Fierz, C.; Satyawali, P. A physical SNOWPACK model for the swiss avalanche warning-Part II. Snow microstructure. *Cold Reg. Sci. Technol.* **2002**, *35*, 147–167. [CrossRef]
30. Colbeck, S. Theory of metamorphism of dry snow. *J. Geophys. Res-Oceans* **1983**, *88*, 5475–5482. [CrossRef]
31. Pinzer, B.R.; Schneebeli, M. Snow metamorphism under alternating temperature gradients: Morphology and recrystallization in surface snow. *Geophys. Res. Lett.* **2009**, *36*, L23503. [CrossRef]
32. Donahue, C.; Skiles, S.M.; Hammonds, K. In situ effective snow grain size mapping using a compact hyperspectral imager. *J. Glaciol.* **2021**, *67*, 49–57. [CrossRef]
33. Bartelt, P.; von Moos, M. Triaxial tests to determine a microstructure-based snow viscosity law. In Proceedings of the International Symposium on the Verification of Cryospheric Models, Zurich, Switzerland, 14 January 2000. [CrossRef]
34. Yue, S.; Yan, Y.; Zhang, S.; Yang, J.; Wang, W. Spatiotemporal variations of soil freeze-thaw state in Northeast China based on the ERA5-LAND dataset. *Acta. Geogr. Sin.* **2021**, *76*, 2765–2779. [CrossRef]

35. Pelosi, A.; Chirico, G. Regional assessment of daily reference evapotranspiration: Can ground observations be replaced by blending ERA5-Land meteorological reanalysis and CM-SAF satellite-based radiation data? *Agric. Water Manag.* **2021**, *258*, 107169. [CrossRef]
36. Flin, F.; Brzoska, J.B.; Lesaffre, B.; Coléou, C.C.; Pieritz, R.A. Three-dimensional geometric measurements of snow microstructural evolution under isothermal conditions. *Ann. Glaciol.* **2004**, *38*, 39–44. [CrossRef]
37. Slaughter, A.E.; Adams, E.E.; Staron, P.J.; Shertzer, R.H.; Walters, D.J.; McCabe, D.; Catherine, D.; Henninger, I.; Leonard, T.; Cooperstein, M.; et al. Field investigation of near-surface metamorphism of snow. *J. Glaciol.* **2011**, *57*, 441–452. [CrossRef]
38. Hansen, A.C.; Brown, R.L. A new constitutive theoiy for snow based on a micromechanical approach. In Proceedings of the Davos Symposium, Davos, Switzerland, 29 January 1987.
39. Izumi, K.; Huzioka, T. Studies of metamorphism and thermal conductivity of snow. I. *Cold Reg. Sci. Technol.* **1975**, *33*, 91–102. [CrossRef]

water

MDPI

Article

Ice Mass Balance in Liaodong Bay: Modeling and Observations

Yuxian Ma [1,2,3], Dewen Ding [1,2], Ning Xu [2], Shuai Yuan [2] and Wenqi Shi [2,*]

1 College of Environmental Science and Engineering, Ocean University of China, Qingdao 266100, China
2 National Marine Environmental Monitoring Center, Dalian 116023, China
3 State Key Laboratory of Coastal and Offshore Engineering, Dalian University of Technology, Dalian 116024, China
* Correspondence: wqshi@nmemc.org.cn

Abstract: During the winters of 2009/2010 and 2020/2021, observations were carried out at an eastern port of Liaodong Bay to examine the variations in sea ice thickness and atmospheric conditions. The daily ice thickness (DIT) and the cumulative ice thickness (CIT) are the two main observation items related to the thickness of sea ice. For DIT, the sea ice thickness gradually decreases as the temperature increases, and the freezing rate a is 1.48 cm/(°C·d)$^{1/2}$. For CIT, when the temperature is −12 °C, the maximum growth rate of ice thickness decreases from 3.5 cm/d to 1.5 cm/d as the ice thickness increases from 0 to 20 cm. The residual method was applied to calculate the oceanic heat flux, which is an important parameter of ice modeling, and both the analytic model (Stefan's law) and numerical model (high-resolution thermodynamic snow-and-ice model) were utilized in this work. It was found that the accuracy of the simulation results was high when the growth coefficient of the analytic mode was 2.3 cm/(°C·d)$^{1/2}$. With an oceanic heat flux of 2 W·m^{-2}, the maximum error of the numerical model approached 60% in 2010 and 3.7% in 2021. However, using the oceanic heat flux calculated in this work, the maximum error can be significantly reduced to 4.2% in the winter of 2009/2010 and 1.5% in 2020/2021. Additionally, the oceanic heat flux in Liaodong Bay showed a decreasing trend with the increase in ice thickness and air temperature.

Keywords: Liaodong Bay; sea ice thickness; Stefan's law; HIGHTSI; oceanic heat flux

check for updates

Citation: Ma, Y.; Ding, D.; Xu, N.; Yuan, S.; Shi, W. Ice Mass Balance in Liaodong Bay: Modeling and Observations. *Water* **2023**, *15*, 943. https://doi.org/10.3390/w15050943

Academic Editors: Zhijun Li, Fang Li, Sasan Tavakoli, Xuemei Liu and Changlei Dai

Received: 2 February 2023
Revised: 26 February 2023
Accepted: 26 February 2023
Published: 1 March 2023

1. Introduction

Global climate change is altering the movement of salts, gases, and nutrients in the ocean–ice–atmosphere system [1–5]. Over the past 70 years, the severity of sea ice in the Bohai Sea has weakened [6–10], and the increased volatility has led to a greater risk of sea ice hazards [11,12]. In order to clarify the relationship between climate change and sea ice in the Bohai Sea and improve the prediction of sea ice thickness, it is important to carry out both direct observations and numerical simulations of sea ice mass balance in this area [13–15].

At present, most observations and numerical simulations of the ice mass balance process focus on the polar and sub-polar regions [16,17]. Researchers observed the ice thickness, ice temperature, and ice salinity of sea ice during both the ice growth and melting periods [18], and the collected data were used to optimize the classical sea ice thermodynamic model [19]. In addition, the data were used to determine the parameters of thermodynamic modes. It was found that the oceanic heat flux plays a significant and important role in the accurate simulation of ice, and the oceanic heat flux has different characteristics depending on region and season [20,21]. However, due to climate change, the Arctic sea ice is gradually shifting from multi-year ice to first-year ice [22], and the research priority will become seasonal sea ice in the future. As the southern boundary of the frozen sea area in the Northern Hemisphere, the research on Bohai Sea ice can be an important reference for Arctic first-year ice. At present, the research works on Bohai Sea ice have mainly focused on the ice extent in recent years; for example, Zhang [23] used machine learning methods to develop a novel empirical model with the aim to predict the sea ice area. However, the thermodynamic process of landfast ice has not been carried out. The last thermodynamic observations of landfast ice can be traced to the winter of 1989/1990, when China and Finland conducted field surveys of meteorology, sea ice temperature, and thickness in the Bayuquan Port, and obtained 8 days of continuous hourly data [24]. Based on these data, Cheng [25,26] carried out the test of the high-resolution thermodynamic snow-and-ice model (HIGHTSI), in which the oceanic heat flux was chosen as a constant 5 W/m^2. Because the oceanic heat flux has its own characteristics in each sea area, Ji and Yue used floating ice thickness and meteorological data collected from the Liaodong Bay JZ20-2 platform and calculated the oceanic heat flux in the 1997/1998 ice season. It was found that the largest value was 200 W/m^2 during the initial ice period, which then decreased to 0 during the melting period [27]. The disadvantage is that the observed data of Ji and Yue were affected by dynamic factors, and we are still lacking knowledge about the oceanic heat flux of landfast ice, although it has been identified to have an important influence on the numerical simulation of sea ice in Liaodong Bay [28–32]. Therefore, the continuous mass balance observation of landfast ice in the Bohai Sea is of great significance for seasonal sea ice research.

In this study, the relationship between sea ice thickness and air conditions was observed in Jiangjunshi Port during the winters of 2009–2010 and 2020–2021. The residual method was applied to calculate the oceanic heat flux and to analyze the factors that affect the flux. Furthermore, based on the observed sea ice thickness and meteorological data as inputs, the sea ice growth and decay were evaluated using both an analytic model and numerical simulation.

2. Observation Area and Data

2.1. Observation Area

The state of sea ice in the Bohai Sea is influenced by global climate change, and Liaodong Bay usually has the maximum ice conditions in terms of ice extent, thickness, concentration, and duration [33]. Meanwhile, ice can be observed every winter season in Liaodong Bay.

Prior to carrying out air–ice–water observations in Liaodong Bay, the primary challenge was to locate a suitable landfast ice site that meets specific criteria. The selected area must allow seawater to flow under the sea ice while remaining stationary under the ocean current. Accordingly, Jiangjunshi Port, which has severe ice conditions, was selected as the observation site. Figure 1 shows the location of Jiangjunshi Port in Liaodong Bay, with coordinates of 39°55′7.72″ N, 121°40′40.77″ E. There is a small port that is approximately 0.16 km^2 in Jiangjunshi Port. The water depth is 8 m, and the seawater salinity is 28‰. This location is ideal for observing the thermodynamic process of sea ice in Liaodong Bay, as it is sheltered from the dynamic factors such as currents and waves. Usually, the ice season begins in December and extends until the end of March [34].

2.2. Sea Ice Thickness and Atmospheric Conditions in the Winter of 2009–2010

Atmospheric data were collected during the winter of 2009–2010 using the automatic meteorological equipment positioned 10 m above the ice surface. The collected elements included wind speed, wind direction, air temperature, humidity, and pressure, with data recorded every 10 min. The measurement accuracies of wind speed, wind direction, and air temperature were 0.01 m/s, 0.01 degrees, and 0.01 °C, respectively. The observation period covered the entire ice season, as presented in Figure 2. On 5 January, the air temperature hit a lowest value of −17.99 °C.

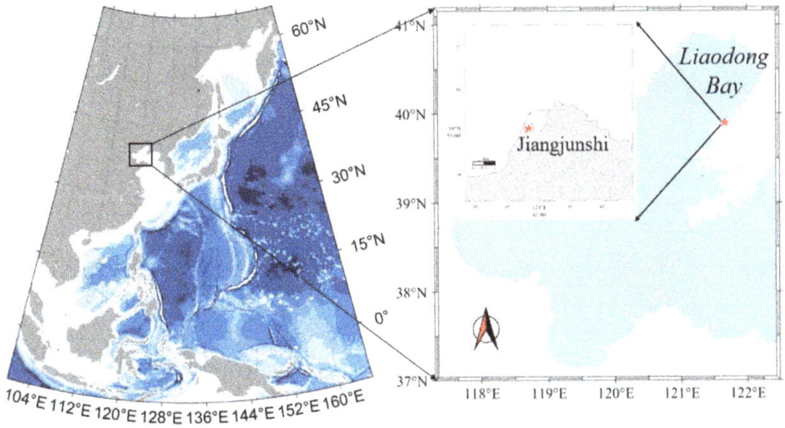

Figure 1. Locations of landfast ice survey site in Jiangjunshi Port.

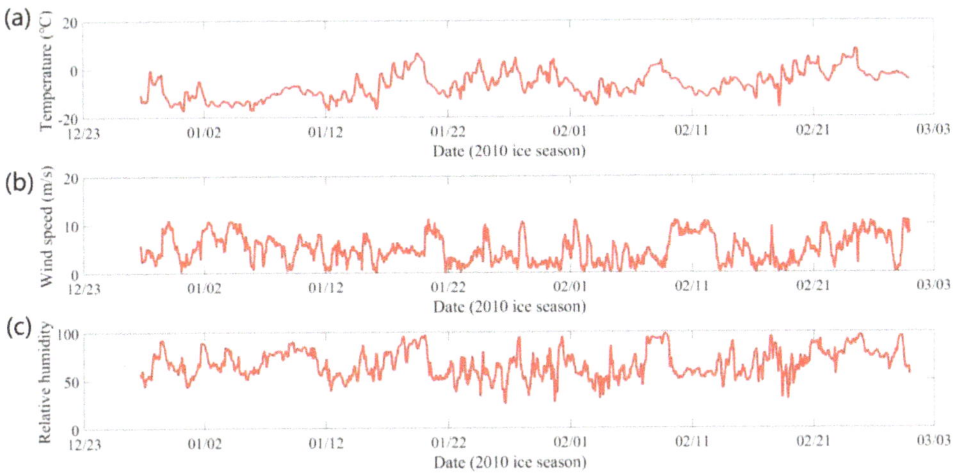

Figure 2. Time series (2009/2010) of (**a**) hourly measured air temperature, (**b**) hourly measured wind speed, (**c**) hourly measured humidity.

During the 2010 ice season, two sea ice observation options were available at the port covered with landfast ice. The following observation steps were taken: (1) Two areas of 2×2 m were cut into the landfast ice, and the ice within these areas was removed. To ensure that the two areas did not affect each other, they were placed 50 m apart and named ice zone 1 (IZ1) and ice zone 2 (IZ2), respectively. (2) The growth thickness was observed in IZ1 for 24 h. After measuring the ice thickness each day, the thickness was recorded as H1 and then all the sea ice in IZ1 was removed. (3) The CIT was observed in IZ2, which was measured at 9 a.m. every day and recorded as H2.

In addition, it was observed that the sea ice occasionally broke due to the tide. The broken sea ice drifted out of the port, and it interrupted the observation of sea ice thickness. According to our observations, the first period of interruption occurred from 11 January to 19 January 2010 for a total of 9 days, with an average air temperature of −7.9 °C. The second period of interruption occurred from 23 January to 15 February 2010 for a total of 24 days, with an average air temperature of −5.8 °C. Throughout the observation period, the temperature remained low, and the sea ice growth was rapid.

2.3. Sea Ice Thickness and Atmospheric Conditions in the Winter of 2020–2021

Atmospheric conditions were also observed during the observation period using automatic meteorological equipment. The installed sensor was the MaxiMet series GMX500, and the observed elements included air temperature, wind speed, wind direction, humidity, and air pressure. The measurement resolution for wind speed was 0.01 m/s, for wind direction it was 0.01°, for humidity it was 1%, and for temperature it was 0.01 °C. The data were recorded every 5 s, and the recording period was from 27 December 2020 to 1 March 2021. Figure 3 presents the time histories of those meteorological data. The wind speed data contain some missing parts, which were caused by the heavy frozen state of the sensor. In the third section of the model calculation, the missing data are replaced by ERA5 reanalysis data.

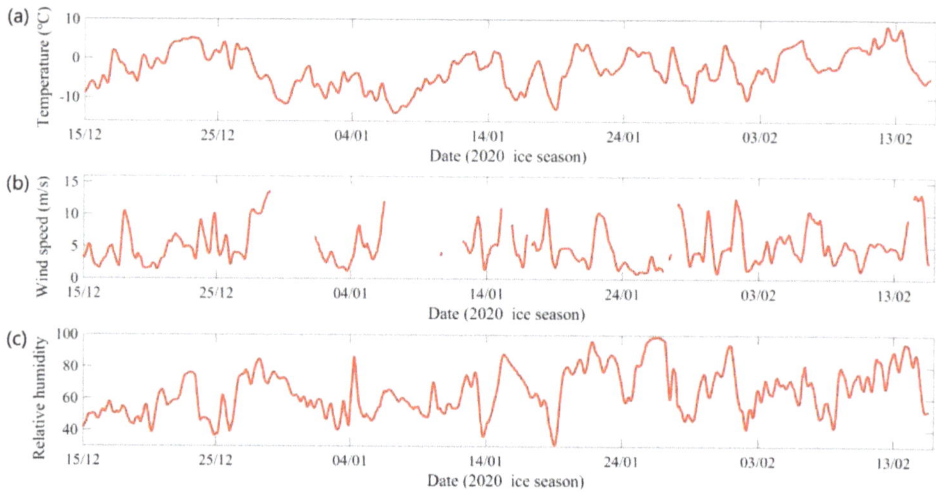

Figure 3. Time series (2020/2021) of (**a**) hourly measured air temperature, (**b**) hourly measured wind speed, (**c**) hourly measured humidity.

On 31 December 2020, the sea ice entered the port and froze completely. On 3 January, the measurement of ice thickness started. The measurement was conducted using an ice ruler, and the thickness was observed at 8:00 and 17:00 during the ice period. On 4 February, the tide caused the sea ice to break and the measurement of ice thickness was completed.

3. Methods

3.1. Thermodynamic Model of Sea Ice

3.1.1. Stefan's Law of Ice Growth

There are two types of thermodynamic models for sea ice: an analytical model and numerical simulation. For the analytical model, Stefan established a formula to calculate ice thickness, which only considers the heat balance at the ice–water interface. The heat released from the freezing occurring at the bottom of the ice layer is transferred to the surface of the ice layer through a constant temperature gradient. This formula is based on 4 basic assumptions: (1) under rapid temperature changes, the lag in ice temperature change is ignored; (2) the solar radiation absorbed by ice is ignored; (3) the heat flux at the bottom of the ice is ignored; and (4) the surface temperature is a function of time [35–37]. The calculation formula is:

$$\rho_i L_f \frac{dh_i}{dt} = \frac{k_i \left(T_f - T_0\right)}{h_i} \tag{1}$$

where k_i is the thermal conductivity of the ice; ρ_i is ice density; L_f is freezing latent heat; T_f is the freezing temperature; and T_0 is the ice surface temperature. According to Formula (1), the ice thickness can be calculated. The initial ice thickness is h_0 at the initial time $t = 0$, and both sides of Formula (1) are integrated simultaneously:

$$h_i = \sqrt{h_0^2 + a^2 FDD} \tag{2}$$

$$a = \sqrt{2k_i/\rho_i L_f} \tag{3}$$

$$FDD = \int_0^t \left(T_f - T_0\right) \tag{4}$$

3.1.2. High-Resolution Thermodynamic Snow-and-Ice Model (HIGHTSI)

Most numerical models of sea ice were developed by combining the principles of energy balance and heat conduction. In this study, the high-resolution thermodynamic snow-and-ice (HIGHTSI) model [38–43] proposed by Maykut and Untersteiner [44] was applied to simulate the growth of sea ice thickness. The HIGHTSI model follows the classical one-dimensional sea ice model, and its core is the partial differential thermal conductivity equation which considers the vertical heat and mass balance through the snow–ice–ocean system. The HIGHTSI model has been widely applied to simulate the ice thermodynamics in various locations, such as, e.g., the Bohai Sea, Antarctic sea, Arctic sea, Baltic Sea, and Finnish lakes. The key processes include the surface heat and mass balance (Equation (5)), the snow and ice temperature (Equation (6)), and the ice bottom heat and mass balance (Equation (7)):

$$(1-\alpha_{i,s})Q_s - I(z)_0 + \varepsilon Q_d - Q_b\left(T_{sfc}\right) + Q_h\left(T_{sfc}\right) + Q_{le}\left(T_{sfc}\right) + F_c\left(T_{sfc}\right) - F_m = 0 \tag{5}$$

where Q_s is the downward solar radiation for all sky conditions; α is the surface albedo; $I(z)$ is the solar radiation penetrating below the surface layer; Q_d and Q_b are the downward and upward longwave radiation under all sky conditions; ε is surface emissivity; Q_h and Q_{le} are turbulent sensible and latent heat fluxes; F_c is the conductive heat flux of the surface layer; F_m is the surface melting of snow or ice; and T_{sfc} is surface temperature.

$$(\rho c)_{i,s}\frac{\partial T_{i,s}(z,t)}{\partial t} = \frac{\partial}{\partial z}\left(k_{i,s}\frac{\partial T_{i,s}(z,t)}{\partial z} - q(z,t)\right) \tag{6}$$

where T is temperature; t is time; z is the vertical coordinate below the surface; ρ is density; c is specific heat; k is thermal conductivity (function of T_i and s_i); $q(z,t)$ is the absorbed solar radiation below the surface layer; and the subscripts s and i are snow and ice, respectively.

$$\rho_i L_i \frac{dh_i}{dt} + F_w = \left(k_i \frac{\partial T_i}{\partial z}\right)_{bot} \tag{7}$$

where h_i is sea ice thickness; L_i is the latent heat of fusion; and F_w is the oceanic heat flux.

The inputs of the mode are wind speed (Va); temperature (Ta); relative humidity (Rh); precipitation; and solar radiation (parameterized). The outputs of the mode are the time series of ice thickness (h_i) and ice temperature.

3.2. Statistical Method

3.2.1. Least Squares Method

The least squares method is commonly used for parameter estimation, system identification, and prediction. In this study, the least squares method was used to estimate the parameters in the relationship between the sea ice thickness and the air temperature data. According to the least squares method, the minimization value is:

$$L = min \sum_{n=1}^{\max(n)} \left(y(n) - y(n)'\right)^2 \quad (8)$$

where $y(n)'$ is the result of sea ice thickness to be fitted, n is air temperature, and $y(n)$ is the observed data of ice thickness.

3.2.2. Coefficient of Determination (R^2)

The coefficient of determination is a statistical measure of the goodness of fit. It represents the proportion of variance in the dependent variable that can be explained by the independent variables included in the model. The value of R^2 ranges from 0 to 1, where a higher value indicates a better fit between the observed data and the model.

$$R^2 = 1 - RSS/TSS \quad (9)$$

where RSS is the sum of squared residuals and TSS is the total sum of squares.

4. Results

Sea ice thickness is an important characteristic of sea ice severity in Liaodong Bay, and its accurate evaluation is crucial for biogeochemical cycle research. The thickness of sea ice is affected by various factors, such as atmospheric conditions, solar radiation, and oceanic heat flux. Previous studies have demonstrated that the temperature is the main factor affecting the sea ice severity in Liaodong Bay. Against this backdrop, this study analyzed the changes in sea ice thickness under meteorological effects.

4.1. *Sea Ice Thickness Analysis Based on Stefan's Law*

4.1.1. Statistical Law of Sea Ice Growth Rate and Temperature

During the growth and melting of sea ice, the air temperature was found to have a significant influence on the sea ice's temperature profile. When the air temperature drops rapidly, the temperature profile of sea ice shows a linear distribution along the ice; meanwhile, when the air temperature becomes high, the ice temperature is low in the middle but high on the surface layer and bottom layer. Therefore, the growth rate of sea ice only shows statistical regularities during the period of rapid growth. Hence, the relationship between sea ice thickness and atmospheric temperature was evaluated only for the winter of 2009–2010.

Based on the observation area IZ1, the analysis focused on the daily ice thickness (DIT), which represents the thickness of sea ice growth starting from 0 cm over a 24 h period. To evaluate the growth law of DIT, Equations (2)–(4) were used to calculate sea ice thickness, where $h_0 = 0$, $h_i = a\sqrt{FDD}$, and a represents the freezing rate. Based on these equations, the relationship between DIT and daily average temperature was plotted (Figure 4). The regression analysis equation is:

$$h_i = 1.48\sqrt{FDD} \quad (10)$$

where FDD is the cumulative temperature (°C·d) and h_i is the thickness of sea ice (cm). The coefficient of determination is 0.38.

The results indicate that as the temperature increases, the DIT of sea ice decreases, with a freezing rate of 1.48 cm/(°C·d)$^{1/2}$. For instance, at an average temperature of −12 °C, the average DIT was 5.3 cm, whereas at an average temperature of −2 °C, the average DIT was about 2.5 cm. Additionally, there is still an increase in ice thickness of 2 cm at temperatures close to 0 °C, which may be attributed to the varying sea temperatures on different days. Notably, the freezing rate of DIT is significantly smaller than the theoretical value of 3.3 cm/(°C·d)$^{1/2}$.

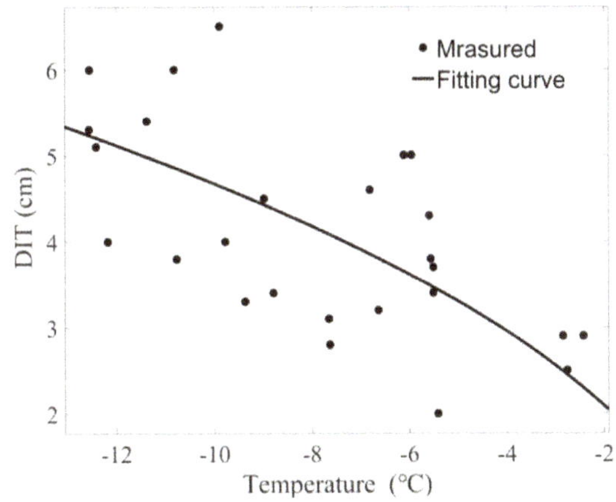

Figure 4. Statistical relationship between DIT and temperature.

In addition to DIT, the cumulative ice thickness (CIT) of sea ice, which was measured every day at 9 am, is another important observational parameter to consider. In Liaodong Bay, it was observed that the sea ice growth is significantly dominated by the cold fronts from Siberia. When the cold air reaches the ice area, the low temperature moves downward in the form of a cold front. It freezes the surface of sea water under the sea ice. As the cold air leaves, both the atmospheric temperature and the ice temperature increase, leading to a stable trend of increasing sea ice thickness. Obviously, the increase in sea ice thickness is affected by the amount of heat consumed by the cold front reaching the bottom of the ice. As the sea ice thickness increases, the same cold front with the same energy intensity can only cause a smaller increase in ice thickness.

To clarify the influence of air temperature and sea ice thickness on the sea ice growth rate, a three-dimensional fitting was performed, and the relationship between the average temperature, sea ice thickness, and sea ice growth rate (cm/d) was plotted, as shown in Figure 5. It can be seen that at a temperature of -12 °C, the maximum growth rate of ice thickness decreases from 3.5 cm/d to 1.5 cm/d as the ice thickness increases from 0 to 20 cm. The regression analysis equation is:

$$dh/dt(T,h) = 1.643 - 0.158T - 0.101h \quad (11)$$

where T is the temperature (°C), h is the thickness of sea ice (cm), and dh/dt is the growth rate of sea ice (cm/d). The coefficient of determination is 0.65.

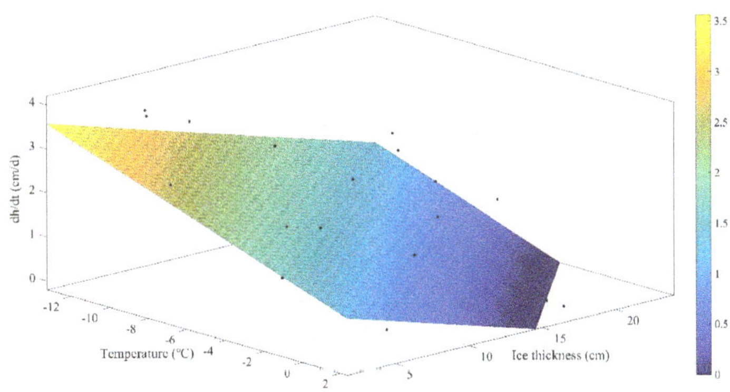

Figure 5. Statistical relationship between growth rate, ice thickness, and average daily temperature.

4.1.2. The Sea Ice Freezing Rate

In Stefan's law, the most important factor is the freezing rate. Based on Equations (2)–(4), the freezing rate is $h_i/\sqrt{FDD}$. For Jiangjunshi Port, the DIT and air temperature were used to calculate the freezing rate of first-day ice growth. The results showed that the freezing rate increased with higher atmospheric temperature and thicker sea ice (Figure 6). The regression analysis equation is:

$$a = 1.459 + 0.170\ T + 0.394\ h \quad (12)$$

where T is temperature and h is DIT. The coefficient of determination is 0.812.

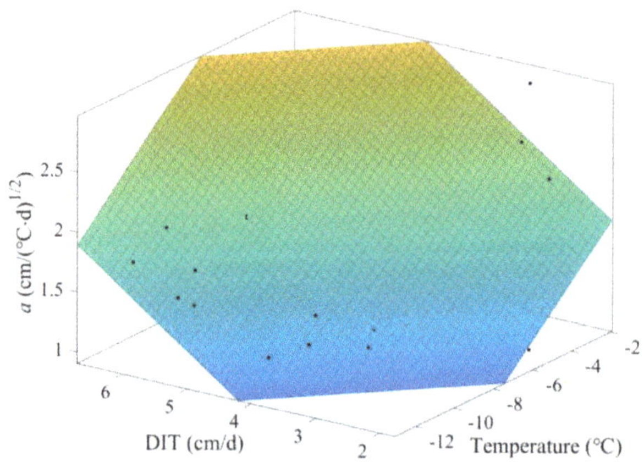

Figure 6. The freezing rate varies with DIT and temperature.

An accurate freezing rate is important for assessing the cumulative growth thickness of sea ice. To calculate the freezing rate of the CIT, the observation data from 23 January to 15 February 2010 were utilized. The data from 12 January to 19 January 2010 were not selected due to a short observation period. The initial ice thickness was 25 cm from 4 January to 4 February 2021. The freezing rate for each day was calculated by using $h_i/\sqrt{FDD}$, and the results are presented in Figure 7. Apparently, except for the two days when the sea ice started to grow, the freezing rate remained stable, with an average of 2.3 cm/(°C·d)$^{1/2}$.

4.1.3. Stefan's Law of Ice Growth

When the integration time step dt = 1 d, FDD represents the cumulative freezing degree days. k_i is 2.03 W/(m·K); ρ_i is 917 kg/m^3; L_f is 333.4 kJ/kg; T_f is the freezing temperature (−1.4 °C); and T_0 is the ice surface temperature, which is approximately equal to the air temperature. According to Equation (3), the theoretical value of a is 3.3 cm/(°C·d)$^{1/2}$, and the unit is calculated from the units of k_i, ρ_i, and L_f.The oceanic heat flux in this study was calculated based on the residual method, which was ineffective to optimize the calculation model based on the same principle. Based on the background, the theoretical values of 1.8 cm/(°C·d)$^{1/2}$, 1.71 cm/(°C·d)$^{1/2}$, 2.3 cm/(°C·d)$^{1/2}$, and 2.7 cm/(°C·d)$^{1/2}$ were used to simulate the thickness of sea ice.

Since Stefan's law assumes a linear ice temperature profile, this method is mainly suitable for the period of rapid sea ice growth caused by low temperatures. Thus, the observation data from the winter of 2009–2010 were used for the simulation, as shown in the results in Figure 8. Comparing with the measured data, the simulation results show that the ice thickness is significantly overestimated when the growth coefficient is 3.3 cm/(°C·d)$^{1/2}$ and 2.7 cm/(°C·d)$^{1/2}$, and significantly underestimated when the growth coefficient is 1.71 cm/(°C·d)$^{1/2}$ and 1.8 cm/(°C·d)$^{1/2}$. Based on the measured data and cumulative freezing days, a sea ice growth rate of 2.3 cm/(°C·d)$^{1/2}$ is recommended for Jiangjunshi Port, and the corresponding calculation results are shown in Figure 8. The calculation errors were analyzed and presented in Table 1. The results indicate that when the growth coefficient is 1.71 cm/(°C·d)$^{1/2}$, the calculation error is over 18%; when it is 1.8 cm/(°C·d)$^{1/2}$, the error is over 13.4%; and when it is 2.7 cm/(°C·d)$^{1/2}$, the error is over 16.4%. However, when the growth coefficient is 2.3 cm/(°C·d)$^{1/2}$, except for in the initial stage, the calculation error is less than 2.8%. The theoretical value of 3.3 cm/(°C·d)$^{1/2}$ yields a calculation error of over 42.2%.

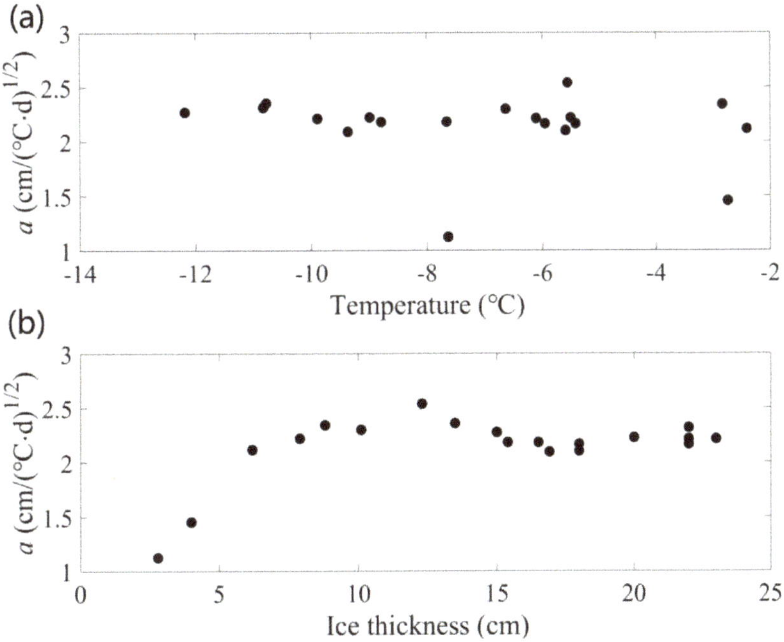

Figure 7. (**a**) The relationship between temperature and freezing rate; (**b**) relationship between ice thickness and freezing rate.

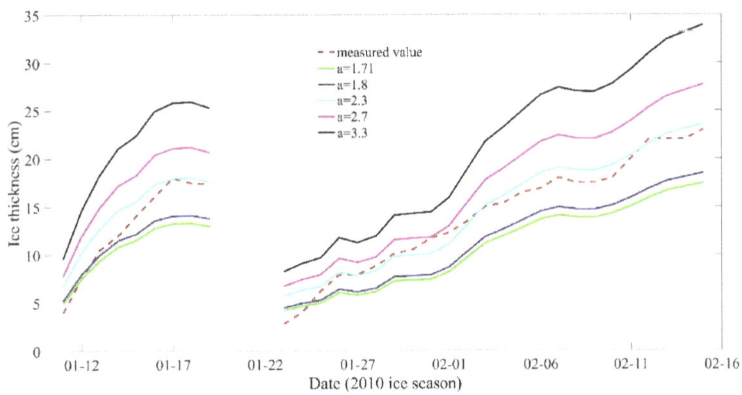

Figure 8. Simulation results under different growth coefficients.

Table 1. Simulation error at different growth coefficients.

	a = 1.71		a = 1.8		a = 2.3		a = 2.7		a = 3.3	
	Difference	Error	Difference	Error	Difference	Error	Difference	Error	Difference	Error
SD1 *-15 January	−2.56	18.1%	−1.88	13.4%	1.51	10.7%	4.22	30.0%	8.30	58.8%
SD1-19 January	−4.37	25.1%	−3.61	20.7%	0.22	1.3%	3.29	18.9%	7.89	45.3%
SD2-27 January	−2.11	26.7%	−1.77	22.4%	−0.07	0.8%	1.30	16.4%	3.34	42.3%
SD2 *-6 February	−3.19	18.9%	−2.38	14.1%	1.65	9.7%	4.87	28.8%	9.71	57.5%
SD2-11 February	−4.93	24.7%	−4.05	20.2%	0.38	1.9%	3.93	19.6%	9.24	46.2%
SD2-15 February	−5.52	24.0%	−4.50	19.5%	0.65	2.8%	4.76	20.7%	10.92	47.5%

Note(s): * SD1 is 12 January–19 January in winter 2009–2010; SD2 is 23 January–15 February in winter 2009–2010.

It was found that the calculation errors were due to two main factors: the negative accumulated temperature in the early stage and the different growth coefficients at different locations. The growth coefficient was measured to be 1.8 cm/(°C·d)$^{1/2}$ in the waters of Huludao on the west coast of Liaodong Bay, 1.71 cm/(°C·d)$^{1/2}$ in the northern waters of Liaodong Bay, and 2.7 cm/(°C·d)$^{1/2}$ on the platform in Liaodong Bay [32]. For the negative accumulated temperature in the early stage, Cao simulated lake ice using Stefan's law and found that predicting ice thickness two weeks in advance was more accurate [36]. In the case of Jiangjunshi Port on the east coast of Liaodong Bay, the measurements showed that the sea ice started to grow from 0 cm, and the growth coefficient given in this study is 2.3 cm/(°C·d)$^{1/2}$.

4.2. *High-Resolution Thermodynamic Snow-and-Ice Model (HIGHTSI)*

4.2.1. Oceanic Heat Flux

In 1982, McPhee and Untersteiner proposed a method for calculating oceanic heat flux using the sea ice energy balance, which relies on the observed data of the sea ice mass balance and temperature profile [45]. This method only requires data of the thermodynamic parameters of sea ice, such as ice bottom position, ice temperature, and ice salinity. In cases where the observation data of the sea ice temperature are not available, the oceanic heat flux can still be calculated by combining the thermodynamic numerical model with the measured thickness data of sea ice. This method is called the residual method and is also known as the method of measuring oceanic heat flux using ice thickness. The residual method has been extensively applied to calculate the oceanic heat fluxes under the pack ice in east Antarctica [46,47], under the pack ice in the Alaskan Beaufort Sea [48], and under the landfast ice in McMurdo Sound [49]. The relationship between the energy balance of ice bottom ablation or accretion can be expressed as:

$$F_w = F_c + F_L \tag{13}$$

where F_w is oceanic heat fluxes; F_c is the conductive heat flux; and F_L is the latent heat flux. The sign convention is that upward and melting are positive, whereas downward and freezing are negative. The calculation formula of conductive heat flux and latent heat flux can be expressed as:

$$F_c = k_i \cdot \partial T_i / \partial z \tag{14}$$

$$F_L = -\rho_i \cdot L \cdot \partial z_i / \partial t \tag{15}$$

where $(\partial T_i/\partial z)$ is the temperature gradient in ice; $(\partial z_i/\partial t)$ is the growth rate of sea ice; and the thermal conductivity, latent heat of fusion, and density of the ice are k_i, L, and ρ_i, respectively, with values of 2.03 W/(m·°C), 333.4 kJ/kg, and 917 kg/m^3.

Oceanic heat fluxes were calculated for three different time periods (twice in 2010 and once in 2021). Sea ice surface temperature was similar to the air temperature, where the freezing point of sea ice is −1.4 °C. The temperature profile in the ice was assumed to be linear from top to bottom. Sea ice thickness was measured daily, and the interval of the sea ice growth rate was set to 1 day. To reduce the errors in the thickness observations, a 4-day running average was used to calculate oceanic heat flux. Figure 9 shows the calculated oceanic heat fluxes for the three time periods (12 January–19 January; 23 January–15 February in the winter of 2009–2010; and 4 January–4 February in the winter of 2020–2021).

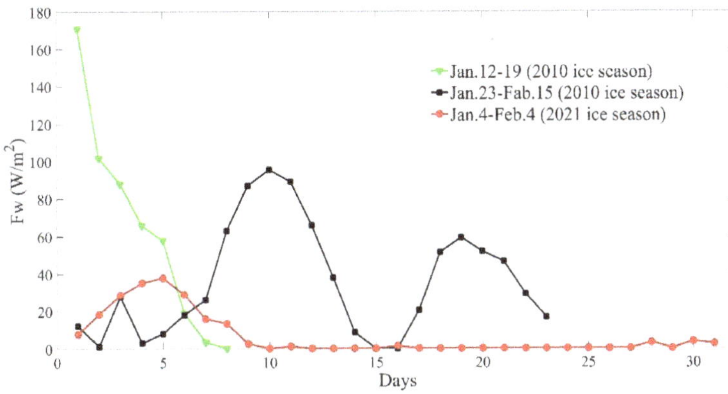

Figure 9. Time series of oceanic heat flux (F_w).

From 12 January to 17 January 2010, the oceanic heat flux decreased rapidly from a peak value of 170.6 W/m^2 to almost 0 W/m^2. From 23 January to 15 February 2010, the oceanic heat flux showed a fluctuating trend with two peaks on the 10th and 19th days, and the maximum value gradually decreased: the first maximum was 95.4 W/m^2, the second maximum was 59.3 W/m^2, and the minimum value of the oceanic heat flux was almost 0 W/m^2. From 4 January to 4 February 2021, the oceanic heat flux was relatively large during the first 10 days of observation and remained low thereafter, with a peak of 37.9 W/m^2 and a minimum value close to 0 W/m^2. The oceanic heat flux was averaged over three time periods. The average oceanic heat flux for the two periods in the winter of 2010 was 60 W/m^2 and 35 W/m^2, respectively. The average oceanic heat flux during the rapid growth of sea ice in the winter of 2021 was 15 W/m^2, and the oceanic heat flux approached 0 W/m^2 after the sea ice thickness stabilized.

The time series of oceanic heat flux were analyzed, which revealed that it was directly affected by both ice thickness and air temperature. When the ice thickness is the same, the intensity of the ice front at the ice–water interface increases with a decreasing air temperature, leading to stronger heat exchange between the ice and water. On the other hand, when the air temperature is the same, thicker ice results in a smaller amount of cold front energy reaching the ice–water interface, leading to a weaker energy exchange between the ice and water.

4.2.2. High-Resolution Thermodynamic Snow-and-Ice Model

The model requires several input parameters, including air temperature (*Ta*), wind speed (*Uz*), relative humidity (*Rh*), and solar radiation (parameterized). The initial input value is the ice thickness measured on the first day, and the calculation step size is 1 h. The values for the physical properties of sea ice are: thermal conductivity of 2.03 W/(m·K), specific heat capacity of 2093 J/(kg·K), latent heat of freezing of 333.4 kJ/kg, ice density of 917 kg/m^3, and freezing point of −1.4 °C. The snow was not included as an input parameter in this study because the snow was very thin in the Bohai Sea, and strong winds blew the snow away from the observation area.

The growth process of sea ice thickness was simulated using the meteorological data from the winters of 2009–2010 and 2020–2021. To investigate the influence of the oceanic heat flux on the model results, two simulations were conducted: one with a fixed oceanic heat flux of 2 W/m^2 and the other with the measured oceanic heat flux. The accuracy of the simulation results was evaluated by comparing them with the observation data of ice thickness.

The sea ice growth process during two time periods in 2010, 11–19 January and 23 January–15 February, was simulated, and the results are presented in Figure 10. Table 2 shows the error between the calculation results and the measured maximum ice thickness. When the oceanic heat flux was assumed to be 2 W/m^2, the ice thickness was significantly overestimated, and the overestimation increased over time. On 19 January, the maximum error of ice thickness was 26.1%. Similarly, when the oceanic heat flux was 2 W/m^2 during the second period, the overestimation of the ice thickness calculation was more severe, with an error of 60% on 15 February 2010. However, by using the oceanic heat flux calculated in Section 4.2.1 in the model calculation, the final errors of the calculation were significantly reduced to 4.2% and 0.8%, as listed in Table 2. This finding indicates that an accurate assessment of the oceanic heat flux is crucial for the accuracy of prediction during the period of rapid sea ice growth.

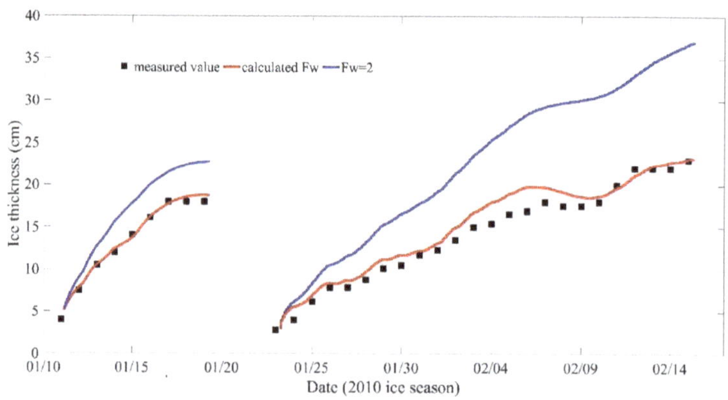

Figure 10. Simulation results under different oceanic heat fluxes (12 January–19 January in winter 2009–2010; 23 January–15 February in winter 2009–2010).

Table 2. Simulation errors at different oceanic heat fluxes.

	19 January (2010 Ice Season)		15 February (2010 Ice Season)		4 February (2021 Ice Season)	
	Difference	Error	Difference	Error	Difference	Error
$F_w = 2$	4.7 cm	26.1%	13.8 cm	60%	1.9 cm	3.7%
Section 3.2 F_w	0.76 cm	4.2%	0.2 cm	0.8%	0.8 cm	1.5%

In the winter of 2010, the initial ice thickness was 0 cm, and the maximum thickness was about 20 cm. However, the 2021 ice season was different from 2010, with an initial ice thickness of 27 cm and a maximum thickness of 51 cm, as observed from 31 December 2020. The model calculations were initiated from 31 December 2020. The sea ice thickness in 2021 was much higher than that in 2010. Figure 11a shows the calculation results of ice thickness from 4 January to 4 February 2021, and the measured maximum ice thickness, along with the calculation error, is presented in Table 2. The difference between the maximum calculated value and the measured value was insignificant when the oceanic heat flux was assumed to be 2 W/m^2. This is because, as the ice thickness increased, the impact of the cold front at the ice–water interface was reduced, leading to a decrease in the oceanic heat flux. However, when the ice thickness was small and growing rapidly, the actual value of oceanic heat flux differed significantly from the assumed value of 2 W/m^2. This indicates that the oceanic heat flux has a significant impact on the accuracy of the model.

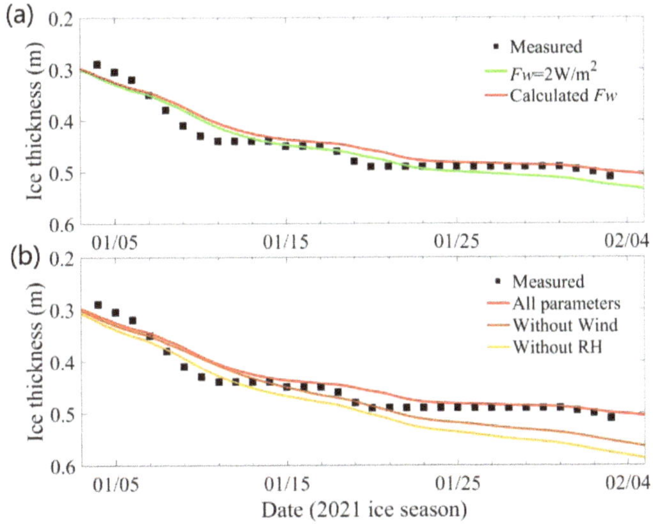

Figure 11. (**a**) Simulation results under different oceanic heat fluxes (4 January–4 February in 2021 ice season); (**b**) sensitivity analysis of wind and relative humidity.

4.2.3. Sensitivity Analysis

Sea ice growth and melting processes are primarily driven by air and hydrological factors, with air temperature being considered the most important driver of sea ice thermodynamic growth. While Stefan's law uses cumulative temperatures for sea ice thickness calculations, the sea ice numerical model adds the heat balance of the sea ice surface, heat conduction within the ice, and heat balance at the bottom of the ice, providing a more detailed understanding of the underlying physical processes. In addition to air temperature, the model also takes into account wind speed and humidity elements at the ice surface, as well as the oceanic heat flux at the bottom of the ice.

The sensitivity analysis of the oceanic heat flux to the sea ice thickness was completed in Section 4.2.2, while this section analyzes the sensitivity of wind speed and humidity. Since wind and humidity affect processes such as turbulent fluxes of sensible and latent heat, the lack of wind and humidity can result in the overestimation of the simulated ice thickness. For instance, on February 4, the ice thickness was calculated as 50.1 cm for all parameters, as 55.7 cm without wind, and as 57.9 cm without humidity, respectively. The corresponding errors compared with the measured data were 1.5%, 9.1%, and 13.6%.

Comparing Figure 11a,b, we observed that the simulated thickness error without wind and relative humidity is greater than the thickness error observed when the oceanic heat flux is set to 2 W/m^2. These results highlight the importance of including wind speed and relative humidity in the sea ice numerical model, especially in cases where their influence is significant.

5. Discussion

The state of sea ice in Liaodong Bay is influenced by both thermodynamic and dynamic factors. Due to the influence of dynamic factors, the thermodynamic observations of sea ice had not been effectively carried out in Liaodong Bay, which has hindered the development of appropriate thermodynamic models for Liaodong Bay. To address this, field observations were conducted on sea ice thickness, temperature, and wind speed in Jiangjunshi Port during the ice seasons of 2009/2010 and 2020/2021. The observed data were used to evaluate the relationship between air temperature and sea ice thickness, and to calculate the oceanic heat flux at Jiangjunshi Port. Additionally, Stefan's law and the HIGHTSI method were applied for evaluating the sea ice growth and melting process based on the observed data.

The average oceanic heat flux during the two ice seasons of 2010 was 60 W/m^2 and 35 W/m^2, while the average oceanic heat flux during the 2021 ice season was 6.1 W/m^2. This trend in oceanic heat flux is consistent with the observations made by Ji [27] during the 1997/1998 ice season in Liaodong Bay on the JZ-20 platform. Ji concluded that the oceanic heat flux gradually decreased throughout the ice season and remained at 0 W/m^2 during the melting period. However, since the observations made in this study do not cover the melting period, calculation of the melting period cannot be provided. It is known that during the melting period, seawater temperature and sea ice temperature gradually increase, and the brine volume fraction (BVF) will increase, resulting in a decrease in latent heat [50–54]. As a result, the heat will gradually decompose the sea ice, and during the melting period, the oceanic heat flux should be greater than 0. Therefore, further studies on oceanic heat flux need to be conducted in combination with the evolution of sea ice temperature and salinity.

To investigate how temperature and ice thickness affect the oceanic heat flux in Liaodong Bay, the analysis of the relationship between ice thickness and oceanic heat flux was carried out, as shown in Figure 12a. Under the same air conditions, the difference in ice temperature distribution between thin and thick ice can affect the oceanic heat flux. It was revealed that the oceanic heat flux was higher during the rapid growth of thin sea ice, but as the sea ice thickness stabilized, the influence of cold fronts on the ice–water interface weakened, resulting in a low oceanic heat flux. Moreover, lower temperatures led to a larger temperature gradient in the ice, causing the sea ice to grow rapidly due to the cold front. It increased the discharge of salt, which in turn, increased the salinity of the seawater and continuously reduced the sea temperature. As a result, the oceanic heat flux became significantly large. Based on these findings, the relationship between air temperature and oceanic heat flux was determined (as shown in Figure 12b), and the heat flux during the rapid growth process of thin ice was calculated. Table 3 shows the statistical results.

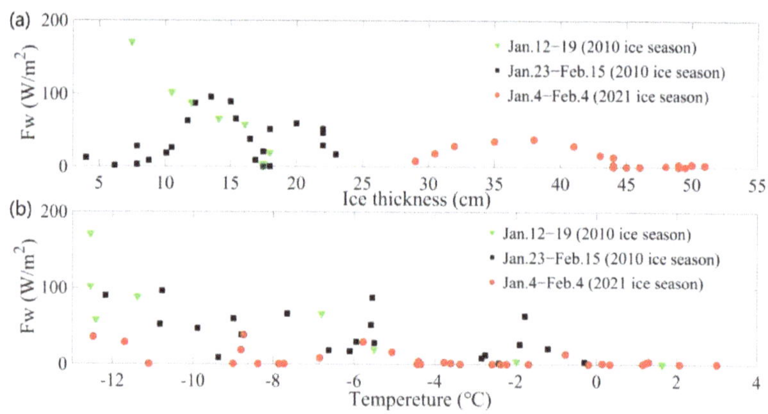

Figure 12. (**a**) The relationship between bottom heat flux and ice thickness; (**b**) the relationship between bottom heat flux and temperature.

Table 3. The oceanic heat flux during the rapid growth process of ice thickness.

Interval Number	**1**	**2**	**3**	**4**	**5**
Temperature (°C)	−6~−4	−8~−6	−10~−8	−12~−10	−14~−12
Average F_w (W·m^{-2})	35.8	41.6	38.2	78.4	104.8

Based on the measured air conditions and ice thickness, the growth of sea ice thickness in Jiangjunshi Port was evaluated using Stefan's law and the HIGHTSI model. The maximum error of Stefan's law was found to be over 50% using the growth coefficients from other sea areas, and thus, a recommended growth coefficient of 2.3 cm/(°C·d)$^{1/2}$ was proposed. The HIGHTSI model was used to calculate ice thickness with an oceanic heat flux of 2 W/m^2, and the maximum errors were 26.1%, 60%, and 5.4% in three different periods. The largest errors occurred with a small initial ice thickness. Using the oceanic heat flux evaluated in Section 4.2.1, the ice thickness errors of the HIGHTSI model were 4.2%, 0.8%, and 0.2% in the same three periods. In 1998, Cheng used sea ice observation data from Liaodong Bay to evaluate the accuracy of the HIGHTSI simulation results and found that the model accurately simulated the growth of sea ice. This paper presents the oceanic heat flux on the east coast of Liaodong Bay and verifies the results of the numerical and analytical models, providing important support for the calculation of ice thickness in the region.

6. Conclusions

Due to the movement of sea ice, there is a serious lack of thermodynamic observations of landfast ice in Liaodong Bay. The insufficient observation data hinder the development of sea ice thermodynamic models in this area. To address this issue, this study presented the observation works carried out in Jiangjunshi Port, with discussions on the oceanic heat flux and the models of the sea ice growth process. The following conclusions can be drawn:

(1) The daily growth of ice thickness from 0 cm decreases with the temperature, and the decrease rate is 0.26 cm/°C. The daily increase in cumulative ice thickness is influenced by air temperature and sea ice thickness. When the sea ice thickness reaches 20 cm, the growth rate decreases to around 1 cm/d.

(2) The error of Stefan's law and the HIGHTSI model was evaluated. It was found that a growth coefficient of 2.3 cm/(°C·d)$^{1/2}$ is more consistent with the measured value. Meanwhile, the HIGHTSI model is strongly dependent on the oceanic heat flux value when the ice is thin.

(3) The residual method was used to calculate the oceanic heat flux of Jiangjunshi Port. The average oceanic heat flux in the first period of the 2010 ice season was 60 W/m^2, and in the second period, it was 35 W/m^2. In the ice season of 2021, the average oceanic heat flux was 6.1 W/m^2.

Nowadays, global climate change has resulted in unpredictable changes in ice conditions in Liaodong Bay, which has a significant impact on human economic activities. Due to the limitations in observation conditions, numerical simulations of sea ice have become an important approach for the continuous evaluation of the ice conditions in Liaodong Bay. Therefore, the work carried out in this study can provide valuable data for assessing the ice thickness here. However, this study also has some limitations, including a lack of observations of sea ice temperature and salinity. Considering more physical and mechanical properties of sea ice in field experiments would provide a more optimized and accurate thermodynamic model for Liaodong Bay.

Author Contributions: N.X. and D.D. initiated this work; Y.M., S.Y. and W.S. conducted field observations; Y.M. and W.S. carried out modeling experiments, analyzed the results, and wrote the original draft; N.X. and S.Y. collected the data. All the authors contributed to the writing and editing of the manuscript. All authors have read and agreed to the published version of the manuscript.

Funding: This research was funded by the National Natural Science Foundation of China, grant number 42206221; Laoshan Laboratory, grant number LSKJ 202203900; National Natural Science Foundation of China, grant number U1806214; State Key Laboratory of Coastal and Offshore Engineering of Dalian University of Technology the Open Fund; and State Environmental Protection Key Laboratory of Coastal Ecosystem.

Data Availability Statement: The data presented in this study are available on request from the corresponding author. The data are not publicly available due to company data secrecy provision.

Conflicts of Interest: The authors declare no conflict of interest.

References

1. Barry, R.G.; Serreze, M.C.; Maslanik, J.A.; Preller, R.H. The arctic sea ice-climate system: Observations and modelling. *Rev. Geophys.* **1993**, *31*, 397–422. [CrossRef]
2. Rind, D.; Healy, R.; Parkinson, C.; Martinson, D. The role of sea ice in 2 × CO_2 climate model sensitivity: Part I. The total influence of sea ice thickness and extent. *J. Clim.* **1995**, *8*, 449–463. [CrossRef]
3. Yuan, X.J.; Martinson, D.G. Antarctic sea ice extent variability and its global connectivity. *J. Clim.* **2000**, *13*, 1697–1717. [CrossRef]
4. Holland, M.M.; Bitz, C.M. Polar amplification of climate change in coupled models. *Clim. Dyn.* **2003**, *21*, 221–232. [CrossRef]
5. Zheng, J.; Ke, C.; Shao, Z. Winter sea ice albedo variations in the Bohai Sea of China. *Acta Oceanol. Sin.* **2017**, *36*, 56–63. [CrossRef]
6. Ma, Y.; Cheng, B.; Xu, N.; Yuan, S.; Shi, H.; Shi, W. Long-Term Ice Conditions in Yingkou, a Coastal Region Northeast of the Bohai Sea between 1951/1952–2017/2018: Modelling and Observations. *Remote Sens.* **2022**, *14*, 182. [CrossRef]
7. Ouyang, L.; Hui, F.; Zhu, L.; Cheng, X.; Cheng, B.; Shokr, M.; Zhao, J.; Ding, M.; Zeng, T. The spatiotemporal patterns of sea ice in the Bohai Sea during the winter seasons of 2000–2016. *Int. J. Digit. Earth* **2017**, *12*, 489–498. [CrossRef]
8. Yan, Y.; Uotila, P.; Huang, K.; Gu, W. Variability of sea ice area in the Bohai Sea from 1958 to 2015. *Sci. Total Environ.* **2020**, *709*, 136164. [CrossRef]
9. Xue, X.; Ren, G.; Sun, X.; Zhang, P.; Ren, Y.; Zhang, S.; Zhao, C.; Yu, X. Change in mean and extreme temperature at YingKou station in Northeast China from 1904 to 2017. *Clim. Chang.* **2021**, *164*, 58. [CrossRef]
10. Yan, Y.; Shao, D.; Gu, W.; Liu, C.; Li, Q.; Chao, J.; Tao, J.; Xu, Y. Multidecadal anomalies of Bohai Sea ice cover and potential climate driving factors during 1988–2015. *Environ. Res. Lett.* **2017**, *12*, 094014. [CrossRef]
11. Gu, W. Research on the sea-ice disaster risk in Bohai Sea based on the remote sensing. *J. Catastrophology* **2008**, *23*, 10–14.
12. Yan, Y.; Huang, K.; Shao, D.; Xu, Y.; Gu, W. Monitoring the characteristics of the Bohai Sea ice using high-resolution geostationary ocean color imager (GOCI) data. *Sustainability* **2019**, *11*, 777. [CrossRef]
13. Ma, Y.; Guan, P.; Xu, N.; Xu, Y.; Yuan, S.; Liu, Y.; Yu, F. Determination of the sea ice parameters for the reliability design of the marine structures in Liaodong Bay. *Ocean Eng.* **2019**, *3*, 136–142. (In Chinese)
14. Ma, Y.X.; Xu, N.; Yuan, S.; Liu, X.; Shi, W.; Zhou, X.; Liu, Y.; Chen, Y. Basic characteristics of sea ice environment on the east coast of Liaodong Bay based on field observations. *J. Glaciol. Geocryol.* **2022**, *44*, 1492–1500. (In Chinese)
15. Ma, Y.X.; Wang, Y.; Yu, F.; Xu, N.; Yuan, S.; Shi, W.Q. Research on the relationship between air temperatures, seawater temperature and ice regime in winter during 2017–2018 at the adjacent sea of Hongyanhe, Liaodong Bay. *J. Glaciol. Geocryol.* **2022**, *44*, 1482–1491. (In Chinese)
16. Vancoppenolle, M.; Fichefet, T.; Goosse, H.; Bouillon, S.; Madec, G.; Maqueda, M.A.M. Simulating the mass balance and salinity of Arctic and Antarctic sea ice. 1. Model description and validation. *Ocean Model.* **2009**, *27*, 33–53. [CrossRef]
17. Vancoppenolle, M.; Fichefet, T.; Goosse, H. Simulating the mass balance and salinity of Arctic and Antarctic sea ice. 2. Importance of sea ice salinity variations. *Ocean Model.* **2009**, *27*, 54–69. [CrossRef]
18. Angelopoulos, M.; Damm, E.; Pereira, P.S.; Abrahamsson, K.; Bauch, D.; Bowman, J.; Castellani, G.; Creamean, J.; Divine, D.V.; Dumitrascu, A.; et al. Deciphering the properties of different Arctic ice types during the growth phase of MOSAiC: Implications for future studies on gas pathways. *Front. Earth Sci.* **2022**, *10*, 864523. [CrossRef]
19. Li, Z.; Zhao, J.; Su, J.; Li, C.; Cheng, B.; Hui, F.; Yang, Q.; Shi, L. Spatial and temporal variations in the extent and thickness of Arctic landfast ice. *Remote Sens.* **2019**, *12*, 64. [CrossRef]
20. Purdie, C.R.; Langhorne, P.J.; Leonard, G.H.; Haskell, T.G. Growth of first-year landfast Antarctic sea ice determined from winter temperature measurements. *Ann. Glaciol.* **2006**, *44*, 170–176. [CrossRef]
21. Perovich, D.K.; Elder, B. Estimates of ocean heat flux at SHEBA. *Geophys. Res. Lett.* **2002**, *29*, 58-1–58-4. [CrossRef]
22. Stroeve, J.; Notz, D. Changing state of Arctic sea ice across all seasons. *Environ. Res. Lett.* **2018**, *13*, 103001. [CrossRef]
23. Zhang, N.; Wu, Y.; Zhang, Q. Forecasting the evolution of the sea ice in the Liaodong Bay using meteorological data. *Cold Reg. Sci. Technol.* **2016**, *125*, 21–30. [CrossRef]
24. Cheng, B. On the Modelling of Sea Ice Thermodynamics and Air-Ice Coupling in the Bohai Sea and the Baltic Sea. Ph.D. Thesis, Helsingin Yliopisto, Helsinki, Finland, 2002.
25. Zhao, J.; Cheng, B.; Vihma, T.; Lu, P.; Han, H.; Shu, Q. The internal melting of landfast sea ice in Prydz Bay, East Antarctica. *Environ. Res. Lett.* **2022**, *17*, 074012. [CrossRef]
26. Zhai, M.; Cheng, B.; Leppäranta, M.; Hui, F.; Li, X.; Demchev, D.; Lei, R.; Cheng, X. The seasonal cycle and break-up of landfast sea ice along the northwest coast of Kotelny Island, East Siberian Sea. *J. Glaciol.* **2022**, *68*, 153–165. [CrossRef]
27. Ji, S.Y.; Yue, Q.J. Determination and analysis of oceanic heat flux under sea ice cover in the liaodong bay. *Mar. Sci. Bull.* **2000**, *19*, 8–14. (In Chinese)
28. Li, R.; Lu, Y.; Hu, X.; Guo, D.; Zhao, P.; Wang, N.; Lee, K.; Zhang, B. Space–time variations of sea ice in Bohai Sea in the winter of 2009–2010 simulated with a coupled ocean and ice model. *J. Oceanogr.* **2021**, *77*, 243–258. [CrossRef]
29. Zhang, N.; Wang, J.; Wu, Y.; Wang, K.-H.; Zhang, Q.; Wu, S.; You, Z.-J.; Ma, Y. A modelling study of ice effect on tidal damping in the Bohai Sea. *Ocean. Eng.* **2019**, *173*, 748–760. [CrossRef]
30. Yan, Y.; Gu, W.; Gierisch, A.M.U.; Xu, Y.; Uotila, P. NEMO-Bohai 1.0: A high-resolution ocean and sea ice modelling system for the Bohai Sea, China. *Geosci. Model Dev.* **2022**, *15*, 1269–1288. [CrossRef]

31. Jia, B.; Chen, X. Application of an ice-ocean coupled model to Bohai Sea ice simulation. *J. Oceanol. Limnol.* **2021**, *39*, 1–13. [CrossRef]
32. Ji, S. Numerical models and applications of engineered sea ice. *Sci. Press* **2011**, *35*, 210–211. Available online: https://kns.cnki.net/KCMS/detail/detail.aspx?dbname=CDFD9908&filename=2001009151.nh (accessed on 1 February 2023). (In Chinese)
33. Cheng, B.; Launiainen, J. A one-dimensional thermodynamic air-ice-water model: Technical and algorithm description report. In *Meri-Report Series of the Finnish Institute of Marine Research*; Finnish Institute of Marine Research: Helsinki, Finland, 1998; Volume 37, pp. 15–35.
34. Yuan, S.; Liu, C.; Liu, X.; Chen, Y.; Zhang, Y. Research advances in remote sensing monitoring of sea ice in the Bohai sea. *Earth Sci. Inform.* **2021**, *14*, 1729–1743. [CrossRef]
35. Lei, R.; Li, Z.; Cheng, B.; Zhang, Z.; Heil, P. Annual cycle of landfast sea ice in Prydz Bay, east Antarctica. *J. Geophys. Res.* **2010**, *115*, C02006. [CrossRef]
36. Lu, P.; Cao, X.; Li, G.; Huang, W.; Leppäranta, M.; Arvola, L.; Huotari, J.; Li, Z. Mass and heat balance of a lake ice cover in the Central Asian arid climate zone. *Water* **2020**, *12*, 2888. [CrossRef]
37. Yang, Y.; Leppäranta, M.; Cheng, B.; Li, Z. Numerical modelling of snow and ice thicknesses in Lake Vanajavesi, Finland. *Tellus A Dyn. Meteorol. Oceanogr.* **2012**, *64*, 17202. [CrossRef]
38. Launiainen, J.; Cheng, B. Modelling of ice thermodynamics in natural water bodies. *Cold Reg. Sci. Technol.* **1998**, *27*, 153–178. [CrossRef]
39. Pirazzini, R.; Vihma, T.; Granskog, M.A.; Cheng, B. Surface albedo measurements over sea ice in the Baltic Sea during the spring snowmelt period. *Ann. Glaciol.* **2006**, *44*, 7–14. [CrossRef]
40. Cheng, B.; Vihma, T.; Pirazzini, R.; Granskog, M.A. modelling of superimposed ice formation during the spring snowmelt period in the Baltic Sea. *Ann. Glaciol.* **2006**, *44*, 139–146. [CrossRef]
41. Cheng, B.; Zhang, Z.; Vihma, T.; Johansson, M.; Bian, L.; Li, Z.; Wu, H. Model experiments on snow and ice thermodynamics in the Arctic Ocean with Chinese 2003 data. *J. Geophys. Res. Ocean.* **2008**, *113*, C09020. [CrossRef]
42. Cheng, B.; Vihma, T.; Launiainen, J. Modelling of the superimposed ice formation and sub-surface melting in the Baltic Sea. *Geophysica* **2003**, *39*, 31–50.
43. Cheng, B.; Vihma, T.; Rontu, L.; Kontu, A.; Pour, H.K.; Duguay, C.; Pulliainen, J. Evolution of snow and ice temperature, thickness and energy balance in Lake Orajärvi, northern Finland. *Tellus A Dyn. Meteorol. Oceanogr.* **2014**, *66*, 21564. [CrossRef]
44. Maykut, G.A.; Untersteiner, N. Some results from a time-dependent thermodynamic model of sea ice. *J. Geophys. Res.* **1971**, *76*, 1550–1575. [CrossRef]
45. McPhee, M.G.; Untersteiner, N. Using sea ice to measure vertical heat flux in the ocean. *J. Geophys. Res.* **1982**, *87*, 2071–2074. [CrossRef]
46. Tang, S.; Qin, D.; Ren, J.; Kang, J.; Li, Z. Structure, salinity and isotopic composition of multi-year landfast sea ice in Nella Fjord, Antarctica. *Cold Reg. Sci. Technol.* **2007**, *49*, 170–177. [CrossRef]
47. Heil, P.; Allison, I.; Lytle, V.I. Seasonal and interannual variations of the oceanic heat flux under a landfast Antarctic sea ice cover. *J. Geophys. Res. Ocean.* **1996**, *101*, 25741–25752. [CrossRef]
48. Wettlaufer, J.S. Heat flux at the ice-ocean interface. *J. Geophys. Res. Ocean.* **1991**, *96*, 7215–7236. [CrossRef]
49. Lytle, V.I.; Massom, R.; Bindoff, N.; Worby, A.; Allison, I. Wintertime heat flux to the underside of East Antarctic pack ice. *J. Geophys. Res. Ocean.* **2000**, *105*, 28759–28770. [CrossRef]
50. Yen, Y.C. *Review of Thermal Properties of Snow, Ice and Sea Ice*; US Army, Corps of Engineers, Cold Regions Research and Engineering Laboratory: Hanover, NH, USA, 1981; Volume 81, pp. 1–27.
51. Sturm, M. Thermal conductivity and heat transfer through the snow on the ice of the Beaufort Sea. *J. Geophys. Res.* **2002**, *107*, 8043. [CrossRef]
52. Notz, D. In situ measurements of the evolution of young sea ice. *J. Geophys. Res.* **2008**, *113*, C03001. [CrossRef]
53. Notz, D. Desalination processes of sea ice revisited. *J. Geophys. Res.* **2009**, *114*, C05006. [CrossRef]
54. Notz, D. A non-destructive method for measuring the salinity and solid fraction of growing sea ice in situ. *J. Glaciol.* **2005**, *51*, 159–166. [CrossRef]

water

MDPI

Article

Recent Advances and Challenges in the Inverse Identification of Thermal Diffusivity of Natural Ice in China

Zhijun Li [1], Xiang Fu [1], Liqiong Shi [2,*], Wenfeng Huang [3,4] and Chunjiang Li [1]

[1] State Key Laboratory of Coastal and Offshore Engineering, Dalian University of Technology, Dalian 116024, China
[2] Department of Basic Sciences, Shenyang Institute of Engineering, Shenyang 110136, China
[3] Key Laboratory of Subsurface Hydrology and Ecological Effect in Arid Region (Ministry of Education), Chang'an University, Xi'an 710054, China
[4] School of Water and Environment, Chang'an University, Xi'an 710054, China
* Correspondence: shilq@sie.edu.cn

Abstract: The ice thermal parameters are the key to reasonably simulating ice phenology, distribution, and thickness, but they have always been a "vulnerable group" in ice research. Technically, it may seem simple to obtain accurate ice thermal property parameters, but in reality, there are numerous impact factors, requiring a rigorous research process. In the 1980s, the thermal conductivity of ice was explored in the field and laboratory, after which there has been no significant progress in China. In this century, mathematics is introduced, after which the inversion identification and analysis with the time-series data of the vertical temperature profiles of ice layers by in situ testing are carried out. The in situ thermal diffusivities of different natural ices were obtained and cross-validated with the inversion identification results. Both natural freshwater ice and sea ice exhibited differences in the thermal diffusivity of the pure ice chosen for the current simulations due to impurities within the unfrozen water among the ice crystals, but the trends are consistent with the results of a small number of laboratory tests on different types of saltwater frozen ice. In this paper, the inversion identification results of the thermal diffusivity of typical ice were selected, and the factors constraining the thermal diffusivities were analyzed. The importance of parameterizing the thermal diffusivity in the phase transition zone of ice under the trend of global warming was illustrated. Future research ideas on the physical mechanism, application value, and parameterization scheme of the thermal diffusivity of natural ice were envisaged.

Keywords: natural ice; thermal diffusivity; inversion identification; vertical temperature profile; research status

Citation: Li, Z.; Fu, X.; Shi, L.; Huang, W.; Li, C. Recent Advances and Challenges in the Inverse Identification of Thermal Diffusivity of Natural Ice in China. *Water* **2023**, *15*, 1041. https://doi.org/10.3390/w15061041

Academic Editor: Pavel Groisman

Received: 30 January 2023
Revised: 6 March 2023
Accepted: 7 March 2023
Published: 9 March 2023

1. Introduction

Thermal conductivity indicates the thermal conduction properties of materials, while thermal diffusivity not only reflects their thermal conduction properties but also takes into account the effects of specific heat and density. In steady-state heat conduction, the thermal conductivity determines the capacity of heat transmission because the inner temperature of the materials does not change with time. Conversely, in unsteady state conduction, the temperature of the materials changes with time, so the thermal diffusivity determines the temperature distribution. The temperature in natural ice changes constantly due to unsteady heat conduction [1,2], so the thermal diffusivity of ice plays an important role. The accuracy of the ice thermal diffusivity improves the water-ice-air coupling model. In most previous studies, especially when simulating large-scale fresh ice, the thermal conductivity, specific heat, and density of pure and clear ice were set as constants [3,4]. For sea ice, only the thermal conductivity is parameterized, while specific heat and density are considered constants [5,6]. These methods are feasible for simulating the evolution of large-scale ice, but there will be a bias for small-scale ice. First of all, the unfrozen water in fresh ice contains trace impurities, and the freshwater ice can be regarded as a three-phase composite material composed of pure ice bubbles and low salinity unfrozen water. Due to the low ice temperature, most of the unfrozen water around the ice crystal is frozen, and a two-phase composite material of pure ice bubbles can be seen. However, if freshwater ice contains sediment (e.g., the Yellow River ice), a four-phase composition of pure ice bubbles unfrozen water sediment depending on the amount of sediment that appears in the ice [7]. Sea ice is recognized as a four-phase composite material composed of pure ice brine cells, bubbles, and solid salt. The distribution, size, shape, and total volume ratio of each phase component control the various properties of the ice for both freshwater ice and sea ice. Secondly, ice-unfrozen water is sensitive to temperature within the temperature range of phase transition, and small temperature changes will alter the ratio of phase components. Research has focused on "low-temperature" ice, where the volume ratios of phases in the ice are stable, particularly the low bubble ratio, and the brine volume ratio has become a key physical indicator for evaluating the properties of sea ice [8]. However, the sea ice temperature in the Bohai Sea is higher than that in the polar regions, with a large bubble ratio in the ice, and it was found in the 1980s and 1990s that the only use of the brine volume ratio would limit the correct evaluation of the mechanical properties of sea ice during the spring ice-melt period [9]. Therefore, porosity, the sum of the brine bubble volume ratio, is a perfect indicator for evaluating its properties [10]. It is also applied in the evaluation of the thermal and mechanical properties of high-latitude ice at present [11,12], but the brine volume ratio is still used as an indicator for evaluating thermal and mechanical properties [13,14]. Finally, as the climate warms, natural ice has increased in temperature in addition to the macroscopic phenomena of thinner ice thickness and shorter ice ages, leading to an increase in the spatial and temporal ratio of ice temperatures in the phase transition zone relative to high temperatures, including polar ice [15]. It may be ineffective to continue to simulate current conditions of ice using the thermal properties of ice as a constant or in a relationship where the relatively low-temperature zone varies only with ice temperature. In particular, the bias will be greater when simulating ice conditions during ice-melting periods.

There are basic models for the thermal conductivity of porous materials in simple structural forms. If the natural ice is viewed as a two-phase composite material, the distribution relationship between the two materials can be normalized and assumed to provide a thermal conductivity model of the two-phase composite material. A report on the application to natural ice and a comparative analysis of different models are shown in the literature [16]. In fact, the composition of other composite materials can be solids containing gases or static or dynamic liquids, but their solid components generally do not undergo a phase transition. Natural ice, however, has a temperature-dependent variation in the volume ratios of the solid, gaseous, and liquid phases, with a significant variation in the phase transition temperature zone. In addition, since the chemical molecular formula of ice is the same as that of water, the total mass of the two remains basically unchanged during the phase transition process, but the volume ratios change, suggesting that natural ice with phase transition is much more complicated than composite materials without phase transition. Although there are few research results on natural ice, the research ideas on the thermal properties of solution-containing composite materials and soil can be used as references [17–19]. In particular, research results on thermal property parameters of frozen soil or ice-containing composite materials with ice phase transition, such as testing technology [20,21], composite material modeling [22], and the evolution process and mechanism of the frozen soil temperature field [23], are presented. For time-series survey data of on-site ice temperature, mathematical inversion identification of the corresponding thermal diffusivity based on the measured temperature field data under ideal conditions is also a basic method for research in similar fields [24–26].

As to the establishment of an accurate thermal property evaluation model of natural ice, the support of measured ice thermal property parameters is needed first. Huang et al. (2013) studied the thermal conductivity of natural freshwater ice using laboratory conditions. Based on the measurements of ice crystals and bubbles, the structure of bubbles in ice is found to be quite complex. At present, a single basic model and empirical equations are difficult to achieve the desired precision, so simple structural models can be improved to form empirical and semi-empirical models [16]. Besides, using the time-series data of the vertical temperature profile within ice layers of the measured natural ice, the inversion identification of the thermal diffusivity of natural ice can be conducted with the finite difference method [27]. Specifically, a one-dimensional ice thermodynamic non-linear distributed parameter system is constructed to identify the thermal diffusivity of ice based on the one-dimensional heat conduction equation of ice. With the theoretical trend and ranges of thermal diffusivity of natural ice as constraints and the bias between the simulated and observed ice temperature as the objective function, an inversion identification model of a discontinuous or non-linear distributed parameter system is established via the small interval refinement stratification of the studied ice layers. Then, the statistical relationship between the inversion-identified thermal diffusivity of ice and freshwater ice temperature or sea ice porosity is established to express the effects of ice temperature, density, and salinity on the thermal diffusivity of natural ice. In this paper, the research results are summarized, especially the problems in the interpretation of physical mechanisms, and the direction of future efforts is reflected.

2. Fundamentals of Physics and Mathematics

2.1. Physical Background of the Thermal Diffusivity of Ice

Ice physics is essential to controlling the thermal diffusivity of ice, and the type of ice crystals only determines which one to choose when applying the composite material model. For freshwater ice, ice temperature and density are necessary, and for sea ice, ice salinity should be added. These physical parameters can describe the thermal conductivity of ice, specific heat, latent heat, etc. [5,28], Thermal diffusivity is the function of thermal conductivity, density, and specific heat. In physical essence, the thermal diffusivity is determined by the ice temperature, density, and salinity as well, i.e.,

$$\lambda = \frac{k}{\rho \cdot c}, \tag{1}$$

where λ is the thermal diffusivity of ice, m^2 s^{-1}, k is the thermal conductivity of ice, W m^{-1} °C^{-1}, ρ is the ice density, kg m^{-3}, and c is the ice specific heat, J g^{-1} °C^{-1}.

The relationships among the thermal conductivity of pure ice, specific heat, and ice temperature, proposed by Yen (1981). Then researchers achieved similar or further development results based on Yen (1981) [29].

$$k_{pi}(T) = 2.0733e^{-0.0057T}, \tag{2}$$

$$c_{pi}(T) = 2096.806 + 7.122T, \tag{3}$$

Equation (4) can be used when the density of pure ice changes with the ice temperature [30].

$$\rho_{pi}(T) = \frac{916.8}{1 + (158T + 0.54T^2) \times 10^{-6}}, \tag{4}$$

where k_{pi} is the thermal conductivity of pure ice, W m^{-1} °C^{-1}, T is the ice temperature, °C, c_{pi} is the specific heat of pure ice, J kg^{-1} °C^{-1}, and ρ_{pi} is the density of pure ice, kg m^{-3}.

Schewerdtfecer (1963) assumed that the bubbles in the sea ice were spherical and applied the thermal conductivity model of classic Maxwell composite materials to derive the equations for the specific heat and thermal diffusivity of sea ice [28], where the effective specific heat of sea ice is Equation (5).

$$c_{si,eff}(z) = \frac{S_{si}(z)}{\alpha T(z)^2} L + \frac{S_{si}(z)}{\alpha T(z)} (c_w - c_{pi}) + c_{pi}, \tag{5}$$

where factor $\alpha = -0.0182°C^{-1}$, z is the sea ice layer depth, m, and $S_{si}(z)$ and $T(z)$ are sea ice salinity (‰) and temperature (°C). L means latent heat of ice freezing, J g^{-1}, and it is L_{pi} = 334 J g^{-1}, and when S_{si} = 8 ‰, L_{pi} = 264 J g^{-1}. c_w = 4.19 J g^{-1} $°C^{-1}$, which is the specific heat of pure water. Equation (3) was not used to calculate the specific heat of pure ice, and c_{pi} = 2.09 J g^{-1} $°C^{-1}$ was taken directly.

The corresponding effective thermal conductivity of sea ice is given by Equation (6).

$$k_{si,eff}(z) = k_{pi} - (k_{pi} - k_b)\frac{S_{si}(z)\rho_{si}}{\alpha\rho_w T(z)}, \tag{6}$$

where ρ_{si} is the density of sea ice, kg m^{-3}, k_b is the thermal conductivity of brine, W m^{-1} $°C^{-1}$, $k_b(T) = 0.52 + 0.023T^2$, and ρ_w = 1000 kg m^{-3}, which is the density of pure water. The thermal conductivity of pure ice was not calculated by Equation (2), and k_{pi} = 2.1 W m^{-1} $°C^{-1}$ was taken.

Therefore, the calculation of freshwater ice is simpler than that of sea ice. However, the thermal diffusivity of pure ice calculated by Equations (1)–(4) is a far cry from that in the phase transition zone of natural ice. According to Chen et al. (2005), the thermal diffusivity of KCl and NaCl saltwater frozen ice in the ice temperature zone of −3 °C to 0 °C varies non-linearly and is the same as the thermal diffusivity of water near the freezing point [31]. Physically, pure ice completes the phase transition at 0°C, with the thermal diffusivity varying directly between 10.0×10^{-7} m^2 s^{-1} and 1.35×10^{-7} m^2 s^{-1} [32], while all others are transitioning in the phase transition zone. This reflects the significant variation in the pure ice, water, and bubble volume ratios in the ice in the phase transition zone, and the thermal diffusivity of ice is mainly determined by the volume ratio of pure ice and water. Only in the low-temperature zone without phase transition, the ice's thermal diffusivity is mainly determined by the volume ratio of pure ice and bubbles. For a unified representation of the effect of the water and bubble volume ratio in the ice, the porosity v is introduced, which is the sum of the bubble volume ratio in the ice and the unfrozen water (brine) volume ratio, expressed as a thousand (‰). As for sea ice, Cox and Weeks (1983) fitted the statistical equation of sea ice porosity in the temperature range of −2 °C to −22.9 °C according to the sea ice phase diagram with the required sea ice physical parameters of temperature, density, and salinity [33]. Later, Leppäranta and Manninen (1988) supplemented the equation for calculating the porosity of low-salinity sea ice at temperatures higher than −2 °C [34]. If the salinity is 0 for freshwater ice, the porosity is the bubble volume ratio of natural freshwater ice, which is only a function of ice density.

2.2. Non-Linearly Distributed Parameter System

The heat transfer in the ice layers is expressed by the classical one-dimensional heat conduction equation [35], as shown in Equation (7).

$$\rho c\frac{\partial}{\partial t}T(z,t) = \frac{\partial}{\partial z}\left(k\frac{\partial T(z,t)}{\partial z} - I(z,t)\right), \tag{7}$$

where z is the vertical coordinate of the ice layer depth, m, with the downward direction defined as the positive direction; t is the time; and I denotes the heat source item, W·m^{-2}, which is equivalent to the radiation component reaching the depth z.

In the inversion identification, the part of the ice surface greatly affected by solar radiation and the bottom layer greatly affected by heat flux from water were removed, or the nighttime data without solar radiation were selected, and the identified ice layer system can be regarded as an ideal situation without external heat sources. In addition, the values of the time and space step lengths used were very small, and interpolation processing was required for the time and space intervals of measured data.

For freshwater ice, its thermal diffusivity decreases with the increase of ice temperature over a range [21]. When the ice temperature changes very little, the change in thermal diffusivity is also small. In the inversion identification, it was assumed that there is a linear relationship between the thermal diffusivity and ice temperature, i.e., $\lambda(T) = a_i + b_i T$, and the heat conduction equation describing the thermal diffusivity of freshwater ice for a small range of ice temperature changes and the corresponding initial-boundary conditions can be expressed as Equation (8).

$$\begin{cases} \frac{\partial}{\partial t}T(z,t) = \lambda(T;a_i,b_i)\frac{\partial^2}{\partial z^2}T(z,t) & (z,t) \in \Omega \times I \\ T(z,0)|_{t=0} = T_0(z) & z \in \Omega \\ T(z,t)|_{z=z_1} = T_1(t) & t \in I \\ T(z,t)|_{z=z_2} = T_2(t) & t \in I \end{cases}, \quad (8)$$

where $\lambda(T;\ a_i,\ b_i)$ represents the thermal diffusivity of ice, which is a function of the ice temperature T as determined by the parameters a_i and b_i. The position of the first temperature probe near the ice surface was taken as the origin of the coordinates, and vertically downward was taken as the positive direction of the oz axis. The starting position of the inversion identified ice temperature range was set as z_1, and the end as z_2, i.e., the inversion identification zone $\Omega = [z_1, z_2]$, time variable $t \in I = (0, t_m]$, and $0 < \text{tm} < +\infty$ is the observation time.

The thermal diffusivity of sea ice decreases with increasing porosity. Similarly, after a variety of simple function calculations [36], in the inversion identification, a simple function $\lambda(v) = a_{si}(1+v)^{b_{si}}$ was used to describe the relation between the thermal diffusivity of sea ice and porosity over a small range of temperature variations, so that the heat conduction equation describing the thermal diffusivity of sea ice and the corresponding initial boundary conditions can be expressed as Equation (9).

$$\begin{cases} \frac{\partial}{\partial t}T(z,t) = \lambda(v;a_{si},b_{si})\frac{\partial^2}{\partial z^2}T(z,t) & (z,t) \in \Omega \times I \\ T(z,t)|_{t=0} = T_0(z) & z \in \Omega \\ T(z,t)|_{z=z_1} = T_1(t) & t \in I \\ T(z,t)|_{z=z_2} = T_2(t) & t \in I \end{cases}, \quad (9)$$

where v is the ice porosity, ‰, and $\lambda\ (v;\ a_{si},\ b_{si})$ represent the thermal diffusivity of sea ice, which is a function of the porosity v determined by the parameters a_{si} and b_{si}.

Based on the research results on the ice thermal diffusivity [4,29,31], the inversion identification model of the non-linear system for the ice thermal diffusivity is Equation (10).

$$\begin{array}{ll} \min & f(z,t;a,b) = \int_{t \in I}\int_{z \in \Omega}\left(T(z,t;a,b) - \overline{T}(z,t)\right)^2 dzdt \\ s.t. & T(z,t;a,b) \in S_{U_{ad}} \end{array} \quad (10)$$

where $f\ (z,\ t;\ a,\ b)$ is the objective function calculating the absolute deviation between calculated and measured ice temperatures, which tries to make the calculated ice temperature close to the measured ice temperature. $T(z,\ t;\ a,\ b)$ is the fitting function for the temperature calculation obtained by the numerical calculation method through Equations (8) or (9), and $T(z,\ t)$ refers to the fitting function of the measured temperature. U_{ad} denotes the allowable set of parameters $(a,\ b)$, which was obtained from the analysis of relevant research [27,37], and S_{Uad} is the set of solutions $T(z,\ t;\ a,\ b)$ of Equations (8) or (9) depending on parameters (a, b) $\in U_{ad}$. Applying the parameter identification theory of the distributed parameter system, the optimal parameters a and b were obtained by computer programming with an improved genetic algorithm. The relevant theoretical analysis of the parameter identification model is detailed in the literature [27,37].

3. Field Survey of the Vertical Temperature Profile of Natural Ice

Since the 1980s, fixed-point temperature vertical profile surveys of natural freshwater ice and sea ice have been carried out, and the survey sites in China are shown in Figure 1. The inversion identification of thermal diffusivity had been conducted for some of the survey data, while some were yet to be carried out. In some of these surveys, both ice crystals and density were measured, while in others, the thermal conductivity was measured. Research of the ice thermal properties was performed in Hongqipao Reservoir, the Thermakarst Lake of Beiluhe on the Qinghai-Tibet Plateau, the estuary of the Yellow River, the Fen River Reservoir II, Bayuquan in the Bohai Sea, and Zhongshan Station in Antarctica, the survey sites shown in Table 1. In addition, the inversion identification of the thermal diffusivity of ice has not yet been carried out at the following sites: Hanzhang Lake, Wuliangsu Lake, Qinghai Lake, Ngoring Lake, and Shisifenzi of the Yellow River.

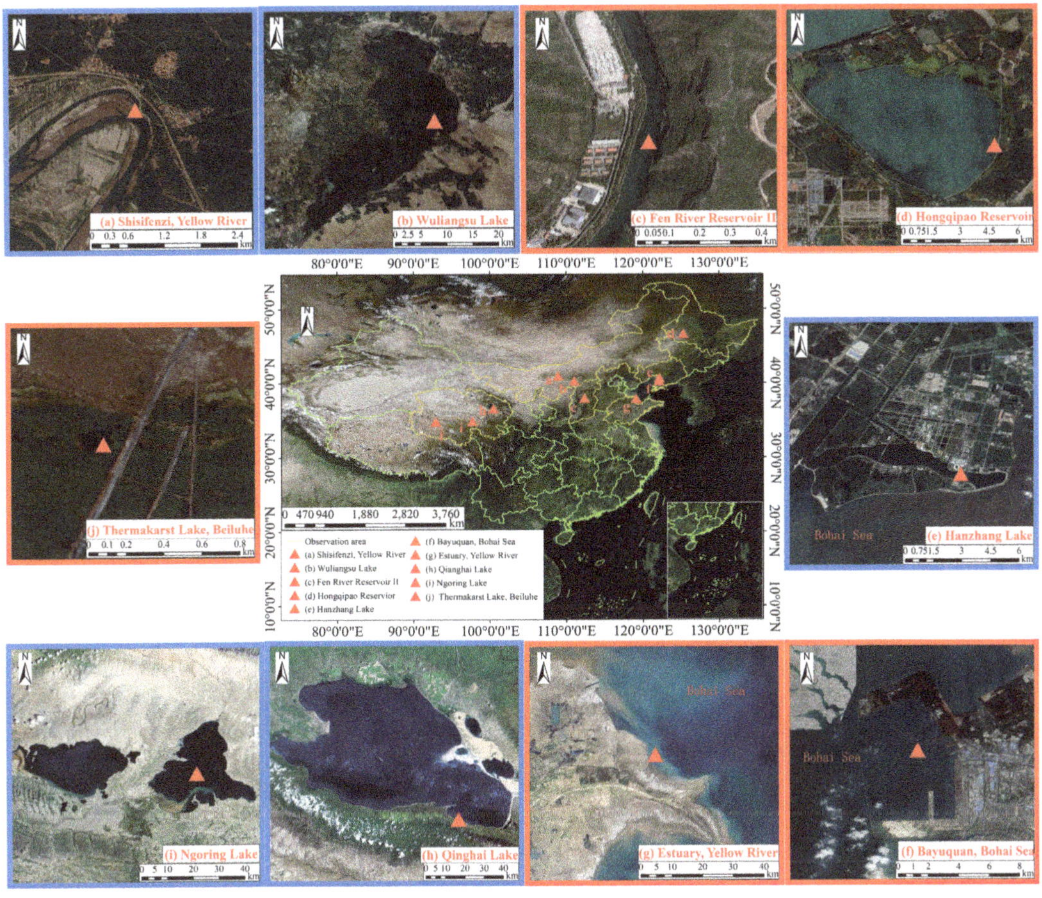

Figure 1. Site locations of a time-series survey of vertical temperature distribution within different ice layers in China (the red boxes indicate that the research on the thermal properties of ice has been carried out, and the blue boxes indicate that the research has not been carried out).

From January to February 2004, a survey of the temperature vertical profile of river ice and the formation and dissipation processes of ice was carried out in a shady place on the Fen River Reservoir II. Specifically, the air temperature at a height of 1.5 m from the ice surface, the ice temperature every 5 cm within 0.55 m vertically downward from the ice surface, and the water temperature at 100 cm and 150 cm below the ice surface were measured. Temperature data were automatically collected every 15 min [38,39]. The PT100 (JUMO, Germany) was used as the temperature probe, with a precision of ±0.1 °C and a resolution of 0.01 °C.

From December 2008 to April 2009 and from October 2009 to April 2010, a field survey on the formation and dissipation process of ice was carried out in Hongqipao

Reservoir [40–42] to obtain the air temperature, ice temperature, and water temperature at 150 cm above the ice surface. The ice or water temperature probes were placed at 2, 7, 12, 17, 32, 47, 62, 77, 92, 107, 122, 152, 182, 212, and 242 cm below the ice surface. Besides solar radiation, wind speed, wind direction, ice thickness, ice crystals, bubbles in ice, and ice density were measured. The PT100 (JUMO, Germany) was adopted as the temperature probe, and the TRM-ZS1 polar wind direction and wind speed gradient monitoring recorder was used for radiation measurement, with a resolution of 1 W m^{-2} and each sampling interval of 1 min.

From October 2010 to July 2011, a set of temperature chains was installed in the Thermakarst Lake of Beiluhe on the Qinghai-Tibet Plateau [16,42], and the PT100 system (Campbell CR10) developed by the State Key Laboratory of Frozen Soil Engineering, Cold and Arid Region Environment and Engineering Research Institute, Chinese Academy of Sciences, was used, with a resolution of 0.1 °C. The temperature sensor probes installed every 5 cm automatically recorded the temperature every 30 min. An array of data, such as air temperature, ice temperature, and water temperature under the ice, was obtained.

From April to October 2006, a field survey of overwintering sea ice physics was carried out near the Zhongshan Station in Prydz Bay, Antarctica. A series of ice core salinity, density vertical profiles, and continuous ice temperature vertical profiles were measured, as well as time-series data on ice thickness. The temperature probe PT100 (JUMO, Germany) was placed every 6 cm, and the sampling interval was 30 min [43].

In the 1980s, the ice thermodynamic properties in Bayuquan in the Bohai Sea and the Yellow River estuary were studied [44]. The ice temperature vertical distribution profile was measured in the Bayuquan harbor basin from 5 to 23 January 1987. There are two sections formed at five temperature measurement points that are parallel and perpendicular to the harbor basin wall, respectively. A total of 12 layers of temperature measurement probes with a copper-composite thermocouple were buried vertically at each temperature measurement position at the depths of 3, 8, 13, 18, 23, 28, 33, 38, 43, 53, 78, and 103 cm below the ice surface. The temperature value was converted based on the thermal potential measured with an UJ33a potentiometer with a precision of 1 microvolt. The precision was ±0.1 °C within the measured temperature range of 0 to −25 °C, calibrated by a secondary standard thermometer [45]. Two CN-9L heat flux sensors (prod. Japan) were inserted at 8 cm and 23 cm, which are highly responsive and allow instantaneous dynamic changes to be measured. The heat flux passing through the heat flux sensor was converted based on the potential values measured at the two poles of the sheet, after which the thermal conductivity was calculated from the calibration coefficient of the heat flux sensor [44]. Meanwhile, samples were taken to measure the salinity (3.2‰), density (876 kg m^{-3}), and crystal (grain size of 2.1–32 mm) vertical profiles of the ice. On January 28, 1988, the ice slabs were collected at the No. 3 Drainage and Irrigation Station of Gudong Oilfield on the north side of the Yellow River estuary, about 1500 m from the shore. Its thickness was 11 cm, the mean salinity of the sea ice was 0.400‰, and the sea ice density was 961 kg m^{-3}. The upper and lower incubators were prepared in the laboratory, and the thermal conductivity of ice samples in the low-temperature zone was tested [44].

Generally, there is less snow accumulation on the ice surface in arid and semi-arid areas of China. Even if some snow falls, it is difficult to remain on the ice surface due to the strong winds. Therefore, snow accumulation has little effect on the ice and the water temperature under the ice. Among the survey sites, only the Hongqipao Reservoir had continuous snow accumulation, which can reduce the effect of radiation on ice and water temperature, but the snow accumulation is not enough to cause changes in ice and water temperature characteristics. The air temperature, some ice temperature, and radiation values of Hongqipao Reservoir from 5 to 10 January 2009 were taken as an example (Figure 2) [46], and the lake ice evolution was analyzed. The ice temperature increases with depth, and the variation decreases and stabilizes in Figure 2. In other words, the influence of ice temperature on air temperature reduces with increasing depth, and the restriction of the water temperature under the ice is generally reflected. Furthermore, the temperature characteristics of river ice and sea ice measured in situ are basically the same as those of lake ice [39,40,42,43,45]. The total solar radiation and reflected solar radiation flux in northern China typically start to increase around 6:30 each day, reach a peak near 12:30, and then start to decrease. There is no total solar radiation or reflected solar radiation between 17:00 each day and 6:30 the next day. In addition, the moment of air temperature/ice temperature peaks shows a certain lag with the radiation peak [40,41].

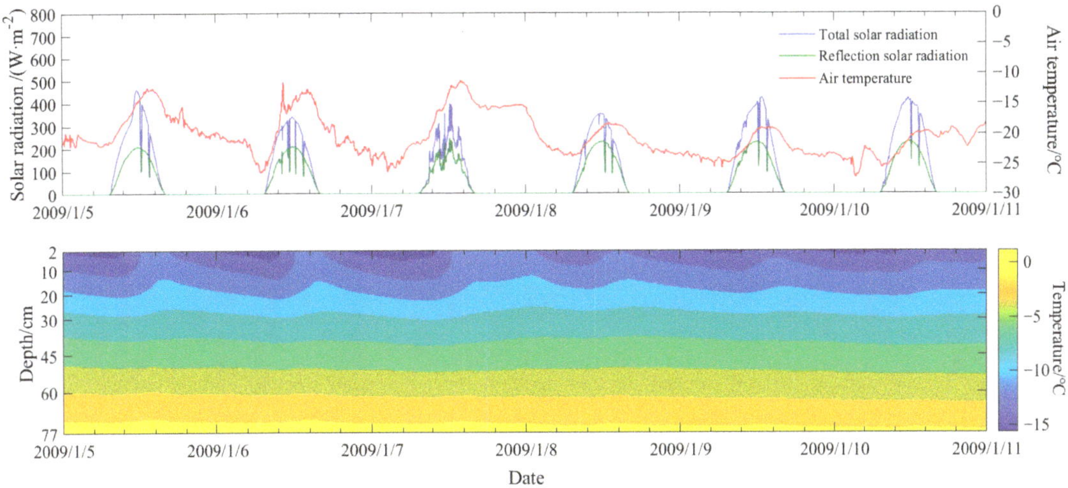

Figure 2. Measured process curves of air temperature, total solar radiation, reflected solar radiation, and ice/water temperature of Hongqipao Reservoir from 1 to 5 January 2009 (modified from Ref. [46]).

The sites selected for the field survey are all flat and stable ice layers with growth caused by thermodynamic processes, so the crystal structure is relatively simple no matter if it is lake ice, river ice, or sea ice. Generally, granular ice is found near the ice surface, followed by columnar ice. If snow falls during the initial freezing period, the granular ice will account for more [42]. A sample of ice crystals from surface to bottom is shown in Figure 3. In the arid area of the Qinghai-Tibet Plateau, granular ice on the ice surface disappears due to sublimation [47]. Granular ice appears crosswise only at the Zhongshan Station, Antarctica, where sea ice is highly dynamic [36].

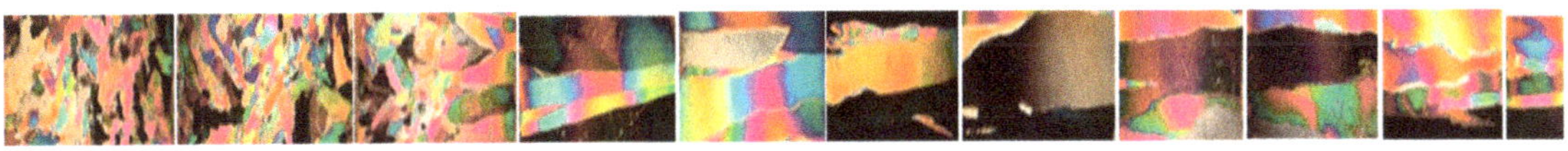

Figure 3. Crystals of reservoir ice from the surface to the bottom in Hongqibao Reservoir (a total length of 85 cm) (modified from Ref. [42]).

Low-temperature ice samples, especially those frozen after sampling, have a low ratio of unfrozen water in the ice. The density of low-temperature ice samples can reflect the bubble volume ratio in natural ice. Bubbles in natural freshwater ice are generally visible but are relatively rare, such as the sword-headed bubbles in Hongqipao Reservoir (Figure 4a) [42]. The lake ice of Beiluhe on the Qinghai-Tibet Plateau has a higher porosity and larger bubbles [47]. Figure 4b,c show spine-shaped and linear granular bubbles in the lake ice of Beiluhe, respectively. The spine-shaped bubbles are larger in size, pancake-like, approximately 1 cm to 5 cm in diameter horizontally, and up to 1 cm to 2 cm thick vertically, with a flat top and bottom and an irregular spherical shape. In contrast, the linear granular bubbles are smaller, generally 0.3–2.5 mm in diameter, with individual bubbles appearing to be spherical, but with some connected in series to form a cylindrical-like shape, and the slenderness ratio (ratio of the width and height of the longitudinal section of the bubble) is 100–200 [47].

Table 1. Basic information about the site for the survey and research on ice thermal properties.

Information	Hongqipao Reservoir	Thermakarst Lake, Beiluhe	Estuary, Yellow River	Fen River Reservoir II	Bayuquan, Bohai Sea	Zhongshan Station, Antarctica
Latitude and longitude/°	112.27° E, 37.60° N	92.92° E, 34.83° N	119.12° E, 37.88° N	112.38° E, 37.98° N	122.07° E, 40.28° N	76.37° E, 69.37° S
Elevation/m	140	4640	2	800–1400	2	11
Duration of freezing period/d	150–180	150–210	80–100	80–100	110–120	300
Ice thickness at severe ice period/m	1.0–1.2	0.7–1.0	0.1–0.2	0.4–0.5	0.3–0.4	1.6–1.8
Mode of Ice Formation	Thermodynamics	Thermodynamics	Thermodynamics	Thermodynamics	Thermodynamics	Thermodynamics
Ice classification	Lake Ice	Lake Ice	Saltwater ice	River ice	Sea ice	Sea ice
Ice crystals	Granular/columnar	Columnar	Columnar		Granular/ columnar	Granular/columnar mixture
Ice physical indicators	Density	Density, bubbles	Salinity, density		Salinity, Density	Salinity, density
Ice thermal indicators	Laboratory thermal conductivity, identified thermal diffusivity	Laboratory thermal conductivity, identified thermal diffusivity	Laboratory thermal conductivity	Identified thermal diffusivity	Thermal conductivity in situ	Identified thermal diffusivity
Literatures	[40,41]	[16]	[44]	[38,39]	[44]	[43]

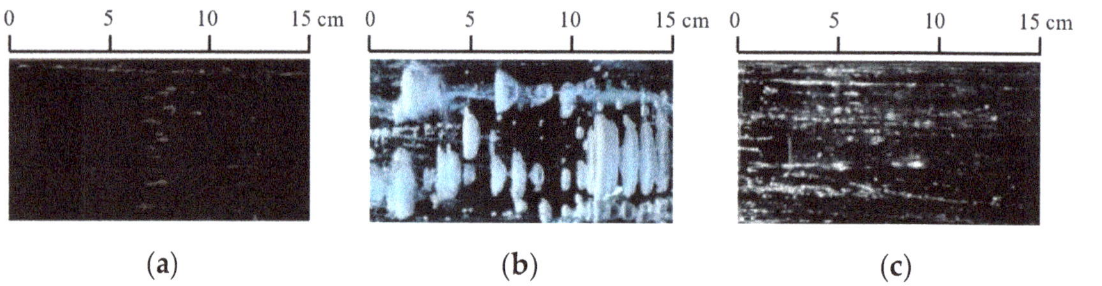

Figure 4. The bubbles in different freshwater ice in China are (**a**) sword-headed bubbles in Hongqibao Reservoir, 28 December 2008 (modified from Ref. [42]); (**b**) spine-shaped bubbles in the Thermakarst Lake of Beiluhe on the Qinghai-Tibet Plateau, 9 December 2010 (modified from Ref. [47]); and (**c**) linear granular bubbles in the Thermakarst Lake of Beiluhe on the Qinghai-Tibet Plateau, 9 December 2010 (modified from Ref. [47]).

4. Thermal Diffusivity Characteristics of Natural Freshwater Ice and Sea Ice

4.1. Thermal Diffusivity Characteristics of Freshwater Ice

The time-series data of the ice temperature vertical profile of Fen River Reservoir II, measured from 0:00 on 23 January to 0:00 on 5 February 2004, were grouped into daily mean temperatures −6 °C, which were recorded −6 to −3 °C and −3 to 0 °C, and then the inversion identification of thermal diffusivity was calculated, respectively. During the inversion identification, the high-temperature section of the river ice was cryptographically divided into several small temperature sections and periods to explore the fine relationship between the thermal diffusivity and temperature near the freezing point. The inversion-identified thermal diffusivity of ice in Fen River Reservoir II is shown in Figure 5a [39].

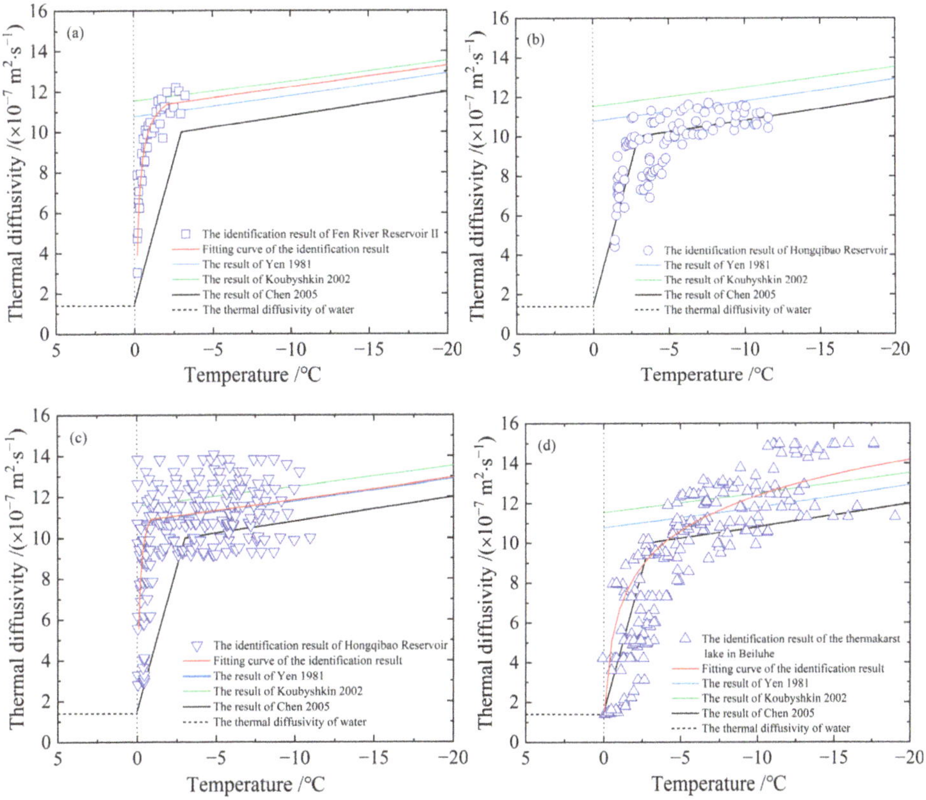

Figure 5. Relationship between inversion-identified thermal diffusivity of different freshwater ice survey data in China and temperature: (**a**) Fen River Reservoir II (modified from Ref. [38]), (**b**) Hongqibao Reservoir (modified from Ref. [46]), (**c**) Hongqibao Reservoir (modified from Ref. [18]), and (**d**) Beiluhe on the Qinghai-Tibet Plateau (modified from Ref. [27]). Some data used in the (**a–d**) are from Refs. [29–31].

According to Figure 5a, when the temperature of natural river ice ranges from 0 °C to −0.76 °C, its thermal diffusivity increases sharply from a small value in a non-linear manner. When the temperature is lower than −0.76 °C, the thermal diffusivity of natural ice is close to the thermal diffusivity of pure ice proposed by Yen (1981) [29]. When the temperature is lower than −2.12 °C, the error between the thermal diffusivity of natural ice and the experimental results of Koubyshkin and Sazonov (1988) [30] is less than 0.035. Therefore, the inversion-identified thermal diffusivity characteristics of freshwater ice in Fen River Reservoir II were summarized as follows: (1) When the temperature of the entire ice layer is low, especially when the ice surface temperature is significantly lower than the melting point, i.e., no phase transition occurs at the ice surface, the results are all closer to the previous experimental ones. (2) When the air and ice surface temperatures are high (above or reach 0 °C), the inversion identification results are smaller than the thermal diffusivity of pure ice calculated by Yen (1981) [29] or Koubyshkin and Sazonov (1988) [30].

Based on the measured ice temperature from the Hongqipao Reservoir, the time and space step length used in the inversion identification were interpolation data of 10 s and 0.5 cm, respectively, and the inversion identification was calculated by layers. The inversion-identified thermal diffusivity of reservoir ice and temperature in the winter of 2008–2009 are shown in Figure 5b, and all the data from the two winters of 2008–2009 and 2009–2010, the thermal diffusivity of ice obtained by inversion identification, are shown in Figure 5c. In the low ice temperature range (−15 °C to −3 °C), the inversion-identified thermal diffusivity of reservoir ice is similar to the results reported in the literature, i.e., the thermal diffusivity reduces slowly with increasing ice temperature. In the high ice temperature range (−3 °C to 0 °C), the inversion-identified thermal diffusivity of reservoir ice decreases significantly with the increasing ice temperature and approaches the thermal diffusivity of water. It suggests that the ice is in the phase transition stage; the ratio of pure ice crystals in the ice decreases and the ratio of unfrozen water increases, ensuring that the thermal diffusivity of ice converges with the thermal diffusivity of water due to the increased water volume ratio [46,48]. On the other hand, the high thermal diffusivity of bubbles cannot be manifested at this stage due to their small ratio.

Based on the measurements of the ice temperature vertical profile in the Thermakarst Lake of Beiluhe on the Qinghai-Tibet Plateau, the inversion identification was calculated by layers with an interpolated time and space step length of 10 s and 0.5 cm, respectively, to obtain the scatter diagram of thermal diffusivity changing with ice temperature, as shown in Figure 5d. In the relatively low ice temperature range (−15 °C to −3 °C), the thermal diffusivity varies slowly with ice temperature. Similarly, in the relatively high ice temperature range (−3 °C to 0 °C), the thermal diffusivity changes drastically, dropping rapidly with the increasing ice temperature and converging to the thermal diffusivity value of fresh water [27].

Researchers have adopted different non-linear fitting methods when establishing the relationship between the inversion identification results and ice temperature in Figure 5. The results of the low-temperature section of the Fen River Reservoir II in Figure 5a are similar to those of previous studies. Bai (2006) suggested using previous results based on actual conditions but introduced a statistical Equation (11) for the high-temperature section [38]. The results supported the inversion-identified thermal diffusivity of other freshwater ice in China [27,36].

$$\lambda(T) = (12.6 + \frac{2.7}{T - 0.12}\Big) \times 10^{-7} \qquad T \in [-0.19, -2.12], \tag{11}$$

Figure 5b shows the inversion identification results of Hongqipao Reservoir in the winter of 2008–2009 without the fitting expression of scattered points (93 points). Figure 5c presents the inversion identification results of Hongqipao Reservoir in the two winters of 2008–2009 and 2009–2010 [46]. The results were used for the ice temperature evaluation in Kanas Lake, Xinjiang, China [49]. The adjustment of the calculation method used in this identification results in 204 identification results for 87 identification temperature intervals. Despite the scattered results, a new piecewise Equation (12) of thermal diffusivity with temperature is given on this basis [48].

$$\lambda(T) = \begin{cases} \left(\frac{T}{0.08T - 0.01}\right) \times 10^{-7} & -0.85 \le T < -0.1 \\ 10.83 \times 10^{-7} \cdot e^{-0.009T} & -15 \le T < -0.85 \end{cases}, \tag{12}$$

The thermal diffusivity of lake ice in the Thermakarst Lake of Beiluhe on the Qinghai-Tibet Plateau varies with ice temperature (Figure 5d) and is calculated as Equation (13) [27]. It was part of a contribution to one of the Norway-China collaboration projects [50].

$$\lambda(T) = [2.61 \times \ln(-T + 0.11) + 6.35] \times 10^{-7} \qquad T \in [-15, 0), \tag{13}$$

The symbols in Equations (11)–(13) are the same as the above ones.

To compare the difference between the inversion identification results of freshwater and the thermal diffusivity calculated from pure ice, the results of Yen (1981) [29], Koubyshkin, and Sazonov (1988) [30] were plotted in Figure 5. These results for pure ice are consistent with those for the thermal diffusivity of pure ice in the laboratory [21]. However, the phase transition of natural freshwater ice is not complete at 0 °C, and there is a phase transition temperature range. The thermal conductivity of any impure ice decreases rapidly from a high value in the phase transition zone as the ice temperature approaches the freezing point [41,44,51]. Chen (2005) used KCl and NaCl saltwater freezing to measure the process of thermal diffusivity change in the phase transition zone (−3 °C to 0 °C) by temperature wave analysis (TWA) and to obtain the thermal diffusivity that rapidly decreases in value for water as it approaches the freezing point [31]. However, the results from the piecewise function are also marked in Figure 5. In comparison, the results of Chen (2005) confirmed the correctness of the inversion identification results of natural ice.

In addition, according to Figure 5d, when the ice temperature ranges from −15 °C to −5 °C, the inversion-identified thermal diffusivity value of lake ice in Beiluhe is significantly higher than other inversion identification results of lake ice and literature reports [31]. While the temperature is higher than −5 °C, the inversion identification results are almost the same as the experimental results [31]. It indicates that when the temperature of a large number of spine-shaped bubbles in the lake ice of Beiluhe (Figure 4b) is lower than −5 °C and the phase transition is stable, the bubbles begin to contribute. Since the thermal diffusivity of gas is much greater than that of pure ice, ice with more bubbles has a greater thermal diffusivity. In most cases, as the size and content of bubbles in river ice and reservoir ice are smaller [42], their impact on thermodynamic properties is often neglected. However, such bubbles in ice as in the Beiluhe cannot be overlooked, and even the combined effects of bubble content, size, and shape on the thermal diffusivity of ice may need to be considered [16].

4.2. Thermal Diffusivity Characteristics of Sea Ice

Sea ice is distinguished from freshwater ice by the presence of salt. The unfrozen water among ice crystals, known as brine, is both saltier and denser than seawater. When the sea ice temperature is high, the edges of the crystalline grains begin to melt, the brine channel widens, and the brine in the ice drains downward by gravity, after which it may be left to fill with gas. It results in both liquids and gases of complex chemical compositions in the ice, and their volume ratios vary with temperature, causing a more complex mathematical expression of the change in the thermal diffusivity of sea ice. Theoretically, instead of establishing a relationship with ice temperature, the thermal diffusivity of sea ice has to be introduced as a function of density and salinity. The porosity of sea ice is a function of temperature, density, and salinity, but when the porosity is introduced and there is a lack of salinity and density vertical profiles for online monitoring, sampling tests will be conducted at different times, resulting in no corresponding porosity even though the inversion-identified thermal diffusivity of sea ice is obtained using ice temperature.

In 2006, during the continuous survey of the formation and dissipation of sea ice and the ice temperature vertical profile at Zhongshan Station, Antarctica, seven ice cores were sampled to test the ice density and salinity. Assuming that the porosity of the ice core did not change for 24 h on the day of sampling [37], the inversion-identified thermal diffusivity of the ice and the porosity at the corresponding location during this period were plotted as a scatter diagram (Figure 6) [36,37] to obtain a statistical relationship Equation (14) between the thermal diffusivity and porosity of Antarctic sea ice. Figure 6 also shows that the thermal diffusivity of each ice core varies significantly.

$$\lambda(v) = 10.8 \times 10^{-7}(1+v)^{-0.302} \qquad v \in (0, 275], \tag{14}$$

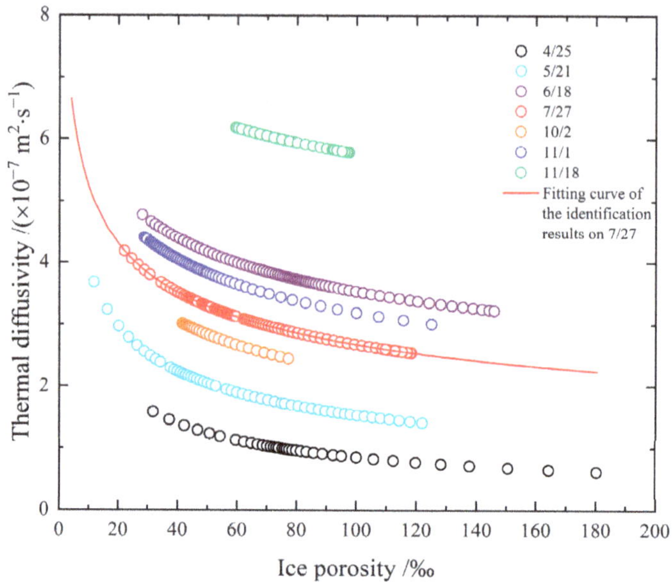

Figure 6. Relationship between inversion-identified thermal diffusivity and porosity of sea ice survey data at Zhongshan Station, Antarctica, in 2006 (modified from Refs. [36,37]).

5. Future Working Directions and Considerations on the Relation between Thermal Conductivity and Ice Physical Parameters

Both in freshwater lake (or reservoir) ice and in river ice and sea ice, the thermal diffusivities of ice are all dominated by three physical parameters that can be quantified: the ice temperature, ice density, and salinity. With regard to pure ice frozen from pure water, its freezing point and melting point are the same, which are 0 °C. Thus, the phase transition is also finished at 0 °C. However, natural water usually contains a certain amount of dissolved chemicals, which will result in a concentration of the unfrozen water's salinity and a lower freezing point. Especially when there are different chemicals at different levels, the freezing and melting points are also diversified [31,52]. This leads to an irreversible phenomenon in the thermodynamics of natural ice as well as in saline frozen soil [53]. In fact, these small changes usually bring deviations from the accurate thermodynamic processes on a small scale. Therefore, researchers must reflect on what exactly the determinants describing the thermal diffusion in ice are and build a parameterization program for the thermal diffusivity of natural freshwater ice or sea ice. Some considerations open for discussion are as follows:

1. The thermal diffusivities in Figure 5 are obtained based on measured data of ice temperature from different fields. The heat conduction equation was solved for numerical solutions during the inversion identification process, which relies on the initial boundary value conditions. Hence, different schemes and the initial conditions will generate various results of thermal diffusivity. The optimal parameters of the inversion identification model of the non-linearly distributed parameter system are not absolutely the best solution. However, comprehending from the measured data's precision, on which the study is based, it will not make a difference in the identified thermal diffusivity. Figure 5a shows the results of a segmented discontinuous approach, which is to divide the temperature into multiple small ranges according to the measured ice temperature. It deems that the thermal diffusivity of ice varies as a linear variation or a power function variation with temperature within small temperature ranges, and the thermal diffusivity was recalculated again in the next temperature

range. Although the results of these two calculations have the same function, their thermal diffusivities are different. Compared with other approaches, this method requires a lot of computation, but the resulting thermal diffusivities perform well in the sensitive high-temperature zone of phase transition, and even the smaller the temperature ranges, the better the results. Consequently, the points in the resulting scatter plot in Figure 5a are rather more concentrated, while the results in Figure 5b–d are relatively dispersed for the expansion of the selected temperature range in identification. Moreover, factors such as the time and space step length, interpolation method, and programming algorithm adopted in the identification calculation process can also influence the identification results in varying degrees. Figure 5b,c show the results from different scholars on account of the measurements taken at the same test site in different years, especially the data in Figure 5c covering the measured temperature data applied in Figure 5b. Nevertheless, the dispersion of the two identification results is apparently different because of the different methods. The steps, including how to unify the step length, interpolation, algorithm, etc., also need to be explored to acquire optimal results;

2. As the global warming develops, lake ice, river ice, and sea ice are all reducing. In the Arctic area, except for the shortening of the freezing period on a macro level, thinner ice thickness, and a decline in the proportion of multi-year ice, there are also phenomena including an increase in ice temperature, a decrease in ice salinity, a reduction in ice density, and a widening of the varying range of ice density [15]. Likewise, the ice conditions in the Bohai Sea and inland China are also decreasing [54,55]. As the spatial and temporal proportions of comparatively "high-temperature" ice are growing worldwide, the simulation effects will be reduced in reality if the previous data on the relation between the thermal diffusivity of ice and temperature with no regard to the phase transition or constants are adopted for numerical simulations. If the thermal diffusivity of ice reduces, the heat storage capacity of the ice body will be strengthened, and it will cause an increase in entropy in the phase transition process from ice to water or from water to ice, which can moderate the melting or freezing rate of ice. In other words, despite the fact that the thermal conductivity of ice in the phase transition zone of the sea ice in the Bohai Sea was also relatively small in the 1980s [44], the ice in the Bohai Sea covers a comparatively small percentage of the global situation of sea ice, so that it is covered by a large amount of other low-temperature ice. However, the spatial and temporal proportions of ice within the phase transition zone have increased. In this case, previously adopted methods may still be fairly feasible on large-scale issues, but they may no longer be proper to describe the thermodynamic behavior of ice on a finite microscale;
3. In the inversion identification of thermal diffusivity, only the time-series data of ice temperature vertical profiles are used, without counting the types of ice crystals at the temperature measurement positions since these types cannot be expressed by numerical values directly. The crystals of ice frozen in calm waters (e.g., reservoir and lake ice) have a pattern of granular ice on the surface and then columnar ice [42]. While this is more complicated in the crystals of ice frozen from rivers and oceans. Granular ice's properties are basically isotropic, while columnar ice exhibits anisotropy. This can result in a difference between the mathematical models and the calculated results, such as the spread velocity of radar waves in the ice as determined by permittivity [7,56]. Research shows that the thermal conductivities of natural columnar lake ice range from 1.60 $W \cdot m^{-1} \cdot {}^{\circ}C^{-1}$ to 2.20 $W \cdot m^{-1} \cdot {}^{\circ}C^{-1}$ in both the vertical and parallel long axes and are slightly higher (about 5%) in the vertical long axis, showing that the thermal conductivity of ice crystals only has weak anisotropy [16]. From the perspective of the dispersion of inversion-identified ice thermal diffusivities, the differences among the fitted curves and the data points are over 5%. The uncertainty in inversion-identified is larger than the difference in the anisotropy of ice crystals. If 5–10% of the error caused by thermal diffusivity can be accepted on the large scale simulated, the influence of ice

crystals can be ignored. In the cases of transformation of ice crystals and overlapping or mixing of granular ice and columnar ice due to dynamics and thermodynamics, the differences of ice crystals can also be neglected. Otherwise, thermal diffusivity models of ice corresponding to various crystal structures should be selected. The ice crystal in the inversion identification of the thermal diffusivity is basically the columnar ice in China;

4. If unfrozen water among ice crystals undergoes a phase transition, its mass will remain the same, but its volume ratio will be different, which is the same as the study of frozen soil [53]. If unfrozen water discharges under gravity, it is likely that the partial space originally occupied by unfrozen water will be replaced by gas. In general, the higher the content of bubbles, the lower the ice density. Therefore, ice density can reveal the content of bubbles [56,57] and is an ideal indicator reflecting the effect of bubbles on the thermal diffusivity of ice. If the content of bubbles is less than 3%, the laboratory-tested thermal conductivity of freshwater ice is close to the value [16] calculated by Hamilton and Crosser's (1962) model [58]. When the content of bubbles is over 16%, any model of the porous medium's thermal conductivity cannot accurately compute the thermal conductivity [16]. A joint computing model of the thermal diffusivity of lake ice must be built by introducing a shape factor that includes the content and shape of bubbles in the ice. In future ice investigations, promoting the ice density test is indispensable for all models. Meanwhile, focusing on ice density can also reflect two potential scientific issues: First, the ice temperature in the phase transition zone is relatively high, and the bubble content is high because of the discharge of unfrozen water. Secondly, as global warming develops, the plants under shallow lakes in mid-latitudes have higher activity, releasing gases under the ice in winter, and greenhouse gases contained in lake bottoms at high latitudes or high altitudes may be released, such as in the thermokarst lake ice of the Qinghai-Tibet Plateau. Since the thermal diffusivity of bubbles is much higher than that of pure ice, the thermal diffusivity of lake ice with bubbles is larger than the theoretical thermal diffusivity of pure ice. Particularly, the thermal diffusivity of bubble-containing lake ice with a relatively low temperature is more obviously higher than the value of pure ice because the content of unfrozen water reduces;
5. Natural freshwater ice contains impurities, and the freezing temperature of unfrozen water is dynamic [31]. Meanwhile, the freezing and melting temperatures of ice with saline water. This shows an irreversible phenomenon in thermodynamics [31,59]. The salinity of sea ice is much higher than that of freshwater ice, and its influence is unmissable. When it comes to freshwater ice, the thermal diffusivity of ice can also be described as the relation between temperature and density if the influence of salinity is ignored, while this is impossible for sea ice because it might need to be an expression of the volume ratio of brine (temperature, salinity, density) and the volume ratio of bubbles (temperature, density). However, the inversion identification result for the thermal diffusivity of Antarctic sea ice (Figure 6) indicates that it is not that simple. It suggests that the refinement of the parameterization for the thermal diffusivity of sea ice is relatively difficult if the thermodynamic irreversible phenomenon is neglected, especially for sea ice in the melting period;
6. The previous results of experimental [31] and inversion identification [48] are expressed as segmented functions instead of continuous functions for ice temperature. Since the thermal diffusivity of natural ice is mainly controlled by the thermal diffusivity of pure ice, bubbles, and saline or pure unfrozen water, with the thermal diffusivity of unfrozen water as the lower limit and the thermal diffusivity of bubbles as the upper limit. It is suggested that future development should be based on the logistic functional form, and the suggested Equation (15) form is as follows:

$$\lambda = \frac{C}{1 + Ae^{BT+D}} + E, \tag{15}$$

where E is the lower limit of Equation (15), and it is close to the thermal diffusivity of unfrozen water (λ_w), which is 1.38×10^{-7} $m^2 \cdot s^{-1}$ at 0 °C. $C + E$ is the upper limit of Equation (15), and it is close to the thermal diffusivity of bubbles (λ_a). Since the bubble volume ratio is related to the ice density, its expression is Equation (16) [56,57].

$$v_a = \frac{\rho_{pi} - \rho}{\rho_{pi}} \times 1000‰, \tag{16}$$

where ρ_{pi} is the density of pure ice, 916.8 $kg \cdot m^{-3}$, and ρ is the density of natural ice. C is a nonlinear relation because the shape factor should be introduced when the bubble in ice is high in content and large in size. According to the model reported in the literature [17], the form of the power function (17) is suggested as follows.

$$C = \alpha(1 + \beta v_a{}^{\gamma}), \tag{17}$$

In (15) to (17), B is the maximum growth rate of the thermal diffusivity of natural ice at a certain temperature, A is also related to salinity, D/B is the ice-water phase transition temperature, which is also relatively complicated for unfrozen saline water not closed in the freezing and melting process, α, β, γ are fitting coefficients.

7. This consideration is not proved by examples yet because of the sparse density data obtained from natural ice in the field. However, there were measured thermal conductivities at different temperatures (−5, −10, −15, −20, and −25 °C) and different densities (300, 350, 400, and 450 $kg.m^{-3}$) of snow samples in the laboratory. Hence, this consideration was utilized to make a fitting (R^2 = 0.906) analysis of 152 groups of data for the thermal conductivity of snow, indirectly proving the feasibility of this consideration. We look forward to continuing to accumulate field density test data on freshwater ice to confirm the validity of this research orientation;
8. The expression of the relation between the thermal diffusivity and porosity of sea ice in Figure 6 is simple, but it exhibits great differences among ice cores. It is hard to explain the physical origin of these differences, either from the aspect of ice ages or bubble volume ratio. Maybe it is incorrect to use temperature and porosity to evaluate the thermal diffusivity of sea ice. The thermal diffusivity probably needs to be expressed as a multi-relation of ice temperature and the volume ratios of brine and bubbles. If this orientation is correct, it will be necessary to collect data on both salinity and density of sea ice. Here, the salinity is computed with electrical conductivity rather than being decided by a chemical analysis of specific substance composition. The laboratory measurements have found that the substance composition also influences thermal diffusivity [31]. The density of sea ice has been an indispensable factor in contemporary physical investigations of sea ice. In the future, we will be developing online measurement technologies for ice salinity and density and discovering refined expressions of the relation between the thermal diffusivity and physical indicators (e.g., temperature, salinity, and density of sea ice).

Author Contributions: Conceptualization, Z.L.; methodology, Z.L.; validation, X.F., L.S. and W.H.; investigation, L.S.; data curation, W.H. and C.L.; writing—original draft preparation, Z.L.; writing—review and editing, W.H. and C.L.; project administration, Z.L.; funding acquisition, Z.L. All authors have read and agreed to the published version of the manuscript.

Funding: This research was funded by the National Key Research and Development Program of China (2018YFA0605901, 2019YFE0197600), the National Natural Science Foundation of China (51979024), and the Open Fund of the State Key Laboratory of Permafrost Engineering (SKLFSE201813).

Institutional Review Board Statement: Not applicable.

Informed Consent Statement: Not applicable.

Data Availability Statement: Not applicable.

Conflicts of Interest: The authors declare no conflict of interest.

References

1. Perovich, D.K.; Elder, B.C.; Richter-Menge, J.A. Observations of the annual cycle of sea ice temperature and mass balance. *Geophys. Res. Lett.* **1997**, *24*, 555–558. [CrossRef]
2. Lu, P.; Cheng, B.; Leppäranta, M.; Li, Z. Modelling on seasonal lake ice evolution in central Asian arid climate zone: A case study. *Adv. Polar Sci.* **2021**, *32*, 356–363. [CrossRef]
3. Magee, M.R.; Wu, C.H. Effects of changing climate on ice cover in three morphometrically different lakes. *Hydrol. Process.* **2017**, *31*, 308–323. [CrossRef]
4. Semtner, A.J. A model for the thermodynamic growth of sea ice in numerical investigations of climate. *J. Phys. Oceanogr.* **1976**, *6*, 379–389. [CrossRef]
5. Pringle, D.J.; Eicken, H.; Trodahl, H.J.; Backstrom, L. Thermal conductivity of landfast Antarctic and Arctic sea ice. *J. Geophys. Res. Oceans* **2007**, *112*, C04017. [CrossRef]
6. Cheng, B.; Vihma, T.; Palo, T.; Nicolaus, M.; Gerland, S.; Rontu, L.; Haapala, J.; Perovich, D. Observation and modelling of snow and sea ice mass balance and its sensitivity to atmospheric forcing during spring and summer 2007 in the Central Arctic. *Adv. Polar Sci.* **2021**, *32*, 312–326. [CrossRef]
7. Li, Z.; Li, C.; Yang, Y.; Zhang, B.; Deng, Y.; Wang, G. Physical scheme and parametrization by using air temperature and ice thickness to improve accuracy of GPR propagation velocity in Yellow River ice. *J. Hydrau. Eng.* **2022**, *53*, 902–913. (In Chinese) [CrossRef]
8. Frankenstein, G.; Garner, R. Equations for determining the brine volume of sea ice from −0.5 °C to −22.9 °C. *J. Glaciol.* **1967**, *6*, 943–944. [CrossRef]
9. Li, Z.; Wu, Z. On the application of ice porosity in the analysis of ice compressive strength. In Proceedings of the 13th International Symposium on Ice, Beijing, China, 27–30 August 1996.
10. Li, Z.; Zhang, L.; Lu, P.; Leppäranta, M.; Li, G. Experimental study on the effect of porosity on the uniaxial compressive strength of sea ice in Bohai Sea. *Sci. China Technol. Sci.* **2011**, *54*, 2429–2436. [CrossRef]
11. Wang, Q.; Li, Z.; Lu, P.; Xu, Y.; Li, Z. Flexural and compressive strength of the landfast sea ice in the Prydz Bay, East Antarctic. *Cryosphere* **2022**, *16*, 1941–1961. [CrossRef]
12. Tarovik, O.; Yakimov, V.; Dobrodeev, A.; Li, F. Influence of seasonal and regional variation of ice properties on ship performance in the Arctic. *Ocean Eng.* **2022**, *257*, 111563. [CrossRef]
13. Zhaka, V.; Bridges, R.; Riska, K.; Cwirzen, A. A review of level ice and brash ice growth models. *J. Glaciol.* **2022**, *68*, 685–704. [CrossRef]
14. Karulina, M.; Marchenko, A.; Karulin, E.; Sodhi, D.S.; Sakharov, A.; Chistyakov, P. Full-scale flexural strength of sea ice and freshwater ice in Spitsbergen Fjords and North-West Barents Sea. *Appl. Ocean Res.* **2019**, *90*, 101853. [CrossRef]
15. Wang, Q.; Lu, P.; Leppäranta, M.; Cheng, B.; Zhang, G.; Li, Z. Physical Properties of Summer Sea Ice in the Pacific Sector of the Arctic during 2008–2018. *J. Geophys. Res. Oceans* **2020**, *125*, e2020JC016371. [CrossRef]
16. Huang, W.; Han, H.; Shi, L.; Niu, F.; Deng, Y.; Li, Z. Effective thermal conductivity of thermokarst lake ice in Beiluhe Basin, Qinghai-Tibet Plateau. *Cold Reg. Sci. Technol.* **2013**, *85*, 34–41. [CrossRef]
17. Saito, T.; Hamamoto, S.; Mon, E.E.; Takemura, T.; Saito, H.; Komatsu, T.; Moldrup, P. Thermal properties of boring core samples from the Kanto area, Japan: Development of predictive models for thermal conductivity and diffusivity. *Soils Found.* **2014**, *54*, 116–125. [CrossRef]
18. Dong, S.; Gao, X.; Ma, Z.; Wang, X.; Gao, C. Ice-templated porous silicate cement with hierarchical porosity. *Mater. Lett.* **2018**, *217*, 292–295. [CrossRef]
19. Arkhangelskaya, T.; Lukyashchenko, K. Estimating soil thermal diffusivity at different water contents from easily available data on soil texture, bulk density, and organic carbon content. *Biosyst. Eng.* **2018**, *168*, 83–95. [CrossRef]
20. Zhao, J.; Sun, S.; Liu, C.; Meng, Q. Thermal conductivity and thermal diffusivity of methane hydrate formed from compacted granular ice. *Heat Mass Transf.* **2018**, *54*, 3287–3295. [CrossRef]
21. Hammerschmidt, U. Thermal transport properties of water and ice from one single experiment. *Int. J. Thermophys.* **2002**, *23*, 975–996. [CrossRef]
22. Calonne, N.; Milliancourt, L.; Burr, A.; Philip, A.; Martin, C.; Flin, F.; Geindreau, C. Thermal conductivity of snow, firn, and porous ice from 3-D image-based computations. *Geophys. Res. Lett.* **2019**, *46*, 13079–13089. [CrossRef]
23. Ling, F.; Zhang, T. Simulating heat source effect of a thermokarst lake in the first 540 years on the Alaskan Arctic using a simple lake expanding model. *Cold Reg. Sci. Technol.* **2019**, *160*, 176–183. [CrossRef]
24. Zhang, T.; Osterkamp, T.E. Considerations in determining thermal diffusivity from temperature time series using finite difference methods. *Cold Reg. Sci. Technol.* **1995**, *23*, 333–341. [CrossRef]
25. Tong, B.; Xu, H.; Horton, R.; Bian, L.; Guo, J. Determination of long-term soil apparent thermal diffusivity using near-surface soil temperature on the Tibetan Plateau. *Remote Sens.* **2022**, *14*, 4238. [CrossRef]
26. Adams, W.M.; Watts, G.; Mason, G. Estimation of thermal diffusivity from field observations of temperature as a function of time and depth. *Am. Mineral.* **1976**, *61*, 560–568.

27. Shi, L.; Li, Z.; Niu, F.; Hang, W.; Lu, P.; Feng, E.; Hang, H. Thermal diffusivity of thermokarst lake ice in the Beiluhe basin of the Qinghai-Tibetan Plateau. *Ann. Glaciol.* **2014**, *55*, 153–158. [CrossRef]
28. Schwerdtfecer, P. The thermal properties of sea ice. *J. Glaciol.* **1963**, *4*, 789–807. [CrossRef]
29. Yen, Y.C. *Review of Thermal Properties of Snow, Ice and Sea Ice*; CRREL Report 81-10; Cold Regions Research and Engineering Laboratory: Hanover, NH, USA, 1981.
30. Koubyshkin, N.V.; Sazonov, K.E. Evaluation of loads due to partial freezing of seawater trapped in enclosed cavities. In Proceedings of the17th International Symposium on Ice, Saint Petersburg, Russia, 21–25 June 2002.
31. Chen, N.J.; Morikawa, J.; Kishi, A.; Hashimoto, T. Thermal diffusivity of eutectic of alkali chloride and ice in the freezing-thawing process by temperature wave analysis. *Thermochim. Acta* **2005**, *429*, 73–79. [CrossRef]
32. James, D.W. The thermal diffusivity of ice and water between −40 and +60 °C. *J. Mater. Sci.* **1968**, *3*, 540–543. [CrossRef]
33. Cox, G.F.N.; Weeks, W.F. Equations for determining the gas and brine volumes in sea-ice samples. *J. Glaciol.* **1983**, *29*, 306–316. [CrossRef]
34. Leppäranta, M.; Manninen, T. *The Brine and Gas Content of Sea Ice with Attention to Low Salinities and High Temperatures*; Finnish Institute of Marine Research: Helsinki, Finland, 1988.
35. Reid, T.; Crout, N. A thermodynamic model of freshwater Antarctic lake ice. *Ecol. Model.* **2008**, *210*, 231–241. [CrossRef]
36. Shi, L.; Li, Z.; Feng, E.; Bai, Y.; Yang, Y. Thermal diffusivity identification of distributed parameter systems to sea ice. *J. Appl. Math.* **2013**, *2013*, 760378. [CrossRef]
37. Shi, L.; Bai, Y.; Li, Z.; Cheng, B.; Leppäranta, M. Preliminary results on relationship between thermal diffusivity and porosity of sea ice in the Antarctic. *Chin. J. Polar Sci.* **2009**, *20*, 72–80. [CrossRef]
38. Bai, Y. *Parameter Identification of Non-Smooth Parameter Systems and Its Application*; Dalian University of Technology: Dalian, China, 2006. (In Chinese)
39. Bai, Y.; Li, Z.; Han, M.; Lu, P. Approaches to revise ice thermal diffusivity from measured time series of temperature in a river. In Proceedings of the 6th International Symposium on Test and Measurement, Dalian, China, 1–4 June 2005.
40. Huang, W.; Li, Z.; Han, H.; Jia, Q. Limit resistive forces from ice frozen to concrete-revetment interface of an inclined dam wall. *Cold Reg. Sci. Technol.* **2017**, *141*, 181–187. [CrossRef]
41. Huang, W.; Li, Z.; Liu, X.; Zhao, H.; Guo, S.; Jia, Q. Effective thermal conductivity of reservoir freshwater ice with attention to high temperature. *Ann. Glaciol.* **2013**, *54*, 189–195. [CrossRef]
42. Li, Z.; Huang, W.; Jia, Q.; Leppäranta, M. Distributions of crystals and gas bubbles in reservoir ice during growth period. *Water Sci. Eng.* **2011**, *4*, 204–211. [CrossRef]
43. Lei, R.; Li, Z.; Zhang, Z.; Chen, Y. Thermodynamic processes of lake ice and landfast ice around Zhongshan Station, east Antarctica. *Adv. Polar Sci.* **2011**, *22*, 143–152. [CrossRef]
44. LI, Z.; Yan, D.; Meng, G.; Zhang, M. Relation of thermal conductivity coefficient with temperature of sea ice in Bohai Sea. In Proceedings of the International Offshore Mechanics and Arctic Engineering, Montreal, QC, Canada, 7–12 June 1992.
45. Li, Z.; Sui, J.; Yan, D.; Meng, G. Temperature, salinity, and density profiles in a fast ice sheet in Liao Dong Bay. In Proceedings of the 10th International Conference on Port an d Ocean Engineering under Artic Conditions, Luleå, Sweden, 12–16 June 1989.
46. Shi, L.; Li, Z.; Lu, P.; Feng, E. Influence of ice temperature on thermal diffusivity of natural freshwater ice. *J. Hydraul. Eng.* **2013**, *44*, 1112–1117. (In Chinese) [CrossRef]
47. Huang, W.; Li, Z.; Han, H.; Niu, F.; Lin, Z.; Leppäranta, M. Structural analysis of thermokarst lake ice in Beiluhe Basin, Qinghai-Tibet Plateau. *Cold Reg. Sci. Technol.* **2012**, *72*, 33–42. [CrossRef]
48. Bai, Y.; Xu, H.; Shi, L. Research on the optimal identification of thermal diffusivity of fresh ice in reservoirs of cold regions. In Proceedings of the 21st IAHR International Symposium on Ice, Ice Research for a Sustainable Environment, Dalian, China, 11–15 June 2012.
49. Nicolaus, M.; Wang, C.; Gerland, S.; Li, N.; Li, Z.; Cheng, B.; Perovich, D.; Granskog, M.; Shi, L.; Lei, R.; et al. Advancing the understanding of variations of Arctic sea ice optical and thermal behaviors through an international research and mobility project. *Adv. Polar Sci.* **2015**, *26*, 179–187. [CrossRef]
50. Cheng, P.; Li, J.; Yu, D.; Hao, Z. Study of the ice cover temperature variation of the Kanas Lake. *J. Glaciol. Geocryol.* **2016**, *38*, 692–698. (In Chinese) [CrossRef]
51. Congé, C.; Andrieu, J.; Laurent, P.; Ferrand, J. Experimental data and modelling of thermal properties of ice creams. *J. Food Eng.* **2003**, *58*, 331–341. [CrossRef]
52. Shum, E.; Papangelakis, V. Water recovery from inorganic solutions via natural freezing and melting. *J. Water Process Eng.* **2019**, *31*, 100787. [CrossRef]
53. Wang, C.; Lai, Y.; Yu, F.; Li, F.; Yang, S. Estimating the freezing-thawing hysteresis of chloride saline soils based on the phase transition theory. *Appl. Therm. Eng.* **2018**, *135*, 22–33. [CrossRef]
54. Ma, Y.; Cheng, B.; Xu, N.; Yuan, S.; Shi, H.; Shi, W. Long-term ice conditions in Yingkou, a coastal region northeast of the Bohai Sea, between 1951/1952 and 2017/2018: Modeling and observations. *Remote Sens.* **2022**, *14*, 182. [CrossRef]
55. Yan, Y.; Uotila, P.; Huang, K.; Gu, W. Variability of sea ice area in the Bohai Sea from 1958 to 2015. *Sci. Total Environ.* **2020**, *709*, 136164. [CrossRef]
56. Li, Z.; Jia, Q.; Zhang, B.; Matti, L.; Lu, P.; Lu, W. Influences of gas bubble and ice density on ice thickness measurement by GPR. *Appl. Geophys.* **2010**, *7*, 105–113. [CrossRef]

57. Gherboudj, I.; Bernier, M.; Hicks, F.; Leconte, R. Physical characterization of air inclusions in river ice. *Cold Reg. Sci. Technol.* **2007**, *49*, 179–194. [CrossRef]
58. Hamilton, R.L.; Crosser, O.K. Thermal conductivity of heterogeneous two-component systems. *Ind. Eng. Chem. Fundam.* **1962**, *1*, 187–191. [CrossRef]
59. Overduin, P.P.; Kane, D.L.; van Loon, W.K.P. Measuring thermal conductivity in freezing and thawing soil using the soil temperature response to heating. *Cold Reg. Sci. Technol.* **2006**, *45*, 8–22. [CrossRef]

water

MDPI

Article

Investigations on Flexural Strength of a Columnar Saline Model Ice under Circular Plate Central Loading

Yukui Tian [1,2,*], Weihang Zhao [2], Chaoge Yu [2], Xuhao Gang [2], Peng Lu [3] and Qianjin Yue [4]

1 State Key Laboratory of Structural Analysis, Optimization and CAE Software for Industrial Equipment, Dalian University of Technology, Dalian 116024, China
2 China Ship Scientific Research Center, National Key Laboratory of Science and Technology on Hydrodynamics, Wuxi 214082, China; tjuf16zwh@163.com (W.Z.); 15225276973@163.com (C.Y.); gxh3865gg@163.com (X.G.)
3 State Key Laboratory of Coastal and Offshore Engineering, Dalian University of Technology, Dalian 116024, China; lupeng@dlut.edu.cn
4 School of Ocean Science and Technology, Dalian University of Technology, Panjin 124221, China; yueqj@dlut.edu.cn
* Correspondence: tianyukui@cssrc.com.cn

Abstract: The properties of ice strength have a significant impact on the design and safety of structures in ice-infested waters. To analyze the flexural strength of columnar saline model ice, we conducted circular plate center loading tests at the Small Ice Model Basin of the China Ship Scientific Research Center (CSSRC SIMB) in China. The tests involved varying the loading rate and ice temperature, and a numerical model was developed using FEM and LS-DYNA for validation and comparison. The results of the tests revealed the crack propagation process, stress distribution, load response, and failure mode of the model ice. The model ice displayed typical brittle failure, and the flexural strength was linearly related to ice temperature but not significantly correlated with loading rate. The porosity of the model ice affected the load response and time of failure but not the failure mode. The model ice with 7% porosity had a 7.8% reduction in load response compared to the nonporous model ice. This study provides a reliable method for measuring and analyzing the flexural strength of model ice. It also serves as a foundation for further research on the interaction between structures and ice sheets.

Keywords: ice model basin; columnar saline model ice; circular plate; flexure strength; crack propagation; porosity

Citation: Tian, Y.; Zhao, W.; Yu, C.; Gang, X.; Lu, P.; Yue, Q. Investigations on Flexural Strength of a Columnar Saline Model Ice under Circular Plate Central Loading. *Water* **2023**, *15*, 3371. https://doi.org/10.3390/w15193371

Academic Editor: Karl-Erich Lindenschmidt

Received: 30 August 2023
Revised: 19 September 2023
Accepted: 21 September 2023
Published: 26 September 2023

1. Introduction

In regions where ice is present, such as beneath a floating object, the ice sheet may fracture under vertical forces. This circumstance has gained significance in the fields of polar exploration and engineering [1]. The durability of the ice plays a crucial role in the design and assessment of structures regarding safety. The upward pressure exerted by the floating entity can trigger a range of failures, including shearing, bending, and flexural failure, which is the most frequently observed [2,3]. Conducting tests to determine flexural strength is essential for accurately predicting ice loads on structures and ensuring their safety. This information is critical in developing and assessing structures for floating objects under ice.

There exist three primary methods to assess ice strength: in situ cantilever tests, simply supported beam tests, and circular plate center loading tests [4–6]. An in situ cantilever test involves obtaining an ice beam directly from the ice layer and conducting the test while the ice remains in its natural environment [7]. Frederking and Timco executed in situ cantilever tests on model ice at the National Research Council Hydraulics Laboratory, discovering no significant relationship between loading rate and flexural strength [8]. Krupina also conducted in situ cantilever tests on sea ice in the Barents Sea, obtaining the distribution of sea ice flexural strength in the region [9]. Gang of CSSRC conducted an in situ cantilever beam test on model ice, revealing that the flexural strength of model ice remains constant as the loading rate changes but decreases with increased rewarming time [10]. The simply supported beam test can be split into three-point bending and four-point bending tests [11]. Ji et al. executed an indoor three-point bending test on the flexural strength of Bohai sea ice and found that the sea ice flexural strength had a linear relationship with the loading rate [4]. Gagnon conducted a four-point bending test on iceberg ice and found that the flexural strength of ice decreases with increased ice temperature [12]. Barrette et al. performed a four-point bending test on iceberg ice and found that the larger the average grain size inside the ice, the smaller its flexural strength [13]. Lastly, in circular plate center loading tests, icicles are cut from the ice layer and then shaped into disks according to predetermined thickness–diameter ratios. Krupina conducted an extensive series of circular plate center loading tests on sea ice situated in both southeastern and northeastern regions of the Barents Sea. The study aimed to investigate the effects of different ice properties, such as temperature, salinity, and brine volume, on flexural strength. Furthermore, the plate test outcomes were juxtaposed with those of the cantilever test, and a suitable correlation was established between the two [9]. Marchenko conducted circular plate center loading tests on sea ice in the northwest Barents Sea and found that flexural strength depends on ice temperature [6].

In conclusion, ice is a complex crystalline material, and its flexural strength is influenced by factors such as the structure, temperature, and external loading conditions of ice crystals [14].

In addition to experimental studies, numerical simulations can also be utilized to investigate the flexural strength of ice. The discrete element method, the finite element method, and the peridynamics method are the most popular simulation methods used today. The discrete element method is suitable for the mechanics of discontinuous media [15], but its application for dynamic failure problems is limited due to result deviation in the continuum stage [16]. On the other hand, the finite element method has advantages in terms of fast calculation efficiency and high accuracy and is widely used to simulate ice–structure interaction [17]. However, the simulation accuracy of the finite element method depends on the material model chosen [18]. The peridynamics method analyzes the mechanics of a continuum with the help of the point method of matter and the idea of molecular dynamics. It constructs the object's motion equation in the form of an integral [19]. However, compared with other numerical calculation methods, the peridynamics method also has the problem of low computational efficiency. When simulating the interaction between sea ice and structure, most scholars consider ice as a continuum, considering its elasticity and plasticity. For example, Li used an isotropic elastoplastic fracture model to simulate ice based on LS-DYNA [20]. Some scholars incorporated the effects of temperature and strain rate and constructed a numerical model of ice based on the multi-surface failure criteria of sea ice. Still, there were some differences between the test and simulation results [21]. Other scholars considered the existence of pores inside the ice, such as Von Bock und Polach et al., who applied LS−DYNA and selected the Lemaitre damage model to simulate ice and analyzed the dependence of the load response on the microstructure of ice [22]. To maintain the quality and energy of the model, some scholars used the node-splitting technique instead of the commonly used elemental erosion technique [23].

Despite the wealth of experimental and numerical data on in situ cantilever and simply supported beam tests, there has been limited research conducted on circular plate center loading tests. The International Towing Tank Conference's mechanical properties test method for model ice also lacks relevant specifications for circular plate center loading tests, indicating the need for improvement in this area. As mentioned earlier, the flexural strength of ice is influenced by various factors, such as temperature and loading rate, and further experimental exploration is necessary. Additionally, there is a need for expanded research based on the material model, particularly regarding the influence of internal porosity on ice strength.

In this paper, the influence of loading rate and temperature on the flexural strength was explored by using a saline model ice under a circular plate center loading test, and the simulation of the porosity of ice was also realized to study the influence of the internal porosity of the model ice. The results of center loading on a circular plate of ice were discussed, and, in particular, the physical mechanisms of crack propagation were analyzed. This study provides a reference for the prediction of the ice's mechanical characteristics.

2. Methods

Both laboratory experiments and numerical simulations were conducted in the present study, and the implementation of the two methods is briefly summarized below.

2.1. Laboratory Experiments

2.1.1. The Columnar Saline Model Ice

The SIMB of the CSSRC is an ice water tank with 8 m in length, 2 m in width, and 1 m in depth, as shown in Figure 1 [24]. The ice-making process used in the SIMB is similar to but partly improved from the method by the Evers [25,26] of the Hamburg Tank (HSVA) in Germany. Cooling fans are suspended from the roof, ceiling panels with tiny holes are installed underneath to exhaust cooling air, and circulating fans with guide plates are set on the side walls to suck the air used for cooling; thus, forced cooling air circulation is formed [27].

Figure 1. Interior scene of CSSRC SIMB.

Model ice is made of sodium chloride solution, and the preparation process is divided into precooling, crystallization, and ice making. To control the density and to lower the strength, an underwater air bubbling system is used to produce tiny bubbles that are trapped in the model ice during the freezing process. Its mechanical properties can be adjusted by rewarming and underwater microbubble generation systems [10]. After rewarming, the temperature of model ice is usually around −0.8 °C.

Through a series of experimental measurements, the mechanical properties of model ice in the SIMB were statistically analyzed. The fresh model ice's main characteristic parameters (−0.8 °C) in the SIMB are summarized in Table 1. With the continuous exploration of the experimenters, it was found that after adding bubbles to the saline model ice, the brittleness of the ice was greatly improved. The flexural fracture mode of the ice and the distribution ratio of the crushed ice flakes of various scales were consistent with natural sea ice, and the similarity could be maintained in terms of physical and mechanical properties and fracture behavior patterns [28].

Table 1. The main physical and mechanical performance parameters of model ice in SIMB [24].

Parameter	Value
Density/g·cm^{-3}	0.89~0.91
Thickness/mm	10~100
Flexural strength/kPa	94.7
Compressive strength/kPa	80~150
Elastic modulus/MPa	250~450
Elastic modulus/flexural strength	800~1500

2.1.2. Circular Plate Center Loading Tests

In a circular ice plate indentation test, vertical loads are exerted on the central area of the disk ice specimen under the circumferentially uniform support, as shown on the schematic diagram in Figure 2.

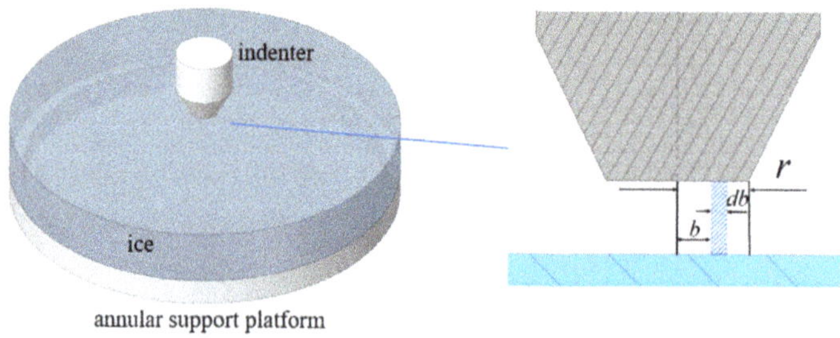

(**a**) (**b**)

Figure 2. Circular plate center loading test: (**a**) schematic diagram; (**b**) partial schematic diagram.

When the plate is thin enough, it belongs to a typical plate–shell structure according to the theory of shell mechanics [29]. The maximum bending moment of the circular plate is at the center of the circular plate with radius R when the load q is uniformly distributed in a circle of radius r. By integrating the deflection caused by the load on the torus plane with radius b and width db (Figure 2b), the deflection of the center of the plate can be obtained, and then the bending moment $M_{\max}$ at the center of the plate can be determined [28]:

$$\begin{aligned} M_{\max} &= q\int_0^r \left(\frac{1-v}{4}\frac{R^2-b^2}{R^2} - \frac{1+v}{2}\ln\frac{b}{R}\right) b\,db \\ &= \frac{F}{4\pi}\left[(1+v)\ln\frac{R}{r} + 1 - \frac{(1-v)r^2}{4R^2}\right] \end{aligned} \tag{1}$$

where F represents the total load $\pi r^2 q$. ν = 0.33, is the Poisson's ratio [30]. The relationship between the bending moment and stress is:

$$\sigma = \frac{6M}{h^2} \tag{2}$$

In the above equation, h is the thickness of the plate, so the maximum tensile stress in the plate, that is, the flexural strength σ_f, is:

$$\sigma_f = \frac{3F}{8\pi h^2}\left[4 - (1-v)\left(\frac{r}{R}\right)^2 - 4(1+v)\ln\left(\frac{r}{R}\right)\right] \tag{3}$$

The relationship between the central bending moment and curvature obtained from the pure bending hypothesis is the basis of the above derivation, and the effect of shear forces on the flexural strength in a plane parallel to the plate surface is not considered. The value of σ_f so far is only an approximate solution, and its accuracy depends on the ratio of the thickness of the disk to the outer diameter [31]. When the thickness–diameter ratio is less than 0.15, the effect of shear force on the bending strength is negligible [6].

2.1.3. Test Conditions

The test was carried out in a small cold laboratory adjacent to the ice water basin. The flexural strength of ice is affected by a variety of factors [1], and the loading rate and ice temperature were selected as the control factors of the present test. Different test conditions are listed in Table 2. The peak force and failure time of the circular plate indentation test are two important measurements. The average temperature of the newly made model ice was about −0.8 °C, the density was 0.91 g·cm^{-3}, and the salinity was 2.75 ppt.

Table 2. Summary of plate specimen size and testing conditions in laboratory experiments.

Diameter/Thickness (mm/mm)	Specimen Number	Loading Rate (mm/min)	Ice Average Temperature (°C)	Laboratory Temperature (°C)	Measurements
140/20	20	100, 150, 200, 250, 300	−0.8	−0.8	Peak force Failure time
140/20	15		−2.0	−2.0	
140/20	15		−4.0	−4.0	
140/20	15		−6.0	−6.0	
140/20	15		−8.0	−8.0	

The test process was divided into three phases: the test preparation stage, the plate specimen sampling and storage freezing stage, and the measurement and data recording stage. In the test preparation stage, the small cold laboratory was first cooled to reach the target air temperature, and the high-speed camera, electronic universal testing machine, thermometer, and other related instruments were precooled. In the plate sampling and storage freezing stage, a total of 80 plate specimens were drilled from the flat model ice sheet. The diameter of the ice plate specimen was 140 mm, and the thickness was 20 mm (Figure 3). It produced a thickness–diameter ratio of less than 0.15, agreeing with the calculation requirements. After sampling, 20 newly made model ice specimens were placed in a small cold laboratory for test measurement, and the rest were stored in a refrigerator for subsequent tests. In the last test and measurement stage, the loading rate of the electronic testing machine was first adjusted according to the requirements of the test conditions. Then the force curve of the ice specimen from the initial bearing to the flexural failure was recorded. The high-speed camera was also used to capture the failure details of the ice specimen. The thickness and temperature of model ice were quickly measured and recorded after failure.

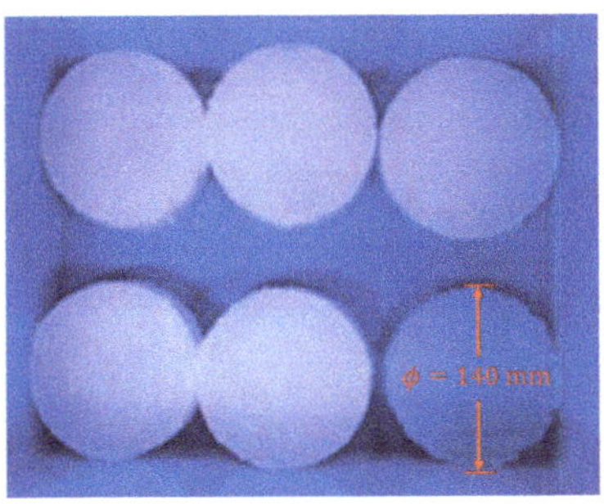

Figure 3. Photo of model ice sample during sampling stage.

In the data recording stage, the force curve of the ice specimen from the initial loading to the flexural failure was recorded with a force measurement system, and the flexural strength was obtained according to the peak force. At the same time, the failure details of the ice specimen were documented with a high-speed video camera. The overall layout scheme of the instruments is shown in Figure 4, and the parameters of the test instrument are shown in Table 3.

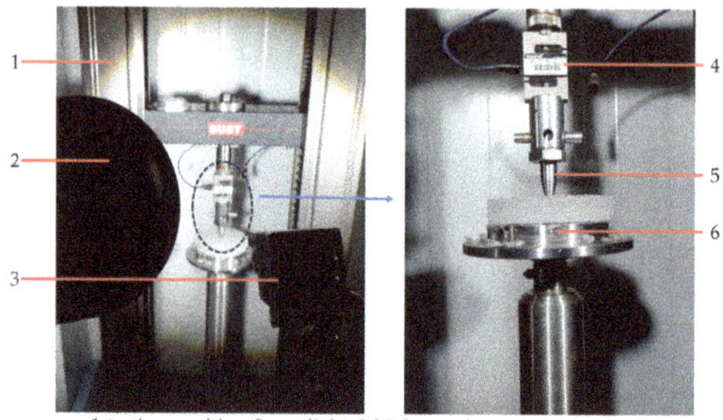

Figure 4. Test device layout.

Table 3. Test instrument parameter table.

Instrument	Accuracy
Thermometer	0.01 °C
Electronic testing machine	1 mm/min
High-speed cameras	Frame rate: 2800
Sensor	0.01 N
Vernier calipers	0.01 mm
Ice density measurement instruments	0.01 g/cm^3
Salinity meter	±3% (FS)

2.2. *Numerical Modeling*

2.2.1. Numerical Model of Ice Material

A big advantage of the finite element method is that many contact algorithms allow the coupling of ice and structural models. Therefore, in this paper, the finite element solver LS-DYNA was used to establish a model ice failure model. The circular plate center loading tests were used to vertically load the central area of the disk ice specimen under the circumferentially uniform support. The contact between the indenter and the ice disk specimen occurred during loading, and the explicit nonlinear finite element method can be used to solve such contact problems. The calculation method used by the nonlinear finite element program is the explicit integration method [32].

When using the finite element method to simulate the model ice mechanical test, it is necessary to determine the material model parameters that match the macroscopic characteristics of the material. Standard model ice constitutive models include the elastoplastic model, elastic brittleness model, etc. According to Karr and Choi [33], model ice materials are considered isotropic in their undeformed state. This assumption was adopted in this paper, so the isotropic elastoplastic fracture model (*MAT_ISOTROPIC_ELASTIC_FAILURE) in LS-DYNA was selected to simulate the model ice. The failure criterion for the material is the Von Mises yield criterion [34].

The material parameters of the model ice are shown in Table 4. The specific parameters of density, plastic hardening modulus, and plastic failure strain were obtained by the model ice mechanical test [1]. According to the relevant ice mechanics numerical simulation, the ranges of other material parameters were obtained, and finally, the specific parameters were given by trial and error.

Table 4. Model ice material parameters.

Material Properties	Value
Density/g·cm^{-3}	0.92
Shear modulus /MPa	76.9
Plastic hardening modulus /MPa	94.7
Yield stress/kPa	83
Bulk modulus/MPa	75
Plastic failure strain	0.05
Failure pressure/kPa	−110

According to von Bock und Polach [22], the air pores inside the model ice can be simulated by deleting elements by using a random algorithm, and the number of deleted elements depends on the porosity. This is another advantage of numerical simulation as compared with the experiments in Section 2.1, because ice porosity is difficult to control according to predetermined values, although an underwater air bubbling system has been placed already. The so-called K file of the model ice numerical model was employed and set the random number function, which was used to delete a part of the model ice elements according to the input value of porosity. The flow chart is shown in Figure 5. There are several key points in this program. First, read the K file, defining each line of information in it as a string. Determine whether each line string represents the coordinate information of the cell; if so, proceed to the next step, and vice versa, and output to a K file. Secondly, enter the sampling rate, combine the random function to obtain a set of unit numbers that need to be deleted, correspond them one-to-one with the read unit string, replace the unit numbers that need to be deleted with spaces, and output them to a new K file.

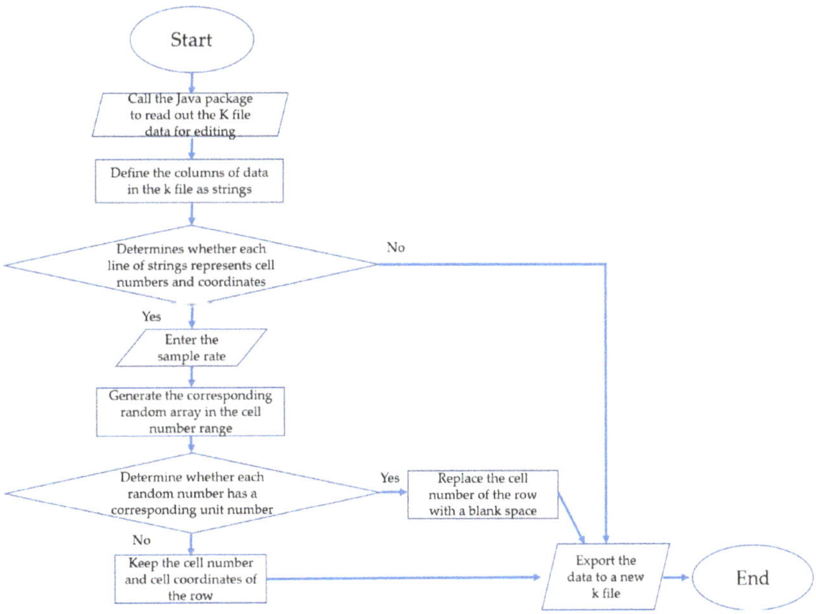

Figure 5. Flow chart of internal defects in construction of model ice.

The corrected ice model with a different porosity is shown in Figure 6. Since the elements used in this model were hexahedral, the resulting pore shape was also hexahedral. In fact, model ice with different porosities differs in the number of pores, but the size of the individual pores does not change, which is the same as Von Bock und Polach [22].

The intervals between pores are random in Figure 6 and may be different from the actual internal pore distribution of ice because the brine channels of actual sea ice tend to be vertical and continuous, but the pores in numerical models cannot guarantee continuity. In addition, the distribution of pores is different from the actual sea ice, the pores of the actual sea ice tend to exist only in the interior, and the surface is continuous. Still, the pores in this paper also exist on the surface. The resulting difference in loading force is hard to evaluate at present, but it is supposed to be ignorable because the given porosity is small (≤7%).

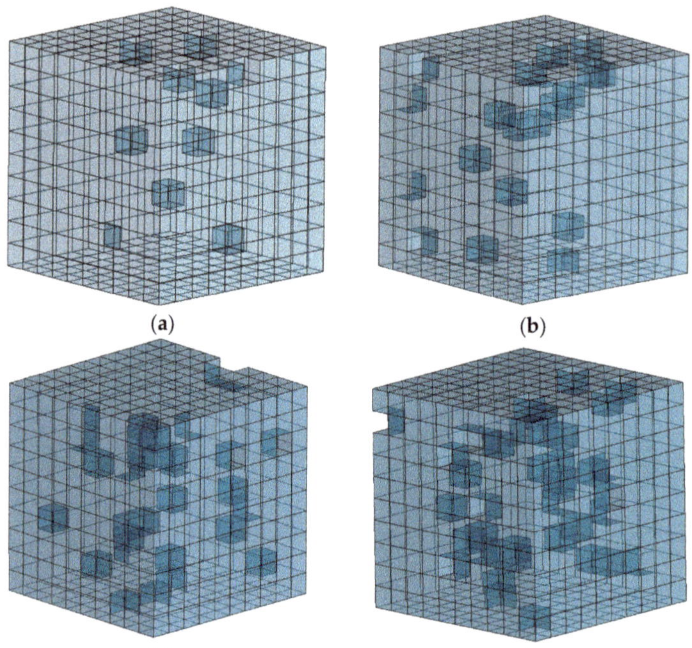

Figure 6. Numerical model of model ice with different porosities of (**a**) 0%, (**b**) 3%, (**c**) 5%, and (**d**) 7%.

2.2.2. Numerical Simulation of Circular Plate Center Loading Tests

In the numerical calculation, the ice specimen was loaded similarly to that shown in Figure 2. During modeling, the annular support below the ice specimen was set as a fixed boundary, the cone indenter had only the degrees of freedom in the Z direction, and the load was applied downward on the upper surface of the ice specimen at a constant loading rate. The main parameters of the numerical model are shown in Table 5. To simulate the fragmentation phenomenon of the model ice, a fine mesh was used in the central area of the model ice, and a large gradient mesh was used in the outer part, as shown in Figure 7a. Through the circular plate center loading tests, the numerical calculation condition was determined. Combining this with the pore simulation principle of Section 2.2.1, a model ice with a porosity of 3% can also be obtained, as shown in Figure 7b.

Table 5. Main parameters of the numerical model of the circular plate center loading test.

Ice Specimen Radius (mm)	Ice Specimen Thickness (mm)	Ring Support Outer Diameter (mm)	Ring Support Inner Diameter (mm)	The Radius of the Lower Surface of the Tapered Indenter (mm)
70	20	70	60	5

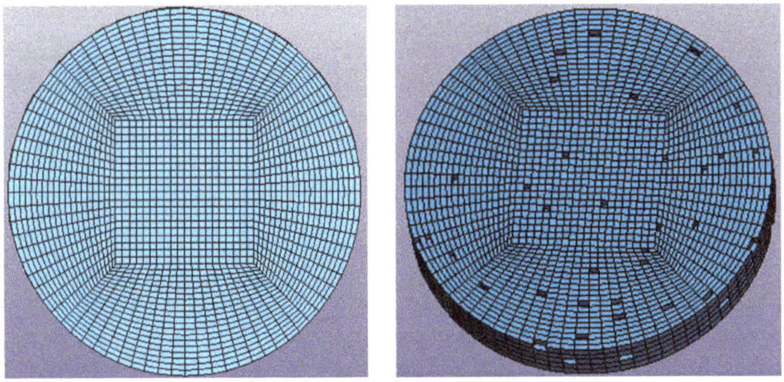

Figure 7. Model ice meshing with different porosities of (a) 0% and (b) 3%.

A rigid body model was selected to simulate the annular support and cone indenter, and the material parameters were defined using the keyword *MAT_RIGID, as shown in Table 6.

Table 6. Ring support and indenter material parameters.

Material Properties	Value
Density/g·cm^{-3}	7.83
Elastic modulus/GPa	207
Poisson's ratio	0.33

LS-DYNA provides a variety of contact algorithms for explicit analysis, divided into single-sided contact, point-to-face contact, and face-to-face contact [35]. The contact between the subglacial surface of the model and the upper surface of the annular support had a large contact area and symmetrical shape, so the surface contact was selected. Since the ring support was modeled using shell elements, the Automatic Contact Algorithm in surface contact was chosen. It can consider the influence of element thickness, allowing contact to appear on both sides of the shell element, making it more accurate when calculating contact forces. The contact between the upper surface of the model ice and the lower surface of the cone table indenter also adopted surface contact, and the erosion contact algorithm in surface contact was selected because the ice breakage effect of the erosion contact simulation model is good. Erosion contact was employed to control time steps. It automatically invoked negative volume failure criteria for all solid elements in the model, which circumvents procedural errors due to negative volumes by removing solid elements that produce negative volumes.

Different test conditions in numerical simulations are listed in Table 7. Compared with the test conditions in laboratory experiments (Table 2), the impact of ice temperature was ignored. In fact, the influence of temperature on the properties of ice materials is difficult to achieve by numerical simulation. Therefore, to reduce the influence of temperature on the results, the input parameters of the ice material in the numerical model were the average mechanical parameters of the model ice at a specific temperature (−0.8 °C). Similarly, in numerical simulation, the flexural strength as an input quantity, and the influence of external parameters' (porosity) changes on the flexural strength cannot be reflected, and the influence of external parameters on bending performance can only be analyzed through the change in load response.

Table 7. Summary of plate specimen size and testing conditions in numerical simulations.

Diameter/Thickness (mm)	Specimen Number	Loading Rate (mm/min)	Ice Porosity (%)	Measurements
140/20	4	150	0	Peak force Failure time
140/20	4		3	
140/20	4		5	
140/20	4		7	

3. Results and Analysis

The test results and the numerical simulation results were compared from the two aspects of the time history curve and damage phenomenon.

3.1. Time History Curve and Failure Mode of Model Ice

The numerically calculated time history curve was compared with the model ice mechanics test time history curve in Figure 8. During the bending process of the ice specimen, both time history curves showed linear changes without an obvious yield stage and increased sharply from 0, reaching the maximum value Pmax within only 0.8 s, and it can be seen that the failure mode of the ice specimen was an obvious brittle failure. The flexural strength σ_f of the model ice can be calculated from Pmax according to Equation (3). The peak force of the two was similar, the loading time difference was 0.01 s, and the absolute error did not exceed 5 N, so the rationality of the calculation model can be verified from the changing trend of the time history curve. However, due to differences in material properties and test conditions, the downward trend of the test curve and the numerical curve was not the same. The downward trend of the test curve was faster because in the model test, when the sensor read the maximum force peak, the indenter stopped pressing down and upward, and the numerical calculation did not simulate this behavior, but this did not affect the numerical model's simulation of the mechanical properties of the model ice.

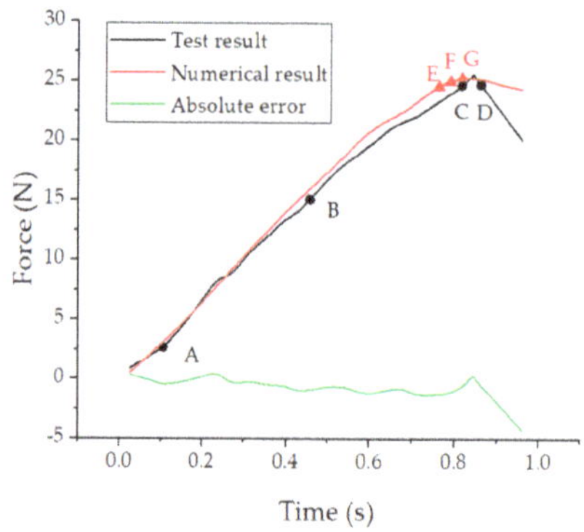

Figure 8. Comparison of time history curves between numerical result and test result (−0.8 °C, 150 mm/min). A–G denote the different loading pictures shown in Figures 9 and 10.

A high-speed camera was used to observe the state of the model ice at the four moments A, B, C, and D in Figure 8, and the flexural failure process of the model ice was captured, as shown in Figure 9. It is clear that the model ice broke along the diameter, and the crack propagation was rapid, with an expansion time of less than 0.06 s. The model ice had no prominent yield stage, and there was no obvious plastic deformation at the failure location. It can be deduced that the failure of the model ice belongs to a brittle failure. Compared with the experimental phenomenon in the breakthrough loads of floating ice sheets carried out by Sodhi [36], the model ice disk did not have holes in the middle area but just failed cracks along the radial direction. This may be because of the different constraints.

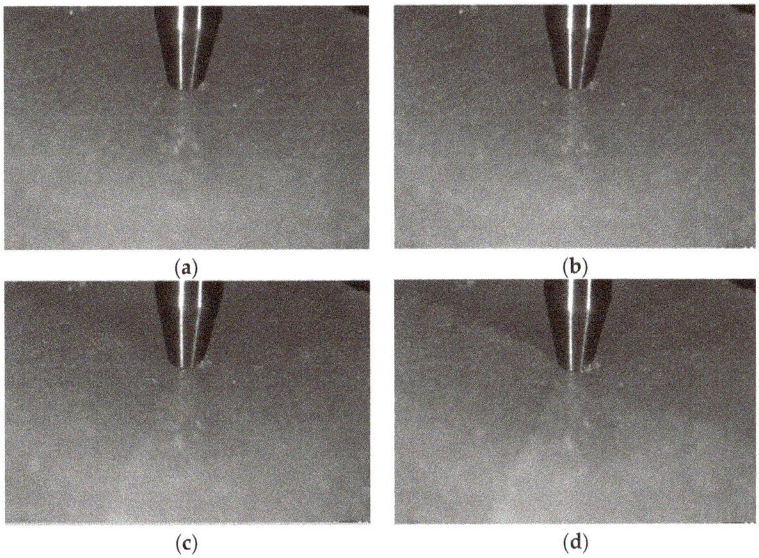

Figure 9. The loading process for model ice at (**a**) 0.11 s, (**b**) 0.45 s, (**c**) 0.79 s, and (**d**) 0.85 s (−0.8 °C, 150 mm/min), corresponding to points A, B, C, and D in Figure 8.

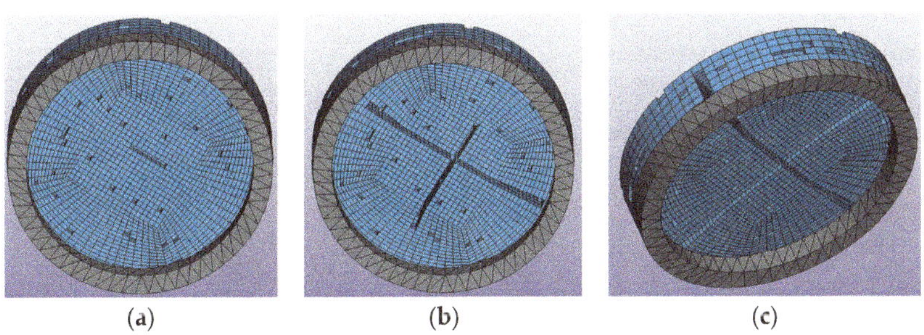

Figure 10. Crack propagation during model ice breakage (**a**) 0.77 s, (**b**) 0.78 s, and (**c**) 0.79 s (−0.8 °C, 150 mm/min), corresponding to points E, F, and G in Figure 8.

The experiment could not observe the failure pattern of the bottom during the model ice fracture failure, but numerical simulations supplemented this part of the study. By observing the bottom of the model ice at the three moments E, F, and G in Figure 8, it was found that the model ice started to crack from the center of the bottom (Figure 10). The crack extended firstly along the radius direction to the bottom boundary, and then extended along the thickness direction to the top surface until the model completely failed, as shown in Figure 10. The flexural failure of the model ice was essentially the tensile failure of the bottom of the model ice, consistent with the flexural failure of the sea ice proposed by Lainey [37]. It indicates that the numerical model of the model ice can better reflect the crack propagation during model ice breakage.

Meanwhile, the damage phenomenon of the model ice was compared between numerical simulations with the picture captured by the high-speed camera in Figure 11. When the ice specimen failed, under the action of the cone indenter, there was a significant deflection change at the center position of the model ice, and the crack propagation was more consistent. It can be considered that the numerical simulation results and the experimental results have good similarity, and the conclusions obtained can confirm each other's results. The impact factors on model ice flexural strength are discussed in two different ways, and these conclusions are considered to apply to both methods.

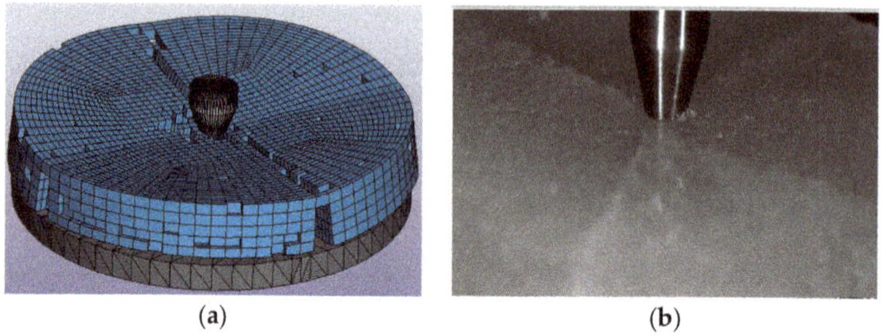

Figure 11. Comparison of failure phenomenon: (**a**) numerical simulation, (**b**) test result (−0.8 °C, 150 mm/min).

The numerical simulation also studied the stress distribution of the model ice at the time of cracking, and the stress distribution is shown in Figure 12. The stress on the top was 249.6 kPa, and the stress on the bottom was 71.3 kPa. The stress on the top was about 3.5 times the stress on the bottom. In fact, under the vertical action of the indenter, the stress on the top was caused by extrusion, and the stress on the bottom was caused by tension. Figure 10 shows that the destruction of the model ice occurred firstly on the bottom surface because the model ice is a brittle material: its tensile strength is much lower than its compressive strength [38]. Thus, the failure usually occurs at the stretch of the bottom surface.

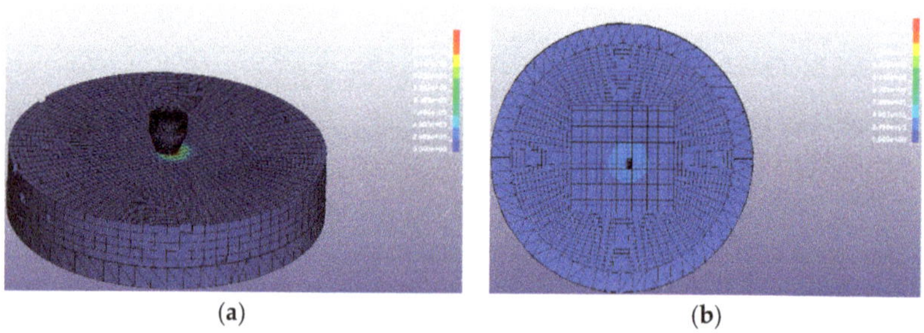

Figure 12. Stress distribution of disk model in failure: (**a**) stress on the top—249.6 kPa; (**b**) stress on the bottom—71.3 kPa (−0.8 °C, 150 mm/min).

3.2. *Impact Factors on Flexural Strength*

3.2.1. The Effect of Loading Rate

Discussions in this section are based on model test results because the behavior of the indenter was not fully represented in numerical simulations, considering that this may interfere with the conclusions drawn for the variable loading rate. Figure 13 shows a typical time history curve for model ice at different loading rates (−0.8 °C). Under different loading rates, the loading force of the model ice failure did not change much. When the loading force increased to the peak point, the time history curve decreased rapidly, and the downward trend was basically the same, which further indicates that the failure of the model ice was a brittle failure. It can be seen from the figure that the time difference between the peaks of the time history curves at different loading rates was obvious, followed by the insignificant change in the peak of the time history curve.

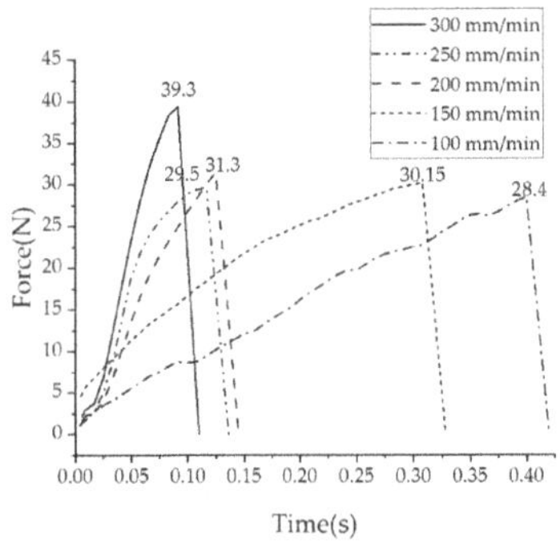

Figure 13. Time history curves of flexural strength experiments (−0.8 °C).

Based on experimental data, the correlation between loading rate and flexural strength was analyzed using the Pearson correlation coefficient and the Spearman correlation coefficient. Among them, the Pearson correlation coefficient can reflect the degree of linear correlation between two random variables; the Spearman correlation coefficient measures the strength of monotonicity between variables. Through calculation, it was found that the Pearson correlation coefficient between the flexural strength and loading rate of the model ice was 0.33, and the p-value was greater than 0.05, indicating that there was no significant linear correlation between the loading rate and the flexural strength of the model ice. The Spearman correlation coefficient between the flexural strength and loading rate was 0.37, and the p-value was greater than 0.05, indicating that there was no significant monotonic correlation between the loading rate and the flexural strength of the model ice, as shown in Figure 14. Therefore, it can be judged that there was no obvious correlation between the ice flexural strength and the loading rate of the model. In fact, in the other four test groups at temperatures of −2 °C, −4 °C, −6 °C, and −8 °C, the relationship between loading rate and flexural strength also showed no significant correlation, and thus was not shown here. This agrees with the conclusion of Frederking et al. in an ice basin [8]. Similarly, in polar sea ice observations, it has been found that the loading rate has less influence on the flexural strength of the ice [39].

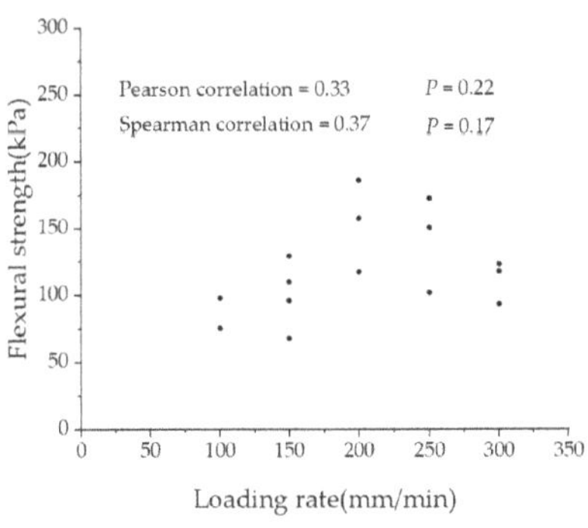

Figure 14. Relationship between flexural strength and loading rate (−0.8 °C).

3.2.2. The Effect of Ice Temperature

As mentioned above, since it is difficult to show the influence of temperature on the mechanical properties of ice materials in numerical simulations, this section is also based on the experiments. This section quantifies the relationship between model ice flexural strength and model ice temperature. The temperature range of the model ice was −0.8 °C to −9 °C, as can be seen from Section 3.2.1.

According to the statistical test data, the Pearson correlation coefficient between ice temperature and flexural strength was −0.89, and the p-value was 3.15 × 10^{-24}. The Spearman correlation coefficient was −0.78, and the p-value was 7.1 × 10^{-13}. Curve fitting to the experimental data was conducted, with a 95% confidence band, as shown in Figure 15, and the regression equation is as follows:

$$\sigma_f = 192.24 - 204.36 \cdot T \quad (4)$$

where σ_f is the model ice flexural strength in kPa, and T is the model ice temperature. It is clear from Figure 15 that in the temperature range from −0.8 °C to −9 °C, the model ice flexural strength decreased linearly with ice temperature, consistent with previous results of sea ice. Ding et al. measured the flexural strength of Bohai sea ice, and the experimental data showed that the flexural strength increased with the decrease in ice temperature, and the relationship between the two was linear [40]. It is worth noting that in the −2 °C to −6 °C range, there are some data points that deviate from the 95% confidence band, possibly due to errors in the measurement of ice temperature. Marchenko [6] tested the flexural strength of sea ice in the northwestern part of the Barents Sea and obtained the relationship between flexural strength in MPa and temperature and salinity, as follows:

$$\sigma_f = 0.236 - 0.095 \cdot S - 0.0134 \cdot T \quad (5)$$

where S is the salinity of sea ice. The empirical formulas for the flexural strength of Barents Sea ice also show ice temperature has an approximate linear relationship with the flexural strength if a constant or average salinity is given. However, it should be especially noted that the model results and the results of the flexural strength measurement in marine conditions in absolute values cannot coincide, as well as other strength characteristics.

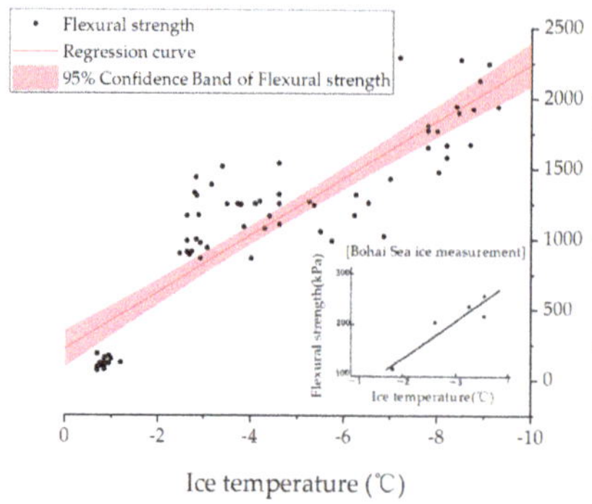

Figure 15. Relationship between flexural strength and ice temperature.

3.2.3. The Effect of Ice Porosity

Since ice crystal observation experiments are labor-intensive and time-consuming, and the expected porosity is difficult to control artificially, this section is based on numerical simulation results. The relationship between porosity, the model ice load response, and the model ice failure mode were studied through four sets of numerical simulations with different porosities, as listed in Table 7. As mentioned earlier, the flexural strength was used as a parameter input to the numerical model, so the influence of porosity on the bending properties of the model ice cannot be discussed in this section through the flexural strength but only by the load response. It was found that the load response of the model ice flexural failure was in the range of 24.44 N~26.5 N, as shown in Figure 16. The greater the porosity, the smaller the load response of the model ice, and the earlier the time for failure to occur. Statistics show that the load response of model ice with 7% porosity was 0.04 s earlier than model ice without porosity. In addition, the peak load decreases gradually with increasing porosity, and this conclusion is similar to Wang's conclusions about sea ice at Prydz Bay, East Antarctic [41].

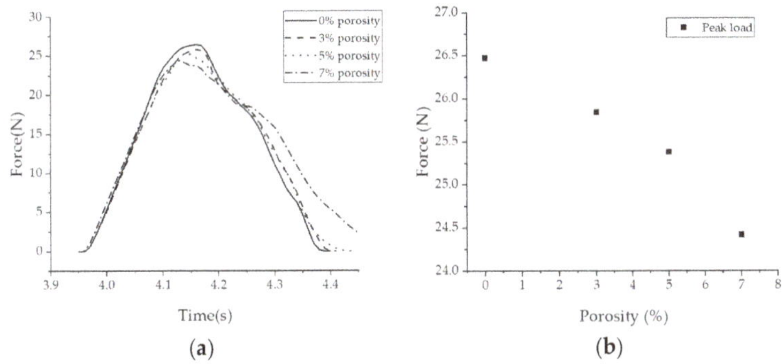

Figure 16. Load response of model ice with different porosities: (**a**) time history curve (**b**) force–porosity curve (−0.8 °C, 150 mm/min).

The final failures of model ice at different porosities are also shown in Figure 17. A crack extending along the diameter appeared on the surface of the model ice under different porosities. There was a significant deflection change at the center of the model ice, and there was a gap in the thickness direction. This indicates that within the range of porosity < 7%, porosity had little effect on the failure mode of the model ice. Compared with nonporous model ice, model ice with pores was more likely to develop microcracks around the pores or further expand the pores. There was a certain crush failure near the indenter, but the failure mode of the model ice was still dominated by flexural failure.

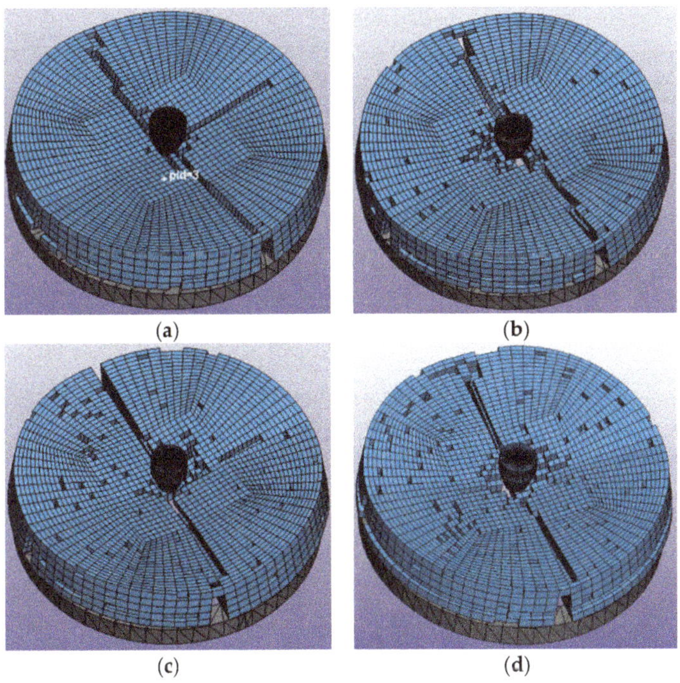

Figure 17. Failure results of model ice with different porosities of (**a**) 0%, (**b**) 3%, (**c**) 5%, and (**d**) 7% (−0.8 °C, 150 mm/min).

4. Conclusions

In this paper, both laboratory experiments and numerical modeling were carried out to investigate the flexural strength of a columnar saline model ice under circular plate central loading. The influence of different factors on the flexural strength of model ice was analyzed and the crack propagation law of the model ice flexural fracture failure was found. The main conclusions are as follows:

1. The experimental and numerical results were compared from two aspects including the time history curve and damage phenomenon, and their results agree well; these could reflect the flexural strength characteristics of the model ice and confirm each other's results.
2. According to the time history curve of the ice specimen from the initial bearing to the flexural failure, it was found that the ice specimen had no obvious yield stage. The high-speed camera observed no obvious plastic deformation at the failure location. The model ice began to crack from the center of the bottom surface, and the crack extended along the radius direction to the lower surface boundary, then extended along the thickness direction to the top surface until complete failure. The failure process of the model ice was judged to be a typical brittle failure.
3. There was no significant correlation between the loading rate and the flexural strength. A significant linear correlation between the model ice temperature and the flexural strength was explored, and the flexural strength of the model ice increases continuously with the continuous decrease in the model ice temperature in the range from −0.8 °C to −9 °C. The larger the porosity, the smaller the load response of the model ice, and the earlier the time of failure. Compared with the nonporous model ice, the load response of model ice with 7% porosity was reduced by 7.8%, and the failure time was 0.04 s earlier. Within the range of 7%, ice porosity had little effect on the failure mode of the model ice.

This paper provided a feasible means to measure and predict the mechanical properties of model ice and realized the simulation of the internal pores of the model ice, forming a preliminary foundation for the research of the mechanism of ice loading under the vertical interaction between the structure and the ice cover. In the future, the results obtained by different methods such as circular plate center loading tests, in situ cantilever tests, and simply supported beam tests will be enriched and compared with each other, to establish a systematic ice flexural strength measurement and analysis technology. In addition, the numerical model will continue to be optimized and improved to reflect the influence of temperature and brine, especially the internal structure and anisotropic characteristics of the model ice.

Author Contributions: Conceptualization, Y.T.; methodology, Y.T. and W.Z.; software, W.Z.; validation, W.Z., C.Y. and X.G.; formal analysis, W.Z.; investigation, Y.T.; writing—original draft preparation, Y.T. and W.Z.; writing—review and editing, P.L. and Q.Y.; project administration, Y.T.; funding acquisition, Y.T. All authors have read and agreed to the published version of the manuscript.

Funding: This work was supported by the Fundamental Research and Development Project of China [Grant No. JCKY2020206B073], the National Natural Science Foundation of China [Grant Nos. 52192690, 52192694], and the Hi-Tech Ship Project of the Ministry of Industry and Information Technology [Grant No. [2021]-342].

Data Availability Statement: The data are not publicly available due to the confidentiality policy of our institution requires.

Conflicts of Interest: The authors declare no conflict of interest.

References

1. Zhao, W.; Tian, Y.; Yu, C. Experimental Investigation on Breaking Strength of Model Ice with Circular Ice Plate Indentation Tests. In Proceedings of the 32nd International Ocean and Polar Engineering Conference, Shanghai, China, 5–10 June 2022.

2. Ye, L.; Guo, C.; Wang, C.; Wang, C.; Chang, X. Peridynamic solution for submarine surfacing through ice. *Ships Offshore Struc.* **2020**, *15*, 535–549. [CrossRef]
3. Meng, X. *Research on the Force of Submarine to Destory the Ice Layer Upwards*; Harbin Institute of Technology: Harbin, China, 2019.
4. Ji, S.; Wang, A.; Su, J.; Yue, S. Experimental studies and characteristic analysis of sea ice flexural strength in Bohai Sea. *Adv. Water Sci.* **2011**, *22*, 266–272. [CrossRef]
5. Weeks, W.; Anderson, D. An experimental study of strength of young sea ice. *Trans. Am. Geophys. Union* **1958**, *39*, 641–647. [CrossRef]
6. Marchenko, A.; Karulina, M.; Karulin, E. Flexural strength of ice reconstructed from field tests with cantilever beams and laboratory tests with beams and disks. In Proceedings of the International Conference on Port and Ocean Engineering under Arctic Conditions, Busan, Republic of Korea, 11–16 June 2017.
7. Li, Z.; Wang, Y.; Li, G. Experimental analysis of flexural strength and elastic modulus of the DUT-1 model ice. *Adv. Water Sci.* **2002**, *13*, 292–297. [CrossRef]
8. Frederking, R.; Timco, G. On measuring flexural properties of ice using cantilever beams. *Ann. Glaciol.* **1983**, *4*, 58–65. [CrossRef]
9. Krupina, N.; Kubyshkin, N. Flexural strength of drifting level first-year ice in the Barents Sea. In Proceedings of the 17th International Offshore and Polar Engineering Conference, Lisbon, Portugal, 1–6 July 2007.
10. Gang, X.; Tian, Y.; Ji, S.; Guo, W.; Kou, Y. Experimental analysis on flexural strength of columnar saline model ice. *J. Ship Mech.* **2021**, *25*, 336–341. [CrossRef]
11. Xu, Y.; Hu, Z.; Chen, G.; Xu, Y. Overview of the investigation methods for ship-ice interaction analysis. *J. Ship Mech.* **2019**, *23*, 110–123.
12. Gagnon, R.; Gammon, P. Characterization and flexural strength of iceberg and glacier ice. *J. Glaciol.* **1995**, *41*, 103–111. [CrossRef]
13. Barrette, P.; Jordaan, I. *Beam Bending and Fracture Behavior of Iceberg Ice*; National Research Council Program on Energy Research and Development: Saskatoon, SK, Canada, 2001; pp. 4–18. [CrossRef]
14. Timco, G.; O'Brien, S. Flexural strength equation for sea ice. *Cold Reg. Sci. Technol.* **1994**, *22*, 285–298. [CrossRef]
15. Gang, X.; Tian, Y.; Yu, C.; Ji, S.; Kou, Y. Numerical simulation analysis of flexural strength of columnar saline ice model. *Chin. J. Ship Res.* **2021**, *16*, 143–149. [CrossRef]
16. Yu, C.; Tian, Y.; Wang, W. Review of research on ice loading of offshore structures in level ice fields. *Chin. J. Ship Res.* **2021**, *16*, 39–53. [CrossRef]
17. Sand, B. *Nonlinear Finite Element Simulations of Ice Forces on Offshore Structures*; University of Technology: Luleå, Sweden, 2008.
18. Wang, C.; Wang, J.; Wang, C.; Guo, C.; Zhu, G. Research on movement of cylindrical structure out of water and breaking through ice layer based on S-ALE method. *Chin. J. Theor. Appl. Mech.* **2021**, *53*, 3110–3123. [CrossRef]
19. Liu, R.; Yan, J.; Li, S. Modeling and simulation of ice-water interactions by coupling peridynamics with updated Lagrangian particle hydrodynamics. *Comp. Part. Mech.* **2020**, *7*, 241–255. [CrossRef]
20. Li, J. *Study on Hydrodynamic Effects on Ice-Structure Interaction*; Dalian University of Technology: Dalian, China, 2020.
21. Wang, J.; Derradji-Aouat, A. Implementation, verification, and validation of the multi-surface failure envelope for ice in explicit FEA. In Proceedings of the 20th International Conference on Port and Ocean Engineering Under Arctic Conditions, Luleå, Sweden, 9–12 June 2009.
22. Von Bock und Polach, R.B.; Ehlers, S. Model scale ice-Part B: Numerical mode. *Cold Reg. Sci. Technol.* **2013**, *94*, 53–60. [CrossRef]
23. Herrnring, H.; Ehlers, S. A finite element model for compressive ice loads based on a mohr-coulomb material and the node splitting technique. In Proceedings of the 40th ASME International Conference on Ocean, Offshore and Arctic Engineering (OMAE), Virtual, 21–30 June 2021.
24. Tian, Y.; Ji, S.; Wang, Y.; Guo, W.; Kou, Y.; Gang, X. Research on sea ice simulation and measurement in small ice model basin of CSSRC. *Mar. Environ. Sci.* **2021**, *40*, 277–286. [CrossRef]
25. Evers, K. Model tests with ships and offshore structures in HSVA's ice tanks. In Proceedings of the 24th International Conference on Port and Ocean Engineering under Arctic Conditions, Busan, Republic of Korea, 11–16 June 2017.
26. Evers, K.; Jochmann, P. An advanced technique to improve the mechanical properties of model ice developed at the HSVA ice tank. In Proceedings of the 12th International Conference on Port and Ocean Engineering under Arctic Conditions, Hamburg, Germany, 17–20 August 1993.
27. Tian, Y.; Ji, S.; Kou, Y.; Guo, W.; Chen, Z.; Gang, X. Characterization of Uniaxial Compression Strength for Columnar Saline Model Ice in CSSRC Small Ice Model Basin. *J. Ship Mech.* **2020**, *24*, 1647–1656. [CrossRef]
28. Yu, C. *Research on Ice Loading of Typical Offshore Ice-Resistant Structure*; China Ship Scientific Research Center: Wuxi, China, 2021.
29. Timoshenko, S.; Woinowsky-Krieger, S. *Theory of Plates and Shells*, 2nd ed.; McGraw-Hill: New York, NY, USA, 1959; pp. 135–146.
30. Timco, G.W.; Weeks, W.F. A review of the engineering properties of sea ice. *Cold Reg. Sci. Technol.* **2010**, *60*, 107–129. [CrossRef]
31. Bassali, W. The transverse flexure of thin elastic plates supported at several points. *Math. Proc. Camb.* **1957**, *53*, 728–743. [CrossRef]
32. Guo, T.; Zhang, A.; Yu, B. Underwater collision simulation of the pressure hull with spherical head based on LS-DYNA. *J. Ship Mech.* **2021**, *25*, 210–219.
33. Karr, D.; Choi, K. A 3-dimensional constitutive damage model for polycrystalline ice. *Mech. Mater.* **1989**, *8*, 55–66. [CrossRef]
34. Qian, Y. *Study on the Calculation Method for the Ice-Load of Vertical Ice-Breaking*; Harbin Engineering University: Harbin, China, 2020.
35. Zhao, H. *LS-DYNA Dynamic Analysis Guide*; The Publishing House of Ordnance Industry: Beijing, China, 2003; pp. 64–71.
36. Sodhi, D.S. Breakthrough loads of Floating Ice Sheet. *J. Cold Reg. Eng.* **1995**, *9*, 4–22. [CrossRef]

37. Lainey, L.; Tinawi, R. The mechanical properties of sea ice a compilation of available data. *Can. J. Civil. Eng.* **1984**, *11*, 884–923. [CrossRef]
38. Ni, B.Y.; Tan, H.; Di, S.-C.; Zhang, C.X.; Li, Z.; Huang, L.; Xue, Y.Z. When does a light sphere break ice plate most by using its net buoyance? *J. Mar. Sci. Eng.* **2023**, *11*, 289. [CrossRef]
39. Barrette, P.; Phillips, R.; Clark, J.; Crocker, G.; Jones, S. Flexural behavior of model sea ice in a centriguge. *J. Cold Reg. Eng.* **1999**, *13*, 122–138. [CrossRef]
40. Ding, D. *An Introduction to Engineering Sea Ice*; Ocean Press: Beijing, China, 2000; pp. 84–88.
41. Wang, Q.; Li, Z.; Lu, P.; Xu, Y.; Li, Z. Flexural and compressive strength of the landfast sea ice in the Prydz Bay, East Antarctic. *Cryosphere* **2022**, *16*, 1941–1961. [CrossRef]

water

Article

Study on the Constitutive Equation and Mechanical Properties of Natural Snow under Step Loading

Hongwei Han [1,2], Meiying Yang [1], Xingchao Liu [1], Yu Li [1], Gongwen Gao [1] and Enliang Wang [1,2,*]

1 School of Water Conservancy and Civil Engineering, Northeast Agricultural University, Harbin 150030, China; hanhongwei@neau.edu.cn (H.H.); 18346271554@163.com (M.Y.); dnlxc@neau.edu.cn (X.L.); 18846771809@163.com (Y.L.); 18045129438@163.com (G.G.)
2 Heilongjiang Provincial Key Laboratory of Water Resources and Water Conservancy Engineering in Cold Region, Northeast Agricultural University, Harbin 150030, China
* Correspondence: hljwel@126.com

Abstract: Snow, as an important component of the cryosphere, holds a crucial role in the construction of polar infrastructure. However, the current research on the mechanical properties of snow is not comprehensive. To contribute to our understanding of the mechanical behaviors of snow in cold regions, uniaxial compression tests under step loading were performed on the snow. With the Maxwell model as the basis, different temperatures, densities, and loading rates were set to establish constitutive equations of snow. The changes in the elastic modulus and viscosity coefficient of snow with respect to three variables were investigated. The results show that the loading rate has no obvious effect on the elastic modulus and viscosity coefficient of snow. Both the elastic modulus and viscosity coefficient of snow follow an exponential function with respect to density, with an increase in density, resulting in a higher value. As temperature decreases, the elastic modulus and viscosity coefficient initially decrease and then increase, whereas no specific functional relationship between them was observed. Additionally, a new constitutive equation considering snow density is derived based on the Maxwell model.

Keywords: snow; step loading; constitutive equation; Maxwell model

Citation: Han, H.; Yang, M.; Liu, X.; Li, Y.; Gao, G.; Wang, E. Study on the Constitutive Equation and Mechanical Properties of Natural Snow under Step Loading. *Water* **2023**, *15*, 3271. https://doi.org/10.3390/w15183271

Academic Editors: Frédéric Frappart and Jueyi Sui

Received: 31 July 2023
Revised: 11 September 2023
Accepted: 13 September 2023
Published: 15 September 2023

1. Introduction

Snow affects the replenishment of water resources, the occurrence of natural disasters, and the changes in air quality [1]. Approximately 98% of the global seasonal snow cover is found in the Northern Hemisphere [2], of which about 60–65% is in Europe and Asia [3]. Extreme snowfall often occurs in the French Alps due to geographical location [4]. In the western United States, snowfall serves as a major source of domestic and agricultural water supply [5]. Similarly, winter snowfall provides water recharge in arid regions of Iran [6]. Snowfall in Southern China increases in winter due to the presence of the Tibetan Plateau, which leads to high relative humidity and discomfort in the south [7]. In recent years, the polar route has attracted significant interest as increasingly more countries have set up scientific research stations in Antarctica for observation and research. Given the extremely cold weather and geographical location of the Antarctic region, the mode of delivering supplies and personnel to these research stations is limited to sea and air transportation. However, sea transportation is time-consuming and subject to seasonal restrictions, leading many countries to initiate airport construction in Antarctica to facilitate research. According to the construction location and runway type, they can be categorized as sea ice runways, blue ice runways, sled runways, and compacted snow runways [8,9]. Challenges faced by the first three types of airport runways include high construction costs, limited available areas, and insufficient snow-layer strength. In contrast, compacted snow runways make full use of Antarctic snow resources and offer low maintenance costs, becoming the preferred choice for airport construction in many countries [10]. Nevertheless, existing research on snow mechanics remains incomplete due to the spatial anisotropy of snow and its continuous metamorphism over time [11,12]. Consequently, analyzing the deformation behavior and mechanical properties of snow is of great interest when it comes to engineering construction.

Snow is composed of three phases of water, air, and ice connected together in the form of particles, and has a certain strength [13]. Subsequently, many experimental studies on snow have been conducted and various constitutive equations have been developed to understand snow deformation under loading conditions [14,15]. At present, the constitutive equations of snow can be divided into two kinds. One is to describe the deformation characteristics of snow from a macroscopic perspective, which is also called the phenomenological method. The other takes a microscopic viewpoint by observing the changes in ice particles and bonds within the snow under the external loads and then deducing the overall deformation.

Mishra and Mahajan [14] considered that snow deformation consists of elastic deformation and creep deformation, ignoring time-hardening and microstructural changes. They described the constitutive relationship of snow based on a complementary power potential, which also predicted volume changes in snow samples. Snow can be regarded as a geotechnical material whose internal microstructure determines the overall stress–strain relationship. Birkeland et al. [16] and Bobillier et al. [17] developed discrete element models to simulate the propagation saw test, aiming to understand the microstructural changes during snow failure. Nicot [18] assumed that the mechanical behavior of snow primarily depends on the mechanical properties of internal bonds. They constructed a probability density function to describe the microstructure of snow. Using a nonlinear Kelvin model and a fabric description as constitutive equations, the overall mechanical behavior is deduced from the local properties of snow. Mahajan and Brown [19] constructed a multi-axial constitutive equation for snow, which pointed out that the deformation of snow was divided into different mechanisms. The deformation of the entire snow sample under external load was deduced based on this determined mechanism. Brown [20] constructed a volume constitutive equation of snow based on the changes in bond size and ice grain diameter. However, the accuracy of the equation is reduced at high snow densities because the interactions between adjacent bonds are not considered. Recent efforts have aimed to combine tomography technology with the discrete element method to build microscopic numerical models of snow and simulate its deformation behavior under load [21–24]. Singh et al. [25] assumed snow to be an orthotropic elastoplastic material and constructed an equation to predict the constitutive relationship of snow with different densities and types. X-ray microtomography combined with the finite element method was used to determine the parameters in the constitutive equation and verify the results. Chandel et al. [26] determined the deformation of ice particles under load based on the damage elastic–plastic constitutive equation of ice. They used X-ray tomography technology and finite element software to simulate the stress–strain relationship of RVE and derived the overall macroscopic constitutive behavior of snow. From the perspective of the relationship between airflow and compaction, a compacted snow constitutive model was proposed under confined compression tests. This model was consistent with laboratory measurements, indicating that air inside the snow sample would be discharged from the pores during testing [27]. During the cold winter months, rivers and lakes begin to freeze and snowfall occurs. The abundant snow and ice resources are often used as building materials. Such as the utilization of ice and snow for constructing airstrips in the Antarctic region [28], creating snow and ice sculptures, constructing distinctive buildings, and the construction of snow and ice roads to facilitate transportation on ice [29]. In construction, snow needs to be crushed and other operations need to be carried out to improve its strength to ensure the safety of buildings and roads. The relevant features of the mechanical behavior of snow need to be fully understood, and constitutive equations should be used to describe the stress–strain relationship of snow.

In previous studies, the constitutive model of snow is frequently investigated using relative theoretical derivations or numerical simulation methods [14,26], and then verified by experimental data. However, the numerous parameters and complex formulas involved in these methods make them difficult to compute. Questions still remain regarding the accurate interpretation of the deformation behavior of snow. In this paper, the constitutive equations of snow obtained by combining laboratory tests and theoretical derivations can more truly reflect the mechanical behavior of snow, which is an extension and supplement to the previous research. In general, the main objective of this paper is to provide a novel constitutive model aimed at representing the most relevant aspects of the overall macro-behavior of snow while retaining adequate simplicity so that a feasible application is achieved in practical (large-scale, long-duration) engineering cases. The derivation method of the constitutive law for solid materials is used to obtain the constitutive law of snow. This study focused on investigating the constitutive equation and mechanical behavior of snow by conducting step-loading uniaxial compression tests on cylindrical snow specimens. Sections 2 and 3 detail the preparation of the snow samples and the uniaxial compression test procedure. After some preliminary notes on the Maxwell model, the derivation process of the elastic modulus and viscosity coefficient through it is introduced. The changes in the elastic modulus and viscosity coefficient of snow in response to changes in snow density, loading rate, and temperatures are studied and discussed in Section 4. A new constitutive equation is proposed to link the elastic modulus and viscosity coefficient of snow to its density, temperature, and loading rate. Section 5 discusses the trends of elastic modulus and viscosity coefficient obtained in this paper with respect to different variables in comparison with the results of other studies in the literature. Finally, the major results and implications of this paper are emphasized in the conclusion.

2. Sample Preparation and Experimental Procedure

2.1. Sample Preparation

The snow samples used in this study are fresh natural snowfall from the top layer of the ground at the Water Conservancy Comprehensive Test Site of Northeast Agricultural University in Harbin, Heilongjiang Province, China. As one of the three main snow areas in China, low temperatures and abundant snow are the primary characteristics of the northeastern region during winter due to the temperate monsoon climate and high latitude. In general, the main snowfall period is from October to March of the following year, and the stable snowfall period is from November to February of the following year. Commonly, fresh snowfall grains are loose and uniform in size, without any bonding for easy compaction. Referring to the snow sample preparation method of [30], from 26 November to 1 December 2022, fresh snow was collected and poured into the compaction equipment by manual layering to achieve uniform density and to minimize errors. The compaction instrument consists of a guard cylinder, a compaction hammer, and a compaction cylinder to compact the snow into a 100 mm diameter, 200 mm high cylinder (height-to-diameter ratio of 2:1), as depicted in Figure 1. It is provided by Suzhou Tuo Testing Instrument Equipment Co., Ltd. located in Suzhou, Jiangsu Province, China. Snow samples with densities of 350 kg/m^3, 400 kg/m^3, 450 kg/m^3, 500 kg/m^3, and 550 kg/m^3 were created using different masses of snow. To prevent the snow samples from sublimating during storage and to maintain integrity, they were wrapped tightly in cling film and stored in a refrigerator at a constant temperature of −25 °C. As the tests were carried out at different temperatures, the snow samples were stored in a refrigerator at the required temperature for 24 h prior to each test to ensure uniform temperature throughout the entire sample. Before the test began, the mass and height of the snow samples were measured using an electronic scale and tape measure to calculate the actual density, ensuring that the error between it and the experimental density was less than ±20 kg/m^3. Through measuring the mass and volume of snow, the natural density of fresh snow was derived to be 128 kg/m^3. The above operations can reduce the error to the tolerable range.

(**a**) (**b**)

Figure 1. Compacting tools and snow sample: (**a**) Compaction equipment, from left to right, the protective cylinder, the compactor hammer, and the compaction cylinder are followed; (**b**) sample.

2.2. Test Procedure

The uniaxial compression tests were carried out on snow samples using a load-controlled step loading method at temperatures of −5 °C, −10 °C, −15 °C, and −20 °C, with loading rates of 5 N/s, 10 N/s, 20 N/s, 30 N/s, 40 N/s, and 50 N/s, respectively. The force difference between the two adjacent steps was set to 300 N, and each step lasted for more than 200 s. As shown in Figure 2, the experimental equipment in this study includes a WDW–100 electronic universal testing machine and two infrared sensors. The WDW–100 machine is made up of a compression device, a control system, and a low-temperature test chamber. It comes from Changchun Kexin Testing Instrument Co., Ltd. located in Changchun city, China. The low-temperature test chamber is used to control the temperature inside the testing machine, so that the snow sample can be tested at a preset temperature, with a minimum temperature of −30 °C. Two infrared sensors were placed inside the testing machine to record the diameter expansion of the snow sample in real-time during compression. One sensor was located near the middle of the snow sample, and the other near the end of the snow sample. Before the test, the upper and lower pressure plates of the testing machine were wrapped with cling film to avoid the end effect impacting the test results during compression. Furthermore, upon setting the internal temperature of the testing machine to the desired test temperature, the sample was put into it when the internal temperature was uniform. During the test, the central axis of the specimen was aligned in a straight line with the center of the upper and lower pressure plates to ensure uniform force distribution. To minimize errors in the recorded diameter growth value, the infrared light emitted by the sensor was directed as closely as possible to the center of the snow sample.

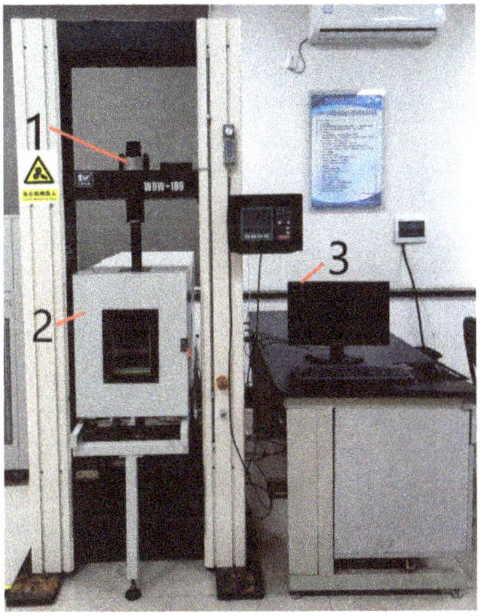

Figure 2. WDW–100 universal testing machine: 1. compression device; 2. cryogenic test chamber; 3. control system.

3. Constitutive Model

3.1. Maxwell Model

The stress–strain curve is a complete macroscopic response to a series of change processes such as the deformation of snow under external force, the rearrangement of internal grains, and the breakage and peeling of the exterior. The constitutive equation is the mathematical expression describing this curve, which is the main basis for studying and analyzing snow's bearing capacity and deformation. There are two main methods to establish the constitutive model of crystalline materials, at present. One of which is to describe the constitutive equation of materials by a series-parallel combination of elastic, viscous, and plastic elements. The other is to synthesize empirical formula to describe the constitutive equation of materials according to the stress–strain change law obtained from experiments. This study adopts the method of component combination to construct the constitutive equation of snow. The widely used classic constitutive models currently include the Maxwell model, Burgers model, and Kelvin model. This study assumes that snow is a kind of viscoelastic material and chooses the Maxwell model as the basis. Three constitutive equations are derived to describe the mechanical behavior of snow to show the effects of temperature, density, and loading rate on the viscoelasticity of snow.

The Maxwell model comprises a Hooke body and a Newtonian body connected in series, as illustrated in Figure 3. When stress is applied to the snow, both elastic strain and plastic strain are generated. The Hooke body represents the instantaneous elastic deformation that occurs upon the application of stress, which can be recovered immediately after the stress is removed. In contrast, the Newtonian body represents the viscous behavior of the material, which produces irreversible deformation under stress. After the stress is unloaded, the deformation remains. A brief description of the Maxwell model is presented as follows.

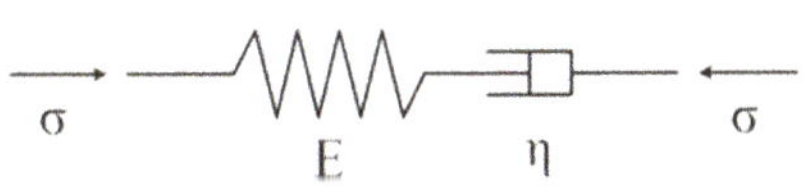

Figure 3. Maxwell model.

The total strain consists of elastic and viscous strains:

$$\varepsilon = \varepsilon_E + \varepsilon_\eta \tag{1}$$

Instantaneous elastic strain is inversely proportional to the elastic modulus, E:

$$\varepsilon_E = \frac{\sigma_E}{E} \tag{2}$$

Viscous strain rate is inversely related to the viscosity coefficient, η:

$$\dot{\varepsilon}_\eta = \frac{\sigma_\eta}{\eta} \tag{3}$$

When the elements are connected in series, they are subjected to the same stresses:

$$\sigma = \sigma_E = \sigma_\eta \tag{4}$$

Substituting these relationships (Equations (2)–(4)) into Equation (1), the constitutive equation is derived:

$$\dot{\varepsilon} = \dot{\varepsilon}_E + \dot{\varepsilon}_\eta = \frac{\dot{\sigma}}{E} + \frac{\sigma}{\eta} \tag{5}$$

Integrating Equation (5) over time, the deformed coordination equation is obtained as:

$$\varepsilon = \varepsilon_E + \varepsilon_\eta = \frac{\sigma_E}{E} + t\frac{\sigma_\eta}{\eta} = \sigma(\frac{1}{E} + \frac{t}{\eta}) \tag{6}$$

3.2. Parameter Calculation

The stress of the snow sample under the uniaxial compression test with step loading remains constant in the horizontal step when the stress rate is 0. Substituting this condition into Equation (5), the expression for the viscosity coefficient of snow can be obtained as:

$$\eta = \frac{\sigma}{\dot{\varepsilon}} \tag{7}$$

The elastic modulus can be obtained using Equation (6):

$$E = \frac{1}{(\frac{\varepsilon}{\sigma} - \frac{t}{\eta})} \tag{8}$$

where σ is the constant stress at the step (MPa); ε is the strain; $\dot{\varepsilon}$ is the strain rate (s^{-1}); t is the time (s); E is the elastic modulus of snow (MPa); and η is the viscosity coefficient of snow (MPa·s).

4. Results

The stress–strain curves of the snow samples obtained by step loading are shown in Figure 4. The uniaxial compression test under step loading is completed alternately by the normal uniaxial compression test and creep test. Indeed, the data of the first and last steps are not involved in the calculation of parameters. Because there had been a pre-pressure adjustment in the first stage, the measured displacement value is greater than the actual deformation, resulting in a relatively small and inaccurately calculated viscosity coefficient. In the final step stage, as the density of the snow sample gradually increases, the bonds between the ice particles begin to deform, break, and rearrange, which cause an increase in deformation of the snow sample under loading. The calculated viscosity coefficient is also relatively small. Due to the low density and the larger internal pores of snow compared to brittle materials such as rocks and ice, the stress of the snow sample under load will rapidly decrease after reaching a set value at the beginning of the step. At this point, the pores inside the snow collapse, and the ice crystals begin to rearrange and combine. When a relatively stable structure is reached, the stress will stop decreasing and enter a creep state as shown in Figure 4. The elastic modulus and viscosity coefficient of compacted snow under step loading are then calculated using Equations (7) and (8) and the variation of these mechanical properties are investigated in relation to temperature, density, and loading rate.

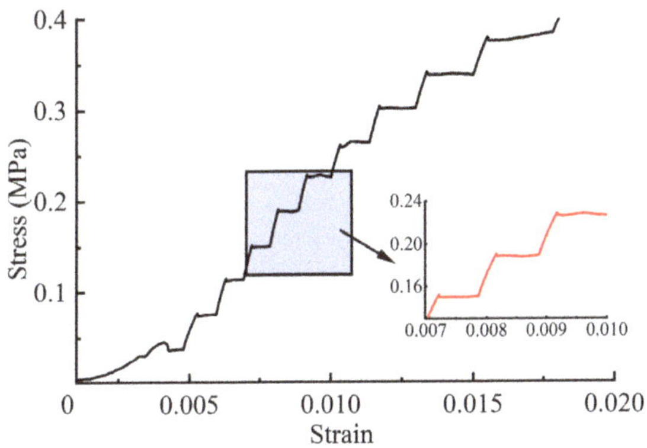

Figure 4. Stress–strain curves of 550 kg/m^3 snow samples at −15 °C and 10 N/s.

4.1. Effect of Loading Rate on Elastic Modulus and Viscosity Coefficient of Snow

Uniaxial compression tests were carried out on 400 kg/m^3 and 500 kg/m^3 snow samples at a constant temperature of −15 °C with step loading rates ranging from 5 to 50 N/s to observe the effect of loading rate on the elastic modulus and viscosity coefficient. As shown in Figure 5a, the elastic modulus is more scattered for the 500 kg/m^3 snow than for the 400 kg/m^3 snow. With increasing loading rate, the value in snow elastic modulus is significantly greater at high densities than at low densities, which indicates that the elastic modulus gradually increases with increasing density. At the same time, it is found that density is a factor that affects the degree of change in elastic modulus, with a greater impact on the elastic modulus with loading rate at higher densities. For example, the elastic modulus of the 400 kg/m^3 snow with a range of 5 to 50 N/s is 3.39 to 9.33 MPa, and the difference is 5.94 MPa. The elastic modulus of the 500 kg/m^3 snow with a range of 5 to 50 N/s is 7.18 MPa to 26.98 MPa, and the difference is 19.80 MPa. However, the results for the elastic modulus of snow show no clear tendency for dependence on the loading rate.

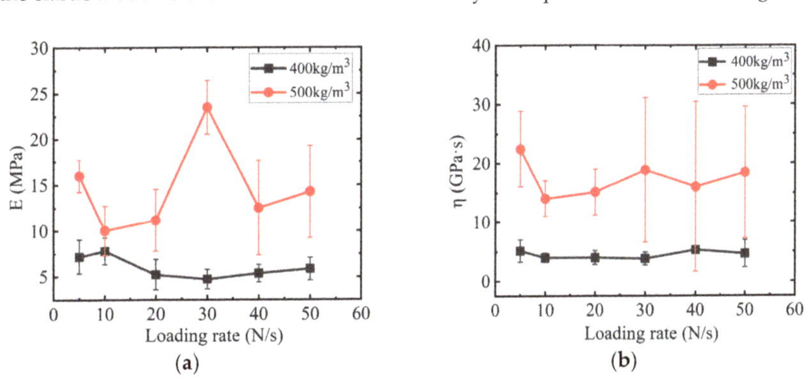

Figure 5. The relationship between the viscoelastic properties and loading rate: (**a**) the elastic modulus; (**b**) the viscosity coefficient.

Figure 5b compares the viscosity coefficient versus loading rate for two density snow samples at −15 °C temperature. The viscosity coefficient varies from 3.26 to 32.98 GPa·s for the 500 kg/m^3 snow sample and from 2.57 to 7.27 GPa·s for the 400 kg/m^3 snow sample, showing that the results are more scattered for the high-density snow. The average value of the viscosity coefficient for 500 kg/m^3 snow is larger than for 400 kg/m^3 snow, which suggests that the viscosity coefficient tends to increase gradually with an increase in density. Density also influences the change degree of the viscosity coefficient from Figure 5b. The difference in viscosity coefficients is 4.70 GPa·s for snow samples at 400 kg/m^3, and 29.72 GPa·s for snow samples at 500 kg/m^3. It is shown that the higher the density, the more intense the change in viscosity coefficient. When the loading rate is less than 10 mm/min, the mean values of the viscosity coefficients at both densities tend to decrease with an increase in loading rate. However, the mean values of viscosity coefficients at the two densities show different trends with increasing loading rate. The mean value of snow viscosity coefficient tends to be a constant at 400 kg/m^3, while the mean value of viscosity coefficient tends to increase, then decrease, and then increase again at 500 kg/m^3.

4.2. *Effect of Temperature on Elastic Modulus and Viscosity Coefficient of Snow*

As shown in Figure 6a, the change in elastic modulus of snow samples with different densities is shown for various temperatures (−5 °C, −10 °C, −15 °C, and −20 °C) under a loading rate of 10 N/s. A trend is observed at temperatures of −15 °C and −20 °C, where the elastic modulus increases as the density increases. However, at −5 °C and −10 °C, the average value of elastic modulus for the 550 kg/m^3 snow samples is lower than that of the 500 kg/m^3 snow samples, which reverses the previous results. It is found that the average value of the elastic modulus first decreases and then increases with decreasing temperature above 500 kg/m^3 and fluctuates in a very narrow band with a density at 400 kg/m^3. This concludes that the temperature has an impact on the change of the elastic modulus in the density range above 500 kg/m^3. However, more experiments need to be conducted at densities below 500 kg/m^3 to uncover the exact density range within which temperature has an impact.

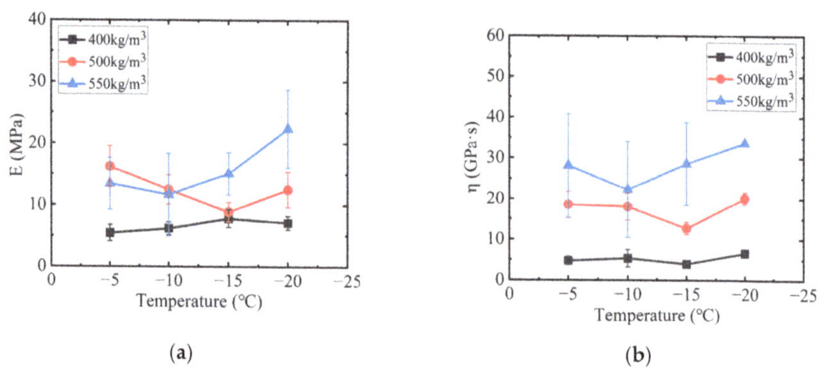

Figure 6. The relationship between the viscoelastic properties and temperature: (**a**) the elastic modulus; (**b**) the viscosity coefficient.

As shown in Figure 6b, the variation of viscosity coefficient with temperature is shown for snow samples at each density. The dispersion degree of viscosity coefficients is lower for low-density snow samples and higher for high-density snow samples. At the density of 400 kg/m^3, the viscosity coefficient of snow varies from 3.36 to 7.77 GPa·s, and the difference is 4.41 GPa·s. At the density of 500 kg/m^3, the viscosity coefficient of snow varies from 11.04 to 21.78 GPa·s, and the difference is 10.74 GPa·s. The mean value of the viscosity coefficient generally decreases and then increases with decreasing temperature. Furthermore, it is noted that the density of the snow samples affects the degree of variation in the mean viscosity coefficient with temperature. We also analyze the changes in the average value of viscosity coefficient under three densities. It is found that the minimum viscosity coefficient of snow at the density of 400 kg/m^3 is 3.98 GPa·s at −15 °C, and the maximum value is 6.64 GPa·s at −20 °C, with a difference of 2.66 GPa·s. The minimum viscosity coefficient of snow at the density of 500 kg/m^3 is 12.71 GPa·s at −15 °C, the maximum value is 20.13 GPa·s at −20 °C, and the difference is 7.42 GPa·s. Similarly, the minimum viscosity coefficient of snow at a density of 550 kg/m^3 is 22.24 GPa·s at −10 °C, and the maximum value is 33.68 GPa·s at −20 °C, with a difference of 11.44 GPa·s. It can be observed that from the above results, the viscosity coefficient of high-density snow samples fluctuates significantly more than that of low-density snow samples as the temperature decreases. Previous research results have shown that the viscosity coefficient can vary with temperature by four to five orders of magnitude, which was concluded via uniaxial compression tests [31].

4.3. Effect of Density on Elastic Modulus and Viscosity Coefficient of Snow

Figure 7 shows the effect of density on the elastic modulus and viscosity coefficient at constant temperature and loading rate. It should be noted that the elastic modulus of compacted snow increases as the density increases, which is consistent with the trend obtained by the authors of [32]. The elastic modulus of snow with the same density is relatively concentrated, ranging from 3.32 to 21.19 MPa. Using the least square method to fit these data in this study, the functional relationship between the elastic modulus and density is obtained, according to Equation (9). It can be concluded that the elastic modulus is an exponential function relationship with the density, and the coefficient of determination (R^2) is 0.76, indicating a good correlation between both.

$$E = 0.396\exp(0.007\rho) \tag{9}$$

where E is the elastic modulus of the snow sample (MPa) and ρ is the density of the snow (kg/m^3).

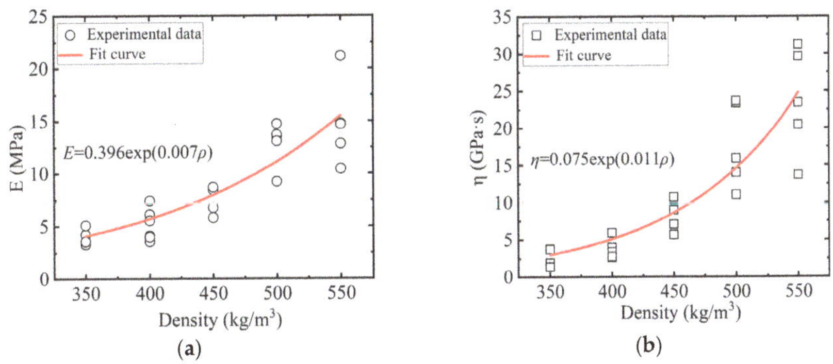

Figure 7. The relationship between the viscoelastic properties and density: (**a**) the elastic modulus; (**b**) the viscosity coefficient.

As shown in Figure 7b, it is observed that the relationship between the viscosity coefficient and density of snow is similar to that between the elastic modulus and density, both exhibiting an increasing trend with increasing density. In the range of 350 ~ 550 kg/m^3, the viscosity coefficient of snow varies from 1.38 to 31.22 GPa·s, which has a relatively poor repeatability compared with the elastic modulus. Furthermore, as the snow density gradually increases, the dispersion of the viscosity coefficient also increases. Taking the results for 350 kg/m^3 and 550 kg/m^3 as examples, the viscosity coefficient of the 350 kg/m^3 snow is from 1.38 to 3.75 GPa·s, and the difference is 2.37 GPa·s. The viscosity coefficient of the 550 kg/m^3 snow is from 13.68 to 31.22 GPa·s, and the difference is 17.54 GPa·s. This indicates that the difference in the viscosity coefficient of snow becomes larger with increasing density, emphasizing the need to consider multiple measurements and calculate the average value for accurate results in practical applications. The functional relationship between the viscosity coefficient and density is obtained by the least square method, according to Equation (10). The relationship between the viscosity coefficient and the density of snow is an exponential function, and the coefficient of determination (R^2) is 0.78, indicating a good correlation between both.

$$\eta = 0.075 \exp(0.011\rho) \tag{10}$$

where η is the viscosity coefficient of the snow sample (GPa·s) and ρ is the density of snow (kg/m^3).

This study has simplified snow as a viscoelastic material and neglected the plastic effect of ice grains, employing a simple Maxwell model to develop constitutive models for snow. The experimental results show that despite the presence of multiple variables such as loading rate, temperature, and density, the elastic modulus and viscosity coefficient of snow exhibits a significant exponential functional relationship only with density. It has incorporated the relationship between elastic modulus and density (Equation (9)) and the relationship between viscosity coefficient and density (Equation (10)) into the Maxwell constitutive equation (Equation (5)). These results are applied in a constitutive model (Equation (11)) for natural snow that considers the influence of density. The R^2 values of both fitted equations are above 0.70, indicating that the fitted curves are closer to the actual data and the derived constitutive equations are more accurate.

$$\dot{\varepsilon} = \frac{\dot{\sigma}}{0.396 \exp(0.007\rho)} + \frac{\sigma}{0.075 \exp(0.011\rho)} \tag{11}$$

where ρ is the density of snow in which 350, 400, 450, 500, and 550 kg/m^3 are taken in the article; σ is the compressive strength of snow (MPa); and $\dot{\varepsilon}$ is the strain rate (s^{-1}).

5. Discussion

5.1. Elastic Modulus

Indeed, the effect of various external environmental variables on the elastic modulus of snow samples has received limited attention in previous research [33,34]. The relationship between the elastic modulus and the loading rate obtained in this study is compared with those in the literature [15,35,36]. It is found to be the same as previous reports from [15,36], yet the opposite in conclusion to that obtained in [35]. Lintzén and Edeskär [36] found that the elastic modulus of both artificial snow materials is independent of the loading rate. They performed uniaxial compression tests on coarse-grained snow and fine-grained snow to calculate elastic modulus at the test temperature of −10 °C and loading rates from 0.5 to 40 mm/min. Scapozza and Bartelt [15] conducted triaxial compression tests at −12 °C on fine-grained dry snow and found the same trend. On the other hand, Kry [35] performed uniaxial compression tests on alpine snow samples with different densities and strain rates (250~450 kg/m^3; 1×10^{-4}~2×10^{-3} s^{-1}). A weak correlation between normalized Young's modulus and strain rate was found, indicating that the elastic modulus of snow increases with strain rate. This result was attributed to the relationship between the elastic modulus of ice and the strain rate. In conclusion, this discrepancy may be attributed to differences in the microstructure of the snow samples, as well as variations in test conditions and methods. To compare and verify these results, standardized test methods for evaluating the viscoelastic properties of snow are required.

Figure 8 compares the elastic modulus versus density for this and other papers [31,37–40]. The results show that the elastic modulus of data F is in the same trend as data B and data D in exponential increase with increasing density. It is worth noting that data B, C, E, and F represent measurements obtained via uniaxial compression test under quasi-static conditions, where the values are relatively concentrated and range from 0.2 to 200 MPa. According to this trend, the value of data F is smaller than that of the other three data, which may be because step loading includes uniaxial compression and uniaxial creep tests. During the creep test, the stress remains unchanged, and there is a certain time for the rearrangement of the snow grains to achieve a relatively dense spatial structure. Compared with the normal uniaxial compression test, greater deformation is generated under the same stress, so that the obtained elastic modulus is smaller. The data A from [37] represent the dynamic modulus obtained by uniaxial compression tests at high strain rates (3×10^{-3}~2×10^{-2} s^{-1}), which is much higher than the elastic modulus value of other data. Köchle and Schneebeli [40] utilize X-ray microcomputer tomography to construct the three-dimensional microstructure of snow and calculate the elastic modulus D using finite element simulation, ranging from 0.3 to 1000 MPa. This value is also higher than the elastic modulus obtained using quasi-static measurements at the same density. This is because the quasi-static measurement of the elastic modulus inevitably takes the viscous strain into account, which leads to the measured elastic modulus not representing the true value.

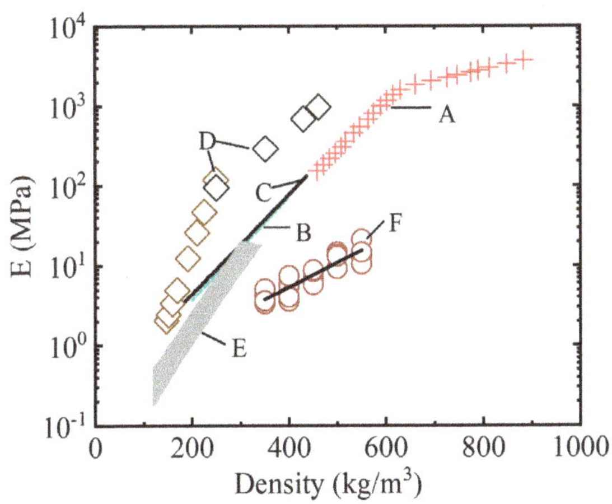

Figure 8. The relationship between the elastic modulus and density from different studies: (A) Uniaxial compression, −25 °C [37]; (B) triaxial compression, −20 °C < T < −2 °C [38]; (C) triaxial compression, −12 °C [41], published in [39]; (D) finite element simulation, the data fit curve [40]; (E) uniaxial compression and tension, −25 °C < T < −12 °C [31]; (F) uniaxial compression under step loading, −5 °C (in this study).

5.2. Viscosity Coefficient

Observations of the relationship between the viscosity coefficient and the density of snow showed an exponential growth trend, which is consistent with the results of other studies. For example, Kojima [42] conducted a quasi-static creep test on natural snow with densities ranging from 150 to 350 kg/m^3 at −7 °C to −10 °C and found a positive correlation between the viscosity coefficient and density. It was concluded that the viscosity coefficient gradually increased with increasing density through triaxial compression tests and uniaxial compression creep tests from [15,35].

The results obtained in this study show an exponential relationship between the elastic modulus and the density of natural snow. Linear regression analysis was used to approximately quantify whether relations existed between elastic modulus and loading rate; however, no tendency for dependence was observed. The functional dependency of the moduli toward temperature was also not observed. Similarly, the viscosity coefficient also increases exponentially with increasing density, and first decreases and then increases with decreasing temperature, but its relationship with loading rate is not clear. Based on these findings, we use the Maxwell model as a framework to present a constitutive equation that take into account the density.

In order to increase and improve the reliability of the constitutive equation, field tests of snow using uniaxial compression equipment may be conducted in future studies. Some field penetration tests have become popular in recent years. Zhao et al. [43] used an Improved Motor-driven Snow Penetrometer to measure the hardness of seasonal snow in Northeast China. The influence of different variables on snow hardness was analyzed using orthogonal tests. Zhuang et al. [44] conducted 74 penetration experiments on seasonal snow in Harbin, China, using a modified Rammsonde. The penetration strength of snow was determined, and its influencing factors were discussed. However, the hardness and the compressive strength obtained from the compression test are two different mechanical parameters and cannot be directly compared.

Snow is made up of countless ice crystals with irregular shapes and sizes. The mechanical properties of snow and ice are similar, but there are also some differences. The elastic modulus of ice increases with decreasing temperature [45]. However, in the present study, the decreasing and then increasing elastic modulus of snow with decreasing temperature was observed only at temperatures greater than 500 kg/m^3. Additionally, the tendency of the elastic modulus to change with density is only indirectly affected at higher temperatures (-5 °C and -10 °C). The effect of temperature on the elastic modulus of snow is relatively complex [46]. Furthermore, the test also shows that the viscosity coefficient of snow increases exponentially with increasing density, first decreasing and then increasing with temperature. Limited by the test conditions, the variable range set in this test was relatively small. For example, the temperature range was only -5~-20 °C, the density range was only 350 kg/m^3~550 kg/m^3, and the loading rate range was only 5~50 N/s. Although the results showed that the elastic modulus and viscosity coefficient varied with temperature and density. A reasonable physical model could not be built to describe the response of elastic modulus and viscosity coefficient to different influencing factors. Therefore, the subsequent temperature and density range should be expanded in an effort to explore its physical phenomenon, which is beneficial for the application of snow in different aspects, for example, in the construction of snow roads in cold regions, runways, and polar infrastructure. The constitutive equations are used to describe the deformation behavior of snow under different loading conditions, which can provide important mechanical parameters and technical support for the design, construction, and maintenance of these snow-related projects.

6. Conclusions

In order to investigate the relationship between the elastic modulus and viscosity coefficient of snow with temperature, density, and loading rate, uniaxial compression tests under step loading were adopted in this study. The constitutive equations of snow were further derived by combining the Maxwell model and the functional relationship between viscoelastic properties and different factors. For practical engineering problems represented by polar airport runways, this study can provide some valuable information for their design, construction, and building. Specifically, the major findings in this study are as follows:

1. The elastic modulus of natural snow increases exponentially with increasing density. Temperature has a certain influence on the elastic modulus of snow. The elastic modulus first decreases and then increases with decreasing temperature, and this relationship is more obvious at high densities. There is no correlation between the elastic modulus and the loading rate.
2. Density is an important factor in the change in the viscosity coefficient of snow. The viscosity coefficient increases exponentially as density increases. The snow viscosity coefficient is affected by temperature, that is, the viscosity coefficient first decreases and then increases as the temperature decreases. The loading rate is weakly correlated with the viscosity coefficient of snow.

3. A new constitutive equation considering snow density is derived by introducing the functional relation between elastic modulus, viscosity coefficient, and the density of snow based on the Maxwell model.

Author Contributions: Conceptualization, H.H. and E.W.; methodology, H.H. and X.L.; formal analysis, H.H. and E.W.; investigation, M.Y., X.L., Y.L. and G.G.; data curation, M.Y., X.L., Y.L. and G.G.; writing—original draft preparation, H.H. and M.Y.; writing—review and editing, X.L. and E.W.; funding acquisition, H.H. All authors have read and agreed to the published version of the manuscript.

Funding: This research was supported by the Natural Science Foundation of Heilongjiang Province of China (No. LH2020E004), the Major Scientific and Technological Projects of the Ministry of Water Resources of China (No. SKS-2022017), and the Project to Support the Development of Young Talent by Northeast Agricultural University.

Data Availability Statement: The data are available upon request.

Conflicts of Interest: The authors declare no conflict of interest.

References

1. Křeček, J.; Šedivá, L.; Palán, L.; Stuchlík, E. Environmental role of snowmelt in headwaters affected by atmospheric acid deposition. *Water* **2023**, *15*, 2458. [CrossRef]
2. Armstrong, R.L.; Brodzik, M.J. Recent northern hemisphere snow extent: A comparison of data derived from visible and microwave satellite sensors. *Geophys. Res. Lett.* **2001**, *28*, 3673–3676. [CrossRef]
3. Parkinson, C.L. Earth's cryosphere: Current state and recent changes. *Annu. Rev. Environ. Resour.* **2006**, *31*, 33–60. [CrossRef]
4. Le Roux, E.; Evin, G.; Eckert, N.; Blanchet, J.; Morin, S. Elevation-dependent trends in extreme snowfall in the French Alps from 1959 to 2019. *Cryosphere* **2021**, *15*, 4335–4356. [CrossRef]
5. Huning, L.S.; Margulis, S.A. Climatology of seasonal snowfall accumulation across the Sierra Nevada (USA): Accumulation rates, distributions, and variability. *Water Resour. Res.* **2017**, *53*, 6033–6049. [CrossRef]
6. Nouri, M.; Homaee, M. Spatiotemporal changes of snow metrics in mountainous data-scarce areas using reanalyses. *J. Hydrol.* **2021**, *603*, 126858. [CrossRef]
7. Wang, L.; Yang, H. Tibetan Plateau increases the snowfall in southern China. *Sci. Rep.* **2023**, *13*, 12796. [CrossRef]
8. Markov, A.; Polyakov, S.; Sun, B.; Lukin, V.; Popov, S.; Yang, H.; Zhang, T.; Cui, X.; Guo, J.; Cui, P.; et al. The conditions of the formation and existence of "Blue Ice Areas" in the ice flow transition region from the Antarctic ice sheet to the Amery Ice Shelf in the Larsemann Hills area. *Polar Sci.* **2019**, *22*, 100478. [CrossRef]
9. White, G.; McCallum, A. Review of ice and snow runway pavements. *Int. J. Pavement Res. Technol.* **2017**, *11*, 311–320. [CrossRef]
10. Sun, B.; Tang, X.; Xiao, E.; Shi, X.; Cheng, X.; Li, L.; Wei, F.; Zhang, T. Ice and snow runway engineering in the Antarctica: Current status and prospect. *Strateg. Study Chin. Acad. Eng.* **2021**, *23*, 161–168. [CrossRef]
11. Li, H.; Zhong, X.; Zheng, L.; Hao, X.; Wang, J.; Zhang, J. Classification of snow cover persistence across China. *Water* **2022**, *14*, 933. [CrossRef]
12. Kabore, B.W.; Peters, B.; Michael, M.; Nicot, F. A discrete element framework for modeling the mechanical behaviour of snow—Part I: Mechanical behaviour and numerical model. *Granul. Matter* **2021**, *23*, 42. [CrossRef]
13. Colbeck, S.; Parssinen, N. Regelation and the deformation of wet Snow. *J. Glaciol.* **1978**, *21*, 639–650. [CrossRef]
14. Mishra, A.; Mahajan, P. A constitutive law for snow taking into account the compressibility. *Ann. Glaciol.* **2004**, *38*, 145–149. [CrossRef]
15. Scapozza, C.; Bartelt, P. Triaxial tests on snow at low strain rate. Part II. Constitutive behaviour. *J. Glaciol.* **2003**, *49*, 91–101. [CrossRef]
16. Birkeland, K.W.; van Herwijnen, A.; Reuter, B.; Bergfeld, B. Temporal changes in the mechanical properties of snow related to crack propagation after loading. *Cold Reg. Sci. Technol.* **2019**, *159*, 142–152. [CrossRef]
17. Bobillier, G.; Bergfeld, B.; Dual, J.; Gaume, J.; van Herwijnen, A.; Schweizer, J. Micro-mechanical insights into the dynamics of crack propagation in snow fracture experiments. *Sci. Rep.* **2021**, *11*, 11711. [CrossRef]
18. Nicot, F. Constitutive modelling of snow as a cohesive-granular material. *Granul. Matter* **2004**, *6*, 47–60. [CrossRef]
19. Mahajan, P.; Brown, R.L. A microstructure-based constitutive law for snow. *Ann. Glaciol.* **1993**, *18*, 287–294. [CrossRef]
20. Brown, R.L. A volumetric constitutive law for snow based on a neck growth model. *J. Appl. Phys.* **1980**, *51*, 161–165. [CrossRef]
21. Bobillier, G.; Bergfeld, B.; Capelli, A.; Dual, J.; Gaume, J.; van Herwijnen, A.; Schweizer, J. Micromechanical modeling of snow failure. *Cryosphere* **2020**, *14*, 39–49. [CrossRef]
22. Hagenmuller, P.; Chambon, G.; Naaim, M. Microstructure-based modeling of snow mechanics: A discrete element approach. *Cryosphere* **2015**, *9*, 1969–1982. [CrossRef]
23. Herny, C.; Hagenmuller, P.; Chambon, G.; Peinke, I.; Roulle, J. Microstructure-based modelling of snow mechanics: Experimental evaluation on the cone penetration test. *Cryosphere Discuss* **2023**. *in review*. [CrossRef]

24. Mede, T.; Chambon, G.; Nicot, F.; Hagenmuller, P. Micromechanical investigation of snow failure under mixed-mode loading. *Int. J. Solids Struct.* **2020**, *199*, 95–108. [CrossRef]
25. Singh, A.K.; Srivastava, P.K.; Kumar, N.; Mahajan, P. A fabric tensor based small strain constitutive law for the elastoplastic behavior of snow. *Mech. Mater.* **2022**, *165*, 104182. [CrossRef]
26. Chandel, C.; Srivastava, P.K.; Mahajan, P. Micromechanical analysis of deformation of snow using X-ray tomography. *Cold Reg. Sci. Technol.* **2014**, *101*, 14–23. [CrossRef]
27. Meyer, C.R.; Keegan, K.M.; Baker, I.; Hawley, R.L. A model for French-press experiments of dry snow compaction. *Cryosphere* **2020**, *14*, 1449–1458. [CrossRef]
28. Shapiro, L.H.; Johnson, J.B.; Sturm, M.; Blaisdell, G.L. Snow mechanics review of the state of knowledge and applications. In *US Army Cold Regions Research and Engineering Laboratory (CRREL) Report 97-3*; Cold Regions Research and Engineering Laboratory: Hanover, NH, USA, 1997.
29. Masterson, D.; Løset, S. ISO 19906: Bearing capacity of ice and ice roads. In Proceedings of the 21st International Conference on Port and Ocean Engineering under Arctic Conditions, Montreal, QC, Canada, 10–14 July 2011.
30. Wang, E.; Fu, X.; Han, H.; Liu, X.; Xiao, Y.; Leng, Y. Study on the mechanical properties of compacted snow under uniaxial compression and analysis of influencing factors. *Cold Reg. Sci. Technol.* **2021**, *182*, 103215. [CrossRef]
31. Mellor, M. A review of basic snow mechanics. In Proceedings of the Snow Mechanics Symposium, Grindelwald, Switzerland, 1–5 April 1974; pp. 251–291.
32. Reuter, B.; Proksch, M.; Löwe, H.; van Herwijnen, A.; Schweizer, J. Comparing measurements of snow mechanical properties relevant for slab avalanche release. *J. Glaciol.* **2018**, *65*, 55–67. [CrossRef]
33. Gerling, B.; Löwe, H.; van Herwijnen, A. Measuring the elastic modulus of snow. *Geophys. Res. Lett.* **2017**, *44*, 11088–11096. [CrossRef]
34. Johnson, J.B.; Schneebeli, M. Characterizing the microstructural and micromechanical properties of snow. *Cold Reg. Sci. Technol.* **1999**, *30*, 91–100. [CrossRef]
35. Kry, P.R. The relationship between the visco-elastic and structural properties of fine-grained snow. *J. Glaciol.* **1975**, *14*, 479–500. [CrossRef]
36. Lintzén, N.; Edeskär, T. Uniaxial strength and deformation properties of machine-made snow. *J. Cold Reg. Eng.* **2015**, *29*, 04014020. [CrossRef]
37. Kovacs, A.; Weeks, W.F.; Michitti, F. Variation of some mechanical properties of polar snow, Camp Century, Greenland. In *US Army Cold Regions Research and Engineering Laboratory (CRREL) Report 276*; Cold Regions Research and Engineering Laboratory: Hanover, NH, USA, 1969.
38. Scapozza, C. Entwicklung eines dichte- und temperaturabhängigen Stoffgesetzes zur Beschreibung des visko-elastischen verhaltens von Schnee. Ph.D. Thesis, Eidgenössische Technische Hochschule ETH Zürich, Zürich, Switzerland, 2004. (In German). [CrossRef]
39. von Moos, M.; Bartelt, P.; Zweidler, A.; Bleiker, E. Triaxial tests on snow at low strain rate. Part I. Experimental device. *J. Glaciol.* **2003**, *49*, 81–90. [CrossRef]
40. Stoffel, M.; Bartelt, P. Modelling snow slab release using a temperature-dependent viscoelastic finite element model with weak layers. *Surv. Geophys.* **2003**, *24*, 417–430. [CrossRef]
41. Köchle, B.; Schneebeli, M. Three-dimensional microstructure and numerical calculation of elastic properties of alpine snow with a focus on weak layers. *J. Glaciol.* **2014**, *60*, 705–713. [CrossRef]
42. Kojima, K. Visco-elastic property of snow. *Low Temp. Sci.* **1954**, *12*, 1–15. (In Japanese)
43. Zhao, Q.; Li, Z.; Lu, P.; Wang, Q.; Wei, J.; Hu, S.; Yang, H. An investigation of the influence on compacted snow hardness by density, temperature and punch head velocity. *Water* **2023**, *15*, 2897. [CrossRef]
44. Zhuang, F.; Lu, P.; Li, Z.J.; Han, H.W.; Li, W. Modified Rammsonde tests in layered compacted snow. *Adv. Polar Sci.* **2019**, *30*, 118–131. [CrossRef]
45. Sinha, N.K. Elasticity of natural types of polycrystalline ice. *Cold Reg. Sci. Technol.* **1989**, *17*, 127–135. [CrossRef]
46. Shenvi, M.N.; Sandu, C.; Untaroiu, C. Review of compressed snow mechanics: Testing methods. *J. Terramech.* **2022**, *100*, 25–37. [CrossRef]

water

MDPI

Article

Observations of Snow–Slush–Snow Ice Transformation and Properties of Brash Ice in Ship Channels

Vasiola Zhaka [1,*], Robert Bridges [2], Kaj Riska [3], Jonny Nilimaa [1] and Andrzej Cwirzen [1]

[1] Department of Civil, Environmental and Natural Resources Engineering, Structural and Fire Engineering, Building Materials, Luleå University of Technology, 97187 Luleå, Sweden; jonny.nilimaa@ltu.se (J.N.); andrzej.cwirzen@ltu.se (A.C.)
[2] TotalEnergies SE, 92400 Paris, France; robert.bridges@totalenergies.com
[3] Formerly TOTAL SA, 92400 Paris, France
* Correspondence: vasiola.zhaka@ltu.se or vzhaka@yahoo.com

Abstract: The thickness and properties of brash ice are usually compared with the properties of the surrounding level ice. The differences between these ice types are important to understand since the consolidated brash ice layer is typically assumed to have the same properties as level ice. Therefore, significant effort in the measurement campaign during the winters of 2020–2021, 2021–2022, and 2023 was made to develop a better understanding of the full-scale brash ice channel development. The channels were located near the shore in the Bay of Bothnia, Luleå, Sweden. The main parameters investigated were the snow, slush, and total ice thicknesses, including ice formed from freezing water and from freezing slush as well as the ice microstructure and strength. To our knowledge, this is the first paper to report the influence of snow in brash ice channels. It was observed that a significant amount of snow covered the brash ice channels between the ship passages. After each ship passage, the snow was submerged and formed slush-filled voids, which thereafter transformed into snow ice (SI) clusters frozen together with columnar ice. The SI content in the brash ice and side ridges was estimated from image analyses. The analyses showed that the snow ice content was 73% in level ice in the vicinity of the ship channel, 58% in the side ridges of the channel, and 21% in the middle of the test channel, whereas in the main channel, the SI contents were 54%, 43%, and 41% in each location, respectively.

Keywords: brash ice; side ridges; snow ice; ship channels; compressive strength

Citation: Zhaka, V.; Bridges, R.; Riska, K.; Nilimaa, J.; Cwirzen, A. Observations of Snow–Slush–Snow Ice Transformation and Properties of Brash Ice in Ship Channels. *Water* **2023**, *15*, 2360. https://doi.org/10.3390/w15132360

Academic Editors: Zhijun Li, Fang Li, Sasan Tavakoli, Xuemei Liu and Changlei Dai

Received: 16 May 2023
Revised: 19 June 2023
Accepted: 19 June 2023
Published: 26 June 2023

1. Introduction

Navigation in fast ice usually occurs in the same ship track by ice breakers and vessels without ice-breaking capabilities [1]. Frequently navigated ship channels are filled with broken ice pieces called brash ice. During navigation, fragments of the brash ice pieces are expelled sideways to form piles of broken ice under the level ice on both sides of a channel, and these side piles are typically called side ridges. The shapes, sizes, and total ice volumes of the broken ice depend on meteorological parameters such as freezing cumulative air temperatures [2–4], the frequencies and speeds of navigation, vessel geometries, and ice strengths [5,6].

The physical and mechanical brash ice properties are often presumed to be similar to level ice properties, for example, when estimating the growth of brash ice (e.g., density) or when simulating a vessel's performance in broken or consolidated brash ice (e.g., ice strength). These assumptions may not be accurate for all situations, and only a few studies have reported full-scale measurements on brash ice properties, e.g., [7,8]. In addition to the most studied effects such as air temperature and frequency of navigation, other effects, including radiation or snowfall, may also influence the brash ice development in ship channels [9,10].

The incoming snow has a dual effect on the ice growth. Firstly, snow has low conductivity and insulates the surface of ice from the atmosphere, thus decreasing the ice growth rate [11]. Secondly, snow transforms to slush when submerged, which freezes to snow–ice [12–14].

Natural flooding of the ice's surface occurs when two main conditions are met. Firstly, the snow mass exceeds the buoyancy of the ice, and, secondly, thermal or mechanical cracks are present [15–17]. During flooding, the bottom of the snow submerges below the water level, forming a fully saturated slush layer. Capillary pressure will force some water to rise in the snow pores and form a partly saturated slush layer above the water level [18–21]. Slush is a mix of fresh or saline water, snow melt, and snow crystals, which have water- and air-filled pores. Slush freezes into snow ice, which has granular equiaxed crystals on a random direction of the c-axis [22].

There are some results on natural snow–slush–snow ice phase changes in the case of level ice [23,24], and, to our knowledge, the effect of snow on deformed ice, particularly in brash ice formed in frequently navigated ship channels, has not been previously addressed [8]. In comparison to fast sea ice, river and lake ice is not always exposed to mechanical deformation; the level ice adjacent to ship channels is exposed to frequent mechanical breaking during navigation. Thus, the cracks formed after each passage can enhance the flooding of adjacent level ice and increase the thickness of the level ice.

The main objective of this study was to develop a better understanding and investigate the influence of incoming snow on the properties of level ice adjacent to ship channels, as well as its contribution to the formation of brash ice and side ridges. This is a descriptive article that aims to further the insight into the processes related to brash ice formation and growth in ship channels, as well as to better understand the impact of navigation on the surrounding level ice. To our knowledge, this is the first paper to report the influence of snow ice in brash ice channels. Different research activities, including field observations and measurements in four different channels, were conducted during three consecutive winters in the fast sea ice in the Bay of Bothnia. These activities provided insight into different aspects of brash ice formation and development, as well as into snow–slush–snow ice transformation phenomena in level ice and brash ice channels. Image analyses of brash ice, ridge ice, and level ice microstructures were carried out to investigate and determine their properties and differences in snow–ice contents. The microstructures, microporosities, and strengths of the ice from the channel and level ice were analyzed and compared. The following chapter details the field investigations, laboratory measurements, and analytical methods applied. The results and analysis are divided into two chapters. The first one (Section 3) discusses the snow–slush–snow ice transformation process in level ice adjacent to the ship channels, and the second chapter (Section 4) investigates and discusses the snow contribution in the ship channels. Finally, the main results and impacts of the paper are summarized in the conclusions.

2. Measurements and Methods

The following subsections present the research location, the in situ research activities, and the laboratory as well as analytical methods that were applied to estimate the snow ice content for different ice types.

2.1. Study Site

This research was carried out in the Bay of Bothnia in the Swedish coastal fast ice between the Icebreaker and Luleå ports, Luleå, Sweden. The Bay of Bothnia consists of brackish water, and, in this particular location, the ice salinity measured in the winter season was zero due to the Luleå River discharge. Luleå Airport's weather station is located approximately 3 km from the research site; see Figure 1.

A channel was created and maintained by Tug Viscaria to be used exclusively for research purposes for two consecutive winters: January–March 2021 and December 2021–March 2022. These channels were located 200 m from the shoreline and are referred to throughout the paper as the test channels TCh01 and TCh02. Another frequently navigated channel, which is established yearly and used by icebreakers and merchant vessels, was located 400 m from the shore and 200 m apart from the test channels. We refer to it as the main channel (MCh), and the brash ice properties from the main channel were studied during 2021 and 2023. Meteorological data, including the air temperature, shortwave radiation, precipitation, snowfall, humidity, and wind, were continuously recorded by the meteorological weather station of SMHI (Swedish Meteorological and Hydrological Institute) in Luleå Airport. The air temperatures and snow thicknesses for the winters 2020–2021, 2021–2022, and 2022–2023 are presented in Figure 2.

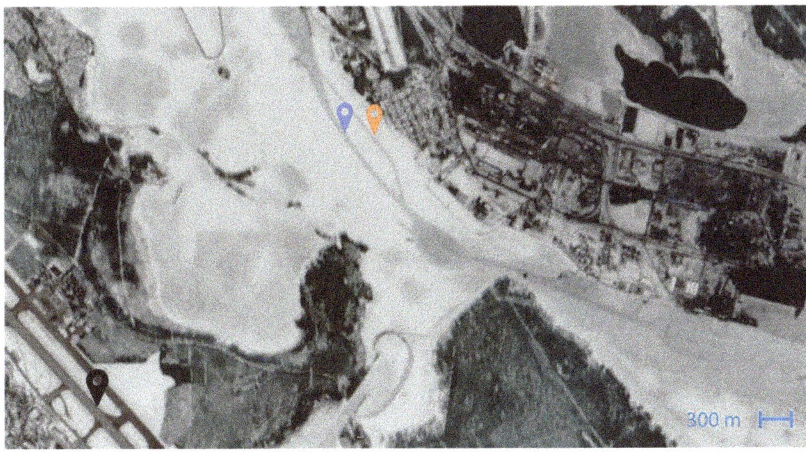

Figure 1. The locations of the research site. The test (TCh) and main channels (MCh) are shown with orange and blue location markers, and the SMHI meteorological station at Luleå Airport is shown with a black location marker. Modified optical (Sentinel) satellite image.

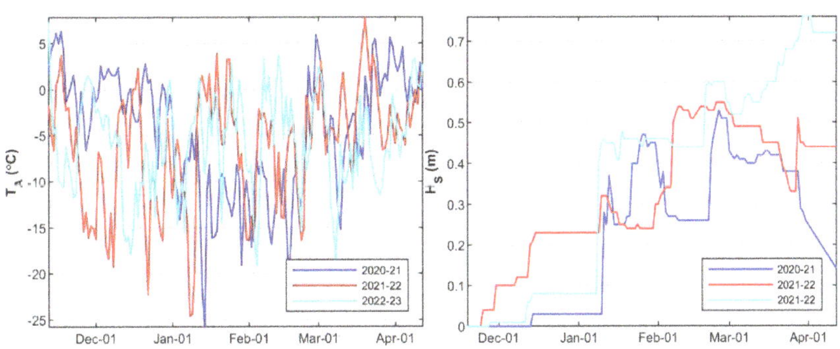

Figure 2. Air temperatures (T_A in °C), and snow thicknesses (H_S in m) were recorded at the SMHIs weather station in Luleå.

2.2. Field Site

The field study was divided into four main activities: (1) observations of the test channels' development instantly after each ship passage through images and video recordings; (2) cross-section measurements of the test channels (TCh01 and TCh02); (3) thickness measurements of the unbroken level ice in the shoreside of the test channels; and (4) mechanical and material property analyses of cores sampled from level ice, brash ice, and side ridges from the first test channel; and the main channels during 2021 and 2023 were investigated in LTU's cold laboratory.

2.2.1. Observation in the Test Channel

The first field activity included brash ice surface and underwater morphology observations after each ship passage in test channels. A GoPro camera was used for underwater observations and recordings. These observations provided insights into the different factors that influence brash ice formation and growth.

2.2.2. Brash Ice Thickness Measurements

The thicknesses of brash ice, side ridges, and level ice up to a distance of 10 m from the channel edge were measured along the cross-section of the channel using a ruler stick with a protrusion in the end. Holes were drilled every 1 m along the cross-section using an auger drill with a diameter of 50 mm. Table 1 summarizes the number of ship passages, dates, and times when the cross-section measurements were carried out. This article discusses only the results of level ice measurements from the vicinity of TCh01. The cross-section results are further analyzed and will be reported in two companion articles. The first article will discuss the brash ice macroporosities and piece size distributions from three brash ice channels located in Luleå. The second article will validate various models for brash ice growth.

Table 1. The date and time of the breaking events and field measurement.

BE	BE Date	BE Time	Cross-Section Measurements	Level Ice Measurements
			Test Chanel #01	
1	2021-01-14	23:00	2021-01-23/26/30	-
2	2021-02-01	10:00	-	-
3	2021-02-02	07:00	-	-
4	2021-02-03	12:00	2021-02-06/08	-
5	2021-02-10	21:00	-	-
6	2021-02-11	11:00	2021-02-13/15/20	-
7	2021-02-21	10:00	2021-02-22/24	2021-02-27
8	2021-03-03	23:00	2021-03-05/06/09/10	-
9	2021-03-12	20:30	2021-03-15/17/18	2021-03/04-13/23/27/01/10
			Test Chanel #02	
1	2021-12-10	21:30	-	2021-12-12/18
2	2021-12-21	17:30	2021-12-23/27	-
3	2021-12-27	16:30	-	-
4	2022-01-05	05:00	2022-01-11	-
5	2022-01-12	10:30	2022-01-16	2022-01-15
6	2022-01-20	00:00	2022-01-22	-
7	2022-01-28	08:30	2022-02-01	2022-01/02-29/05
8	2022-02-06	09:45	2022-02-16	2022-02-12/19
9	2022-02-19	18:00	2022-02-23	-
10	2022-03-06	05:00	2022-03-09	2022-03/04-06/12/19/03/10

2.2.3. Level Ice Measurements

The thicknesses of the level ice on the shoreside of channels TCh01 and TCh02 were measured in two consecutive winters: February–April 2021 and December 2021–April 2022; see Table 1. During the first winter, the level ice thickness measurements were placed in a grid of 30 m along the channel's length and 20 m towards the shore, starting from a minimum distance of 5 m from the channel edge. Approximately 15 cores were drilled each time, as shown in Figure 3. In the second winter, the measurements were carried out in a grid of 60 m along the channel's length and 40 m towards the shore. Twelve cores were drilled each time.

On-site measurements included the snow thickness, followed by drilling an ice core of 200 mm in diameter. The snow ice, congelation ice, and total level ice thicknesses were investigated and recorded for each core. In addition, the freeboard, slush, and ice thicknesses were measured from all drilled holes. The thickness measurements were carried out with a millimeter-scale ruler. The measurement technique's error was below 5 mm, whereas discrepancies could be present due to spatial thickness variations in slush, snow, or ice. This field study aimed to detect any possible slush formation due to flooding or melting and its subsequent freezing to snow ice.

Figure 3. The study location for the investigation level ice was adjacent to the test channel (TCh01 27 March 2021), where cores were examined from a test grid, starting approximately 5 m from the channel edge, and extending up to 30 m towards the shoreline. Cores were also collected along the channel direction for a distance of 30 m. The spacing between cores was 5 m.

2.2.4. Brash Ice and Level Ice Cores

The final group of activities involved sampling ice cores to investigate and analyze the fractions of snow ice and congelation ice. Two brash ice cores were sampled from TCh01 on the 8 February 2021, and two side ridge cores were sampled on the 23 March 2021. On the 6 March 2021, five cores were sampled at different locations along half of the cross-section of TCh01. The first core was drilled in the middle of the channel, the second was drilled between the mid-channel and the edge, the third core was drilled at the edge of the channel, and the fourth was drilled in the ridge between the edge and the level ice. The fifth core was drilled in the level ice near the ship channel.

On the 7 March 2021, nine cores were drilled between the test and main channels, aiming to observe the thickness of snow ice between the two channels. Additionally, cores were sampled along the side ridge of the main channel. Three cores were sampled in the level ice (LI) between the test and main channels, three cores were sampled in the vicinity of the ridge, and the remaining three cores were sampled along the ridge of the main channel. In each position, the samples were spaced 6 m apart from each other and equidistant from the edge of the main channel. Immediately after sampling, holes were drilled through the thickness of each ice core, and a calibrated thermocouple was used to measure the ice temperature. The temperature profiles of these nine cores are shown in Figure 4.

In addition, 10 cores were sampled along half the cross-section of the main channel on the 25th and 26th of February 2023. The cores were transported to Luleå University of Technology and stored in a freezing box with a constant temperature of −20 °C. They were investigated in the cold laboratory during the summer of 2021 and the spring of 2023.

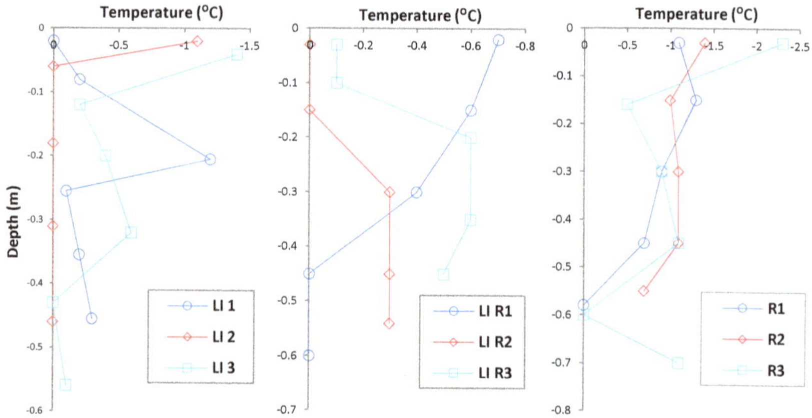

Figure 4. Temperature profiles of the ice cores were sampled on the main channel on 7 March 2021. "LI" indicates level ice samples between the test and main channels; "LI R" indicates the cores from the vicinity of the main channel's ridge; the ridge cores are denoted with "R". Three ice cores were sampled at each location, spaced 6 m from each other (numbers 1–3).

2.3. Methods

This section describes the methods used to determine the ice sample's snow ice content, microporosity, density, and uniaxial compressive strength.

2.3.1. Snow Ice Content

Thin ice sections ranging from 2 to 10 mm were observed under cross-polarized light as well as against dark and bright backgrounds using plain transmitted light. The crystal structure of ice can be identified under cross-polarized light, whereas the pores and pore sizes can be observed against a dark or bright background [25]. A total of 28 cores from the test and main channels were analyzed to determine the quantity of granular ice (snow ice) based on their microstructure. In the case of level ice cores, the thickness ratios between snow−ice and the total core thickness were used to quantify both snow−ice and columnar ice. Snow−ice (SI) resulting from the freezing wet snow (slush) is opaque compared to the transparent ice formed by freezing water [25]. Different terms are used to characterize ice formed from freezing, such as columnar ice, which refers to its crystallinity; congelation ice, which refers to freshwater ice [22]; and clear ice, which refers to the transparency of freshwater ice with low porosity. The notation CI is used for all these terms.

The SI could not be accurately estimated as a ratio of thicknesses in the ridge and brash ice samples due to their mixed characteristics. Instead, thin sections of ice were used to quantify the snow−ice content. Images of thin sections recorded under cross-polarized light, and under dark and bright backgrounds, were analyzed with ImageJ software. The boundaries of the thin ice sections and the boundaries of the snow−ice clusters were manually marked, and their surface areas were examined by the software. The SI content was calculated by determining the ratio between the surface area of the SI and the total surface area of the thin ice section. When analyzing images captured under cross-polarized light, snow ice was identified by its fine granular crystals [22]. On the other hand, for images captured with a dark and bright background, snow ice was distinguished by its fine round air pores [26]. Figure 5 provides an example of image analysis, showing a thin ice section from the channel's middle photographed under plain-transmitted light and cross-polarized light. Trapped round air micropores, as depicted in Figure 5, typically form when slush freezes to snow ice and indicate the presence of SI. Transparent ice with a low pore concentration, as seen in Figure 5, can be defined as CI.

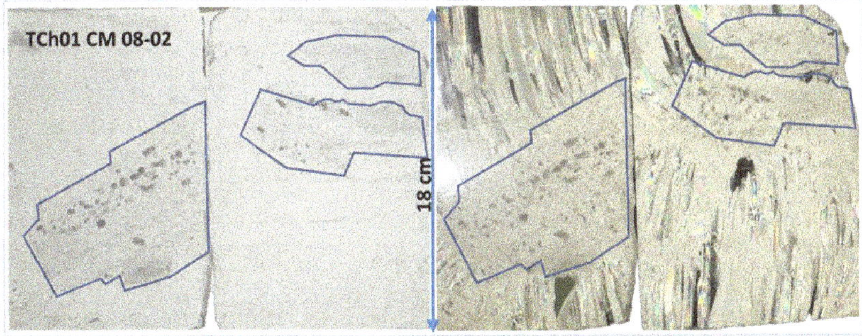

Figure 5. Microstructure of brash ice sampled in the middle of the test channel (TCh01) on 8 February 2021 under plain transmitted light and cross-polarized light. The surface area of snow ice is underlined with a solid blue line.

2.3.2. Physical and Mechanical Properties

Cylindric samples with maximum sizes of 70 mm in diameter and 170 mm in length were horizontally sampled from the 200 mm ice cores. The volume (V) of each cylinder was determined using the measured length and diameter. The densities of the cylindrical samples were calculated using the volume–mass method [27]. The microporosity (p) of each cylinder, with a volume V and a measured mass m_m, was estimated by comparing the measured and the theoretical mass (m_{th}) assuming zero salinity and a density (ϱ_{pi}) of pure ice equal to 917 kg m^{-3} [28].

$$p = 1 - \left(\frac{m_m}{V\varrho_{pi}}\right) \quad (1)$$

A total of 23 cylinders from TCh01 and 22 cylinders from MCh 2021 were subjected to unconfined uniaxial compressive strength tests at nominal strain rates of 10^{-3} s^{-1}. Additionally, 32 cylinders from MCh 2023 were tested at nominal strain rates equal to 10^{-4} s^{-1}. The setup for the unconfined uniaxial compression tests and the calibration of the deformation system were previously described in [8,29]. Immediately after carrying out the compressive strength tests, the temperature of each ice cylinder was measured using a calibrated thermocouple.

3. Level Ice Results and Discussion

Observations and results from the snow–slush–snow ice transformation in level ice adjacent to the test channels are presented and discussed in the following sub-sections.

3.1. Flooding of Level Ice

Earlier studies have indicated that natural flooding does not occur if the ice does not have thermal or mechanical cracks or high porosity, even if the snow cover is thick enough to allow the water to rise at the snow/ice interface [15–17,30]. However, the flooding of the level ice adjacent to ship channels was influenced by both the cracks formed from the ship passages and the water pushed onto the ice during each ship passage. In the first channel (TCh01), instances of flooding were observed after the second, third, fourth, seventh, and eight BEs.

Figure 6 illustrates a set of photographs that clearly demonstrate the formation of slush on the level ice next to TCh02. For example, the first and second images show that a 9 cm snow layer transformed into a slush layer with 3 cm of slush below the water level and the snow thickness significantly decreased after the slush froze into SI. In TCh01, after the second BE, the submerged slush layer was 5 cm, and the slush formed above the water level due to capillarity was 3 cm, resulting in a total slush thickness of 8 cm. Previous laboratory studies have indicated that the capillarity is influenced by the snow permeability, crystal size, and snow density [21,31]. It was found that the capillarity increase was higher for fine pores composed of high-density snow and small round crystals [19].

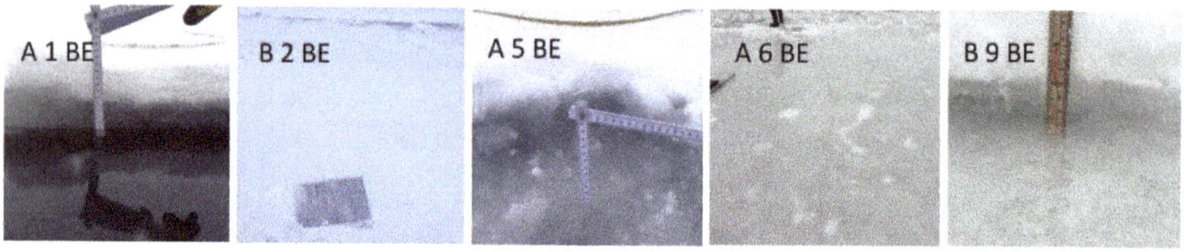

Figure 6. The level ice in the vicinity and adjacent to the channel (TCh02) was flooded after the ship passages. A 1 BE shows the adjacent level ice flooded instantly after the first ship passage, whereas B 2 BE shows the level ice before the second ship passage, with the slush frozen to snow ice.

Figure 7a shows the measurements of the individual slush thicknesses on the level ice near and adjacent to both ship channels (TCh01 and TCh02). This graph aims to illustrate the potential for slush formation and the action of capillarity. The total thickness of slush (HSL) and the thickness of slush above the water level (HSLa) are plotted against the water level on top of the ice (WL), which, in this case, is positive if the ice is flooded and the freeboard (FB) or ice top is negative (underwater). The increase in HSL correlates well with the increase in WL, as shown by the linear regression function. However, there is no correlation between HSLa, formed due to capillarity, and the WL. On average, HSla was 4 cm with a standard deviation of 1.2 cm. The scatter in the data may be attributed to measurement errors, such as measuring the slush thicknesses before reaching equilibrium between WL and HSLa.

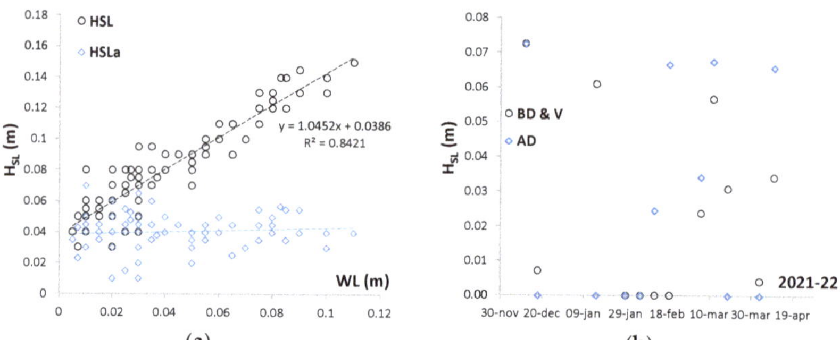

Figure 7. (**a**) The total slush thickness (HSL) and the slush thickness formed above the water line (HSLa) are plotted against the water level (WL). (**b**) The plot shows the slush thickness before drilling or slush thickness in voids (BD and V) and the slush thickness after drilling (AD).

The saturation of snow above the water level differs from the submerged snow and varies depending on the height of the capillarity rise [19]. As a result, the slush below and above the water line has different physical properties. It is also anticipated that the freezing of slush above and below the water line will form snow–ice with different porosities and densities [12].

Figure 7b differentiates between the sum of the slush layer before drilling, including unfrozen slush voids measured during drilling (BD), and the total slush formed after drilling (AD). In TCh02, average slush layer thicknesses of 7 cm, 6 cm, and 6 cm were measured on the adjacent level ice after the first, fifth, and tenth BEs. These slush thickness values specifically refer to the slush layer observed on level ice before drilling or to unfrozen slush voids measured after drilling. After the eighth and ninth BEs, additional slush was formed on level ice after drilling compared to the slush naturally formed before drilling. This suggests that snow mass and ice buoyancy were not in equilibrium, even in the presence of mechanical cracks. One reason could be the rapid freezing of the cracks compared to reaching an equilibrium between the ice buoyancy and snow mass. Additionally, the speed at which flooding progresses spatially at the snow/ice interface may be another factor [20].

The mass balance calculations between snow mass and ice buoyancy identified four flooding occasions in the first winter, which lasted from one to several days; see Figure 8. Two of these flooding events coincided with the measured slush thickness, occurring between the first and second BEs and after the seventh BE. Slush was also measured after the fifth, sixth, and eight BE, even though flooding was not detected in the buoyancy calculations. Seven flooding instances were detected in the second winter. According to the model, the sixth flooding event following the seventh ship passage lasted for 10 days and resulted in an 11 cm-thick slush layer.

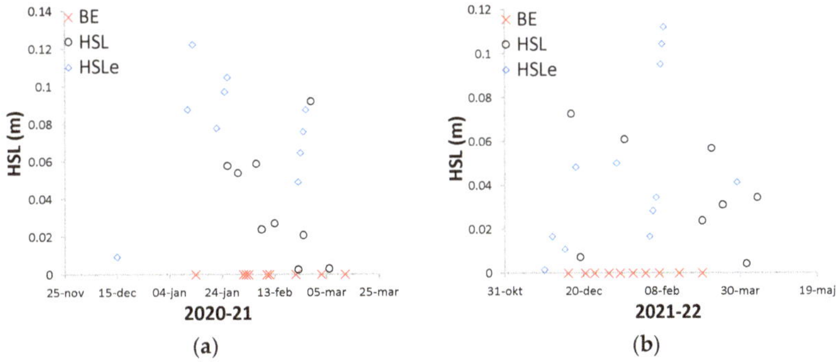

Figure 8. The measured and estimated slush thicknesses were obtained (HSL and HSLe) from the mass balance equation between snow load and ice buoyancy for TCh01 (**a**) and TCh02 (**b**). The time of the breaking events (BE) is illustrated with red crosses.

3.2. *Snow–Slush–Snow Ice Transformation*

The measured snow thicknesses in both winters did not comply with the land snow thickness measured by SMHI, which was assumed to represent the incoming snowfall. Figure 9 illustrates the development of the difference between the measured snow thicknesses on level ice and the land snow thickness ($\Delta H_S = H_{S\,SMHI} - H_{Sm}$) over time. Several factors were observed that could contribute to this difference. Firstly, the varying wind actions on the measurement locations may lead to the redistribution of snow, causing uneven snow thickness compared to the original snowfall position [32]. However, no extreme wind actions such as dune formations were observed in this study location. Secondly, the flooding of the snow/ice interface can result in partial melting of snow due to the water temperature being slightly above the freezing point. Consequently, snowmelt and slush formation can create a snow–ice layer with a reduced thickness compared to the submerged snow thickness. Lastly, superimposed ice can form at the snow/ice interface during spring when the snow melts due to incoming radiation during the day and refreezes at night. Rain precipitation can also form superimposed ice. We observed the melting of 7 cm of snow and the formation of 3.5 cm of superimposed ice after March 20th (TCh01). Previous studies on fast sea ice in Svalbard have reported a lower superimposed ice thickness, such as 23 cm of snow transformed to 6 cm of ice [33]. In the Baltic Sea, 15 cm of snow transformed into 7 cm of ice [34]. This literature result is consistent with our observations.

During the first winter, the SI measurements were taken after snow melting had already occurred, as indicated by a decrease in ΔH_S, as the snow was melting on both land and ice. Snow ice reached its maximum thickness (38 cm) on the 27th of March and thereafter started to melt. In the second winter, the ΔH_S remained higher than the H_{SI} until the melting of snow and SI began after the 3rd of April.

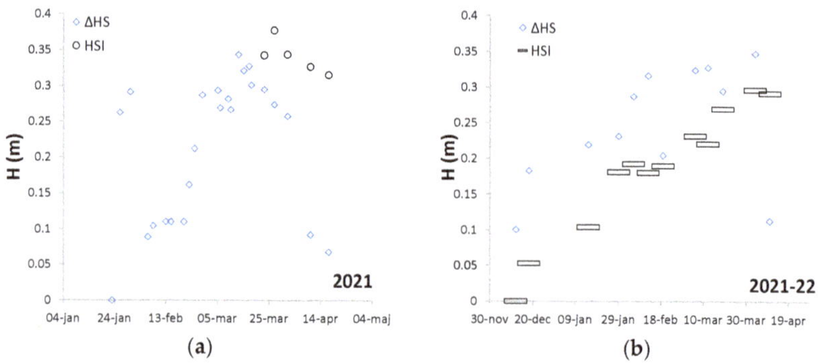

Figure 9. The difference in the snow thickness (ΔH_S) between the incoming snow thickness ($H_{S\,SMHI}$) and measured snow thickness on ice (H_{Sm}) during the first (**a**) and the second winters (**b**).

Figure 10a illustrates the relationship between the snow–ice thickness and the reduction in the snow thickness (ΔH_S). The slope of regression line, 0.09, indicates that from the total snow thickness reduction, about 9 cm of snow did not transform into snow–ice. Figure 10b shows that on average, 70% of ΔH_S and 50% of the incoming snowfall transformed into snow–ice. The remaining part of snow either melted when submerged or was blown away by the wind. However, these processes were not investigated in this study. A previous study on snow–ice formation on lake ice reported similar results [24], whereas another study on lake ice showed that a third of the initial snow layer transformed into snow–ice [23].

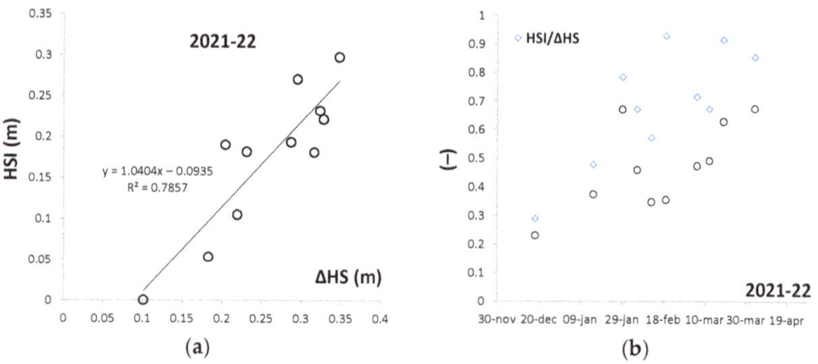

Figure 10. (**a**) The difference (ΔH_S) in the incoming ($H_{S\,SMHI}$) and measured (H_S) snow thicknesses vs. the measured snow ice thickness (H_{SI}); (**b**) the time-dependent ratios of $H_{SI}/\Delta H_S$ and $H_{SI}/H_{S\,SMHI}$.

In the first winter, the measured thickness of SI was twice the CI, with a maximum measured thicknesses of 38 and 19 cm, respectively. The SI fraction ranged from 64 to 75% of the total level ice thickness on 10 April 2021, as shown in Figure 11a. In the second winter, the SI fraction reached 50% of the total level ice thickness on April 3rd, as shown in Figure 11b. Previous studies on snow ice formation in lakes in Finland have reported SI content ranging between 10 and 43% during the winters of 1993–1999 [23], and between 30 and 50% in the winters of 2009–2012 [16].

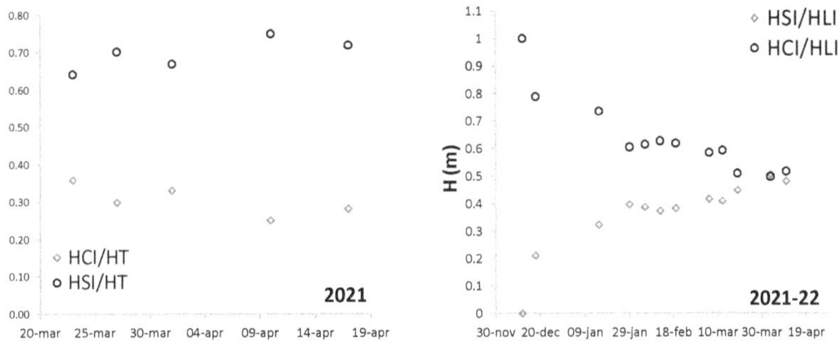

(**a**) (**b**)

Figure 11. Congelation ice (CI) and snow ice (SI) fractions of total level ice adjacent to the ship channels TCh01 (**a**) and TCh02 (**b**).

4. Brash Ice Results and Discussion

In this section, we present and analyze a series of photographs that exemplify the snow's influence on the formation of the brash ice channel. Furthermore, we examine the snow ice content, the brash ice growth history, and physical and mechanical properties associated with it.

4.1. *Snow–Slush–Snow Ice Transformation*

A series of photographs from TCh02 are used to illustrate and discuss the development and transition in the test channels throughout an increasing number of breaking events, as shown in Figure 12. Various snow-related phenomena were observed, including snow accumulation between ship passages, snow submergence between ice pieces, and snow compression at the edges of the channel. The average snow thickness on the cross-section of the channel (H_S BI), excluding the snow accumulated on the side ridges, the snow accumulated on level ice in the vicinity of the test channel (HS LIv), and the average snow thickness on the adjacent level ice (H_S LIa), were measured from one to several days before any breaking event. The results of these measurements are summarized in Table 2.

Table 2. The date and time of the breaking events and field measurement.

	Test Chanel #01		
Nr. BE	H_S BI (cm)	H_S SR (cm)	H_S LIv (cm)
2	14	16	16
5	1	16	15
7	14	18	16
8	29	34	29
9	5	16	9
	Test Chanel #02		
2	-		-
3	4	6	7
5	5	11	9
6	0	9	6
7	0	1	1
8	5	5	5
9	14	18	18
10	6	28	23

In the first channel (TCh01), the average thickness of snow submerged in the channel during any breaking event varied from 1 to 29 cm. Despite receiving a significant amount of snowfall after the seventh BE, as much as 29 cm, the surface of the channel after the eighth BE was filled with both slush and open water pools. During the eighth ship passage, the slush appeared to be in a lower content compared to the slush observed after the seventh or ninth BEs. This reduction in the slush content was likely due to melting caused by air temperatures above 0 °C.

In TCh02, the first breaking event occurred on the 10th of December. After the ship's passage, the channel was predominantly filled with slush and water and the broken ice pieces were not visible on the surface of the channel, as shown in Figure 12. The channel was snow covered before the third, fifth, eighth, ninth, and tenth BEs. After the eighth BE, there was a significant accumulation of snow that acted as insulation and slowed the freezing of the channel, despite low air temperatures.

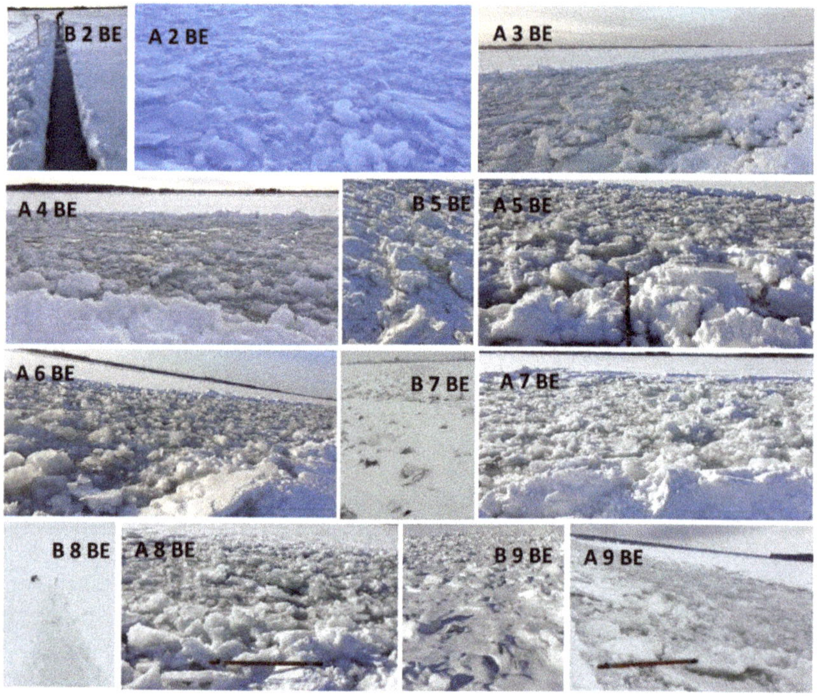

Figure 12. An overview of the ship channel (TCh02) development with breaking events (BE). The notation "B 2 BE" implies before the second breaking event and "A 1 BE" refers to after the first breaking event.

4.2. Snow-Ice in the Ship Channels

A series of photographs captured from the surface and underwater in the ship channels after different breaking events are illustrated in Figures 13 and 14. These images demonstrate that the brash ice pieces within the channel and along its edges consisted of both congelation ice and snow ice. Additionally, the images reveal submerged slush and snow clusters that were partly submerged and partly above the water line. This indicates the role of snow in the ice formation, as it transforms into slush and subsequently freezes to snow–ice. Similar types of snow–ice clusters were previously observed in a brash ice channel with similar characteristics and specifications [35]. The SI pieces merged into a cohesive layer of brash ice as water froze in between. For distances of several centimeters between the brash ice pieces the ice grew in a lateral direction, see Figure 13c. This lateral growth is likely initially driven by the internal heat stored within the ice pieces, as previously hypothesized for the pressure ridge consolidation [36,37]. The earlier models of brash ice growth neglected this lateral growth [9,38,39].

Figure 13. A set of photographs illustrating snow–ice (SI), congelation ice (CI), and slush on the surface of the first channel (TCh01). The orange arrow highlights the SI, and the blue arrow highlights the CI. In the first and second figures (**a**,**b**) pools of submerged slush between pieces are present, whereas lateral growth is also evident in the last image (**c**).

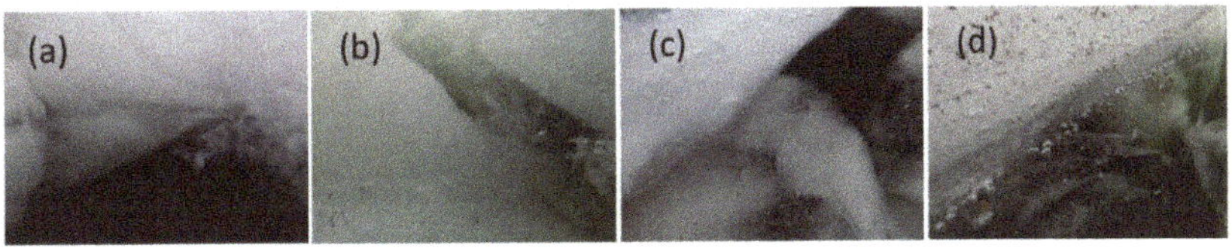

Figure 14. Sub-surface images were recorded at 30 cm depth in the side ridge of TCh01 after the sixth BE (**a**,**b**); after the ninth BE (**c**,**d**).

Figure 14 displays a series of underwater photographs captured on the edge of TCh01. The figures show that instantly after the submergence, the original ice blocks can consist of CI and SI. However, as CI grows around the original ice blocks, it becomes challenging to discern the interface between SI and CI from underwater images. Clusters of submerged snow ice frozen together with CI are evident in Figure 14a,c. Figure 14a illustrates ice crystal clusters in the form of slush, which were also previously observed and reported by [35]. Snow ice originating from submerged level ice is given in Figure 14b,d.

4.3. The History of Brash Ice Formation

The microstructure of ice depends on the ice development process and can indicate different formation phenomena. In the current section, the brash ice microstructure is analyzed and the history of the brash ice formation is discussed.

4.3.1. Ship Channel (TCh01)

Figure 15 presents the microstructure of two cores sampled from the side ridge of the first channel (TCh01) denoted as R1 and R2. The top of the first core consists of approximately 26 cm of snow−ice. The snow−ice having different layers distinguished by the transitional zones may indicate various stages of snow−ice formation. The SI on the CI of the first core may have been formed during four separate flooding occasions, as indicated by the buoyancy estimations detecting four flooding occasions. In addition, two SI clusters were mixed with CI. In Figure 15, the SI layers or clusters are highlighted with orange arrows, and the spaces between them represent either a transition zone or thicker layers of CI. The second core initially consisted of SI, followed by CI with horizontal growth directions, which is common growth pattern observed in lake ice [40]. Similar big columnar crystals were earlier observed in a similar brash ice channel type and may be attributed to a slow growth rate [35].

Figure 16 shows the cross-section profile of TCh01 on 6 March 2021, along with the microstructure of three ice cores sampled on the same day. TCh01 (R) was sampled at the side ridge, TCh01 (M−E) was sampled between the mid-channel and the edge, and TCh01 (M) was sampled from the middle of the brash ice channel. Both brash ice cores, measuring 40 cm and 62 cm in thickness, consisted of SI clusters mixed with congelation ice. The random orientations and varying crystal sizes of CI were a result of the frequent freezing and breaking processes. An attempt was made to identify the boundaries between different merged brash ice pieces in the first 30 cm of the middle−edge core by considering the direction of the crystal growth. The presumed boundaries are shown with solid blue lines. However, it is harder to analyze and define the boundaries between ice pieces from the second brash ice core, as the CI fraction is higher, with inconsistent growth directions, which ultimately indicates the presence of small brash ice pieces that have been frozen together. Previous studies on other deformed ice structures have also observed and reported similar mixed granular and columnar crystalline structures [8,29,41,42].

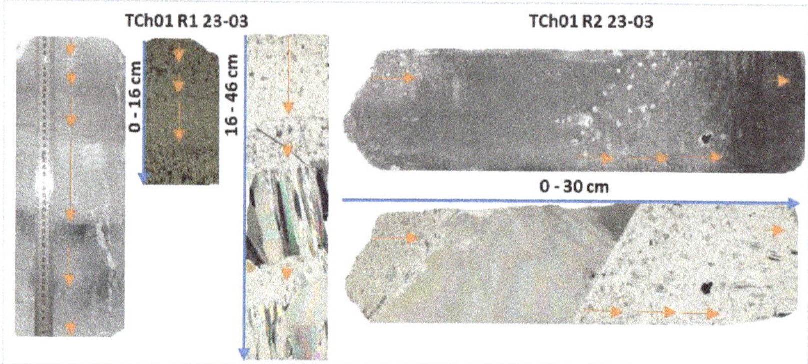

Figure 15. The microstructure of two cores (R1 and R2) sampled on the side ridge of TCh01 on 23 March 2021. The orange arrows distinguish different snow ice layers, and the blue arrows indicate the thickness of the thin sections.

4.3.2. Main Channel (MCh2021)

Nine cores were sampled on 7 March 2021: three cores in the level ice (LI) between the test and main channels, three cores at the very beginning of the ridge (LI-R), and three additional cores at the side ridges (R). The distance between LI-R and R was 8 m. The sampling positions and the microstructure of one core from each position are illustrated in Figure 17.

The level ice cores sampled between the two channels consisted of CI with vertically elongated crystals, similar to freshwater ice seeded by granular ice [43]. A thick layer of SI had formed due to extensive flooding, made possible by mechanical cracks formed at each ship passage in both channels. Eight different SI layers were identified, suggesting eight flooding incidents during the winter season. The orange arrows in the level ice microstructure distinguish the different SI layers, see Figure 17. However, the presence of eight layers may also indicate four flooding scenarios. This hypothesis is supported by the capillarity rise above slush after each flooding event, resulting in two SI layers from one flooding [12,19]. This can be explained by considering that the slush below and above the water line have different porosities and ice crystal concentrations, leading to distinguishable microstructures and SI characteristics. A detailed laboratory study is required to fully understood this process. The mass balance calculations of snow load and ice buoyancy identified four possible flooding events.

The ice core sampled at the beginning of the ridge (LI R1) consisted of a layered structure of SI and CI. Surprisingly, the submerged SI layers were homogenous, but at different depths. Approximately 23 cm of SI had formed on top of CI, originating from surface flooding. The subsequent 32 cm of CI has formed from bottom growth. This two-layered structure is believed to be the original level ice. However, beneath this formation, there was likely a brash ice piece consisting of 4 cm of SI, followed by 1 cm of CI, 6 cm of SI, and a macropore of 1 cm. This pore suggests the presence of a new ice block that had been displaced by the ship passages and consolidated beneath the level ice.

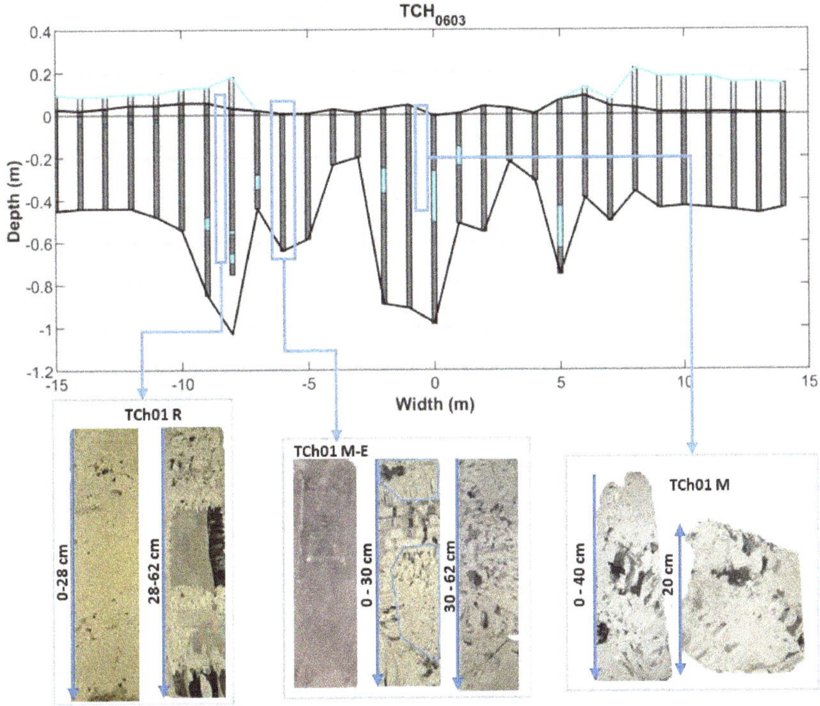

Figure 16. Cross-section profile of the first channel TCh01 measured on 06/03/2021, including the microstructure, and the location of three cores sampled in TCh01. R indicates the side ridge sample, M−E is the sample between the middle and the edge of the channel, and M indicates the middle of the channel.

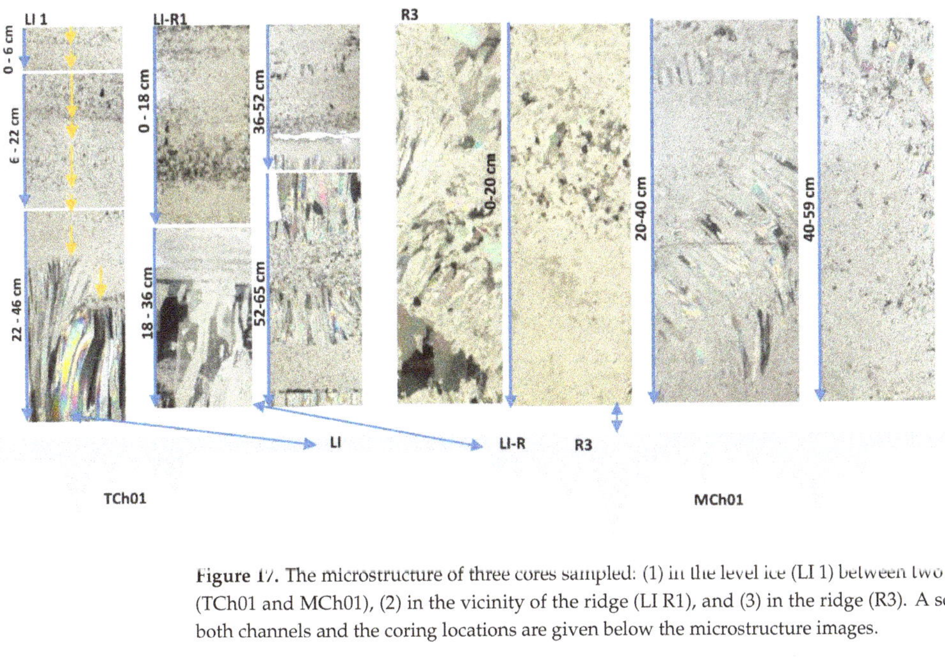

Figure 17. The microstructure of three cores sampled: (1) in the level ice (LI 1) between two channels (TCh01 and MCh01), (2) in the vicinity of the ridge (LI R1), and (3) in the ridge (R3). A scheme of both channels and the coring locations are given below the microstructure images.

This ice block consisted of about 7 cm of CI, followed by 2 cm of SI, 5 cm of CI, 2 cm of SI, and, finally, 1 cm of CI. This structure suggests the freezing of two ice pieces, originating from the same parental level ice. Once the ice pieces merged, the bottom of the ice grew by 1 cm. Considering the presence of a macropore between the LI and the ice piece beneath it, the bottom ice growth was probably driven by the heat stored in ice at the time of submergence. Thus, in this sampling position, the vicinity of the side ridge had three layers of consolidated broken ice.

The microstructure of R3, sampled from the ridge of the main channel, does not represent the entire thickness of the ridge but only the top part that could be cored. This core consisted mainly of a mixed microstructure of granular and columnar ice. The lateral growth of CI observed from 0 to 40 cm illustrates the consolidation of brash ice pieces. Additionally, two different thin sections of the first 20 cm show the horizontal variations in the microstructure of the ridge.

4.4. *Brash Ice Physical and Mechanical Properties*

The snow ice content, uniaxial compressive strength, microporosity, and density results are discussed in the following subsections.

4.4.1. Test and Main Channels (2021)

Level ice in the vicinity of TCh01 had the highest SI content, equal to 73%. This core had an average microporosity of 8% and a density of 846 kg m^{-3}; see Figure 18. The lowest SI content of approximately 21% was found in the channel's middle, accompanied by a density of 897 kg m^{-3} and a porosity of 2.3%. It is likely that a part of the submerged snow melted or displaced sideways, resulting in lower SI content in the channel compared to the ridge and the adjacent level ice.

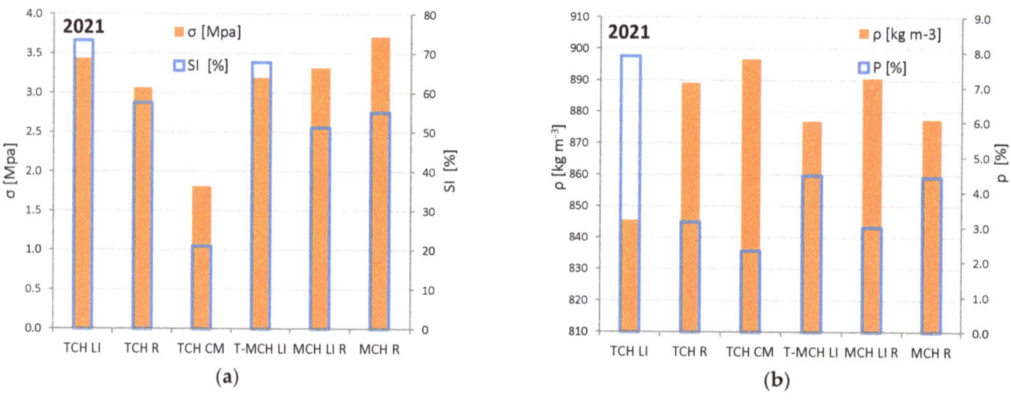

Figure 18. (**a**) The average compressive strength (σ in MPa) and the average snow ice content (SI in %). (**b**) The average microporosity (p in%) and ice density (ϱ in kg m^{-3}). Samples from the level ice in the vicinity of the ship channel are denoted as TCH LI; the brash ice cores are denoted with TCH M, and the ridge samples with TCH R. T-MCH LI indicates level ice samples between TCH and MCH; MCH LI R and MCH R indicate samples from the ridge vicinity and the ridge in the main channel.

The SI fraction in the ridges consists of both snow ice clusters formed in the channel and displaced under the level ice, as well as SI formed on the surface from flooding. However, the proportion of SI was lower in the ridge compared to the adjacent level ice, with values of 58% for the TCh and 53% for MCh. This difference in SI content between the ridges and level ice may be attributed to snow melting in the side ridges due to water being pushed sideways during each ship passage.

During the compressive tests, the temperature of the ice cylinders varied from −5 to −12 °C, with an average value of −8 °C. The brash ice samples taken from the middle of

TCh01 displayed the lowest strength (1.8 MPa), whereas the ridge samples from the main channel exhibited the highest strength. This can be attributed to the continuous breaking of brash ice, resulting in weaker freezing bonds between brash pieces compared to ridge structures. The ridge structures, being less affected by breaking, are exposed to a longer freezing period, thus resulting in stronger bonds. It has been previously observed that the bonding between two ice pieces depends on factors such as the freezing time, thickness, shape, temperature of the ice blocks, and surrounding medium [44–47].

On average, the compressive strength in the horizontal direction for level ice and ridges varied from 3 to 4 MPa for a strain rate of 10^{-3} s^{-1}, see Figure 18. All the ice samples exhibited brittle behavior. A previous study showed that deformed ice with a mixed microstructure, loaded horizontal has higher strength compared to pure columnar or granular ice [29]. This finding may also explain the difference in strength between samples of brash ice and ridge samples.

4.4.2. Main Channels 2023

The thicknesses of level ice, side ridge, and brash ice were measured in February 2023 along the half-cross-section profile of the main channel. The results are shown in Figure 19, where the dark grey bars represent the thicknesses of ice pieces, and the cyan bars in between show the water- or slush-filled macropores. It should be noted that only the top layer of the brash ice was cored in the positions where the macropores were present, limiting the results to the properties of this specific layer. During uniaxial compressive tests, the ice temperature varied between −6 °C and −10 °C with an average value of −8.2 °C. Table 3 summarizes the results.

Table 3. The width in m shows the locations where the ice cores were sampled along the brash ice channel's cross-section. The width zero represents the channel's middle and "−45 m" is the position of the level of ice adjacent to the channel. The core thicknesses, average snow ice content, compressive strength, standard deviation of the compressive strength, density, and microporosity of the ice cores in the sampling location are noted as HT, SI, σ, Std, ϱ, and P, respectively.

Width (m)	H_T (m)	SI (%)	σ (Mpa)	Std (Mpa)	ϱ (kg m^{-3})	p (%)
−45	0.53	44.4	5.2	0.4	883.1	3.8
−35	0.5	64.5	3.3	0.9	829.5	9.6
−25	0.48/0.50	34.8	4.6	0.5	882.9	3.8
−15	0.35	45.6	2.8	0.2	891.2	2.9
−10	0.65	50.0	3.2	1.1	886.2	3.5
−5	0.2/0.23	33.5	2.8	1.2	872.8	4.9
0	0.55/0.24	48.9	4.1	0.7	887.4	3.3

This channel was not always navigated only in the mid position; the side ridges were often broken by ice breakers. However, based on field observations, the last passages before the measurement were carried out in the middle of the channel. Assuming that the main channel was predominantly navigated in the middle (20 m), we have categorized the local results into three main categories: level ice (LI), ridge (R), and brash ice (BI), as shown in Figure 19. The completely solidified ice in the vicinity of the channel was classified as level ice, while the ice between the mid-channel and level ice were considered part of the ridge. The average compressive strength, snow ice content, density, and microporosity for each ice group are illustrated in Figure 20. The snow ice content and compressive strength slightly decreased in the following order: level ice–ridge–brash ice. Similar trends were also yielded in the test and main channels sampled in 2021 discussed in the previous section.

The average snow ice content was found to be 54%, 43%, and 41% for LI, R, and BI, respectively. The lowest density (0.78 kg m^{-3}) and highest microporosity (15%) were recorded in the side ridge, whereas the highest density (0.91 kg m^{-3}) and lowest microporosity (0.85%) were recorded in level ice samples consisting only of columnar ice. The horizontal compressive strength varied from 2 MPa to 5.7 MPa for LI, 2 MPa to 5.3 MPa for

R, and 1.2 MPa to 4 MPa for BI locations. Except for the pure columnar ice samples from the LI cores, all the other samples with a pure snow ice or mixed microstructure showed ductile behavior. In a previous study conducted on a test channel type with lower navigation activity, the average horizontal strength of brash ice and level ice from the same location was between 4 and 6 MPa for a similar strain rate of 10^{-4} s^{-1}, with most of the samples displaying a ductile behavior [35]. In January 2013, the average compressive strength of brash ice was 4.3 MPa, whereas the average compressive strength of level ice was 6.2 MPa.

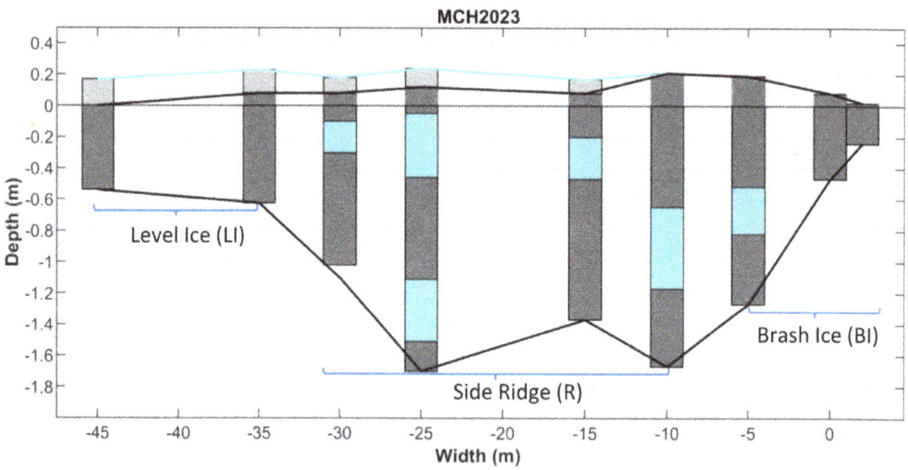

Figure 19. Half-cross-section profile of the main channel was measured on 25/26–02/2023. Each measurement represents the location where the ice cores were sampled. The thickness of each ice core was equal to the ice thickness until the first water void, as shown in this cross-section.

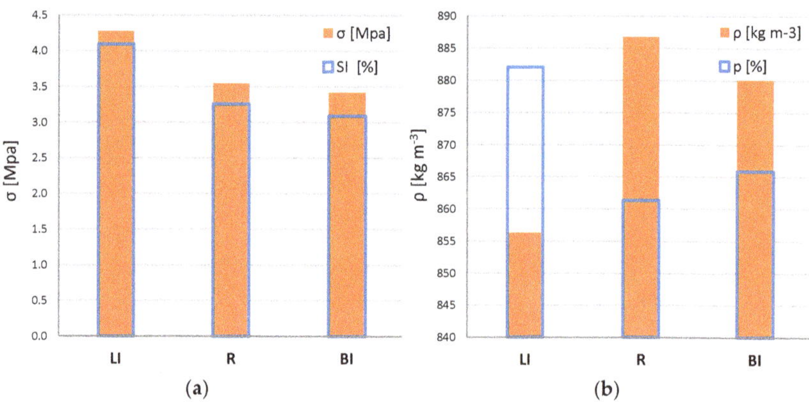

Figure 20. (**a**) The average compressive strength (σ in MPa) and the average snow ice content (SI in %). (**b**) The average microporosity (p in %) and ice density (ϱ in kg m^{-3}). LI, R, and BI imply level ice, side ridges, and brash ice.

Different results were reported for the average strength of brash ice sampled from a refrozen ship channel in the Bay of Bothnia, Finland [8]. The compressive strength of brash ice at a strain rate of 10^{-4} s^{-1} was 4.2 MPa, which is relatively similar to the current test results. However, at a strain rate of 10^{-3} s^{-1}, the average strength was 6.3 MPa [8], which was higher than the results obtained from the test and main channel discussed in the previous section (ranging from 2 to 4 MPa). Similar trends in compressive strengths for the same strain rates reported for the refrozen brash ice in [8] were also reported for pressure ridges located in the Barents Sea [29].

5. Conclusions

This study focused on four main research activities carried out on fast sea ice in the winters of 2020–2023. Two ship channels were specifically created for research purposes and maintained by an ice-breaking tug of Luleå port. Thickness measurements were carried out in the channel's cross-section and adjacent level ice, together with visual observations of the channel's development instantly after each ship's passage. Cores of level ice, brash ice, and ridges were collected from both the test channel (TCh01) and an existing frequently navigated channel (MCh). These cores were further analyzed in LTU's cold laboratory during the summer of 2021 and spring of 2023. The main conclusions of this study are:

1. Four flooding events were estimated to have occurred in the ice adjacent to TCh01. The snow ice microstructure from the level ice in the vicinity of the channel indicated these flooding events as it consisted of four snow–ice layers. However, the ice sampled between the test and main channels had eight different snow ice layers. This difference can be attributed to two possible causes. Firstly, the MCh was navigated more frequently than TCh01, which can cause additional flooding incidents. Secondly, considering that the slush formed from capillary action has different properties than those from four flooding events, eight snow–ice layers could be formed.
2. The total slush thickness measured on level ice adjacent to the ship channels increased linearly with the water level, whereas the slush thickness formed due to capillarity showed no correlation with the water level and had an average thickness of 4 cm. In the second winter, approximately 50% of the incoming snowfall transformed into slush. About 70% of the difference between the measured and incoming snowfall transformed into snow–ice. The rest of the snow (30%) was displaced by the wind or melted when the snow submerged in water. The thickness of snow–ice measured on the level ice adjacent to the first channel was twice as thick as the columnar ice, whereas in the second winter, the snow–ice fraction reached 50% of the total ice thickness.
3. After each ship passage, the snow that accumulated on the channel became submerged and formed a slush layer. Snow–ice clusters between ice blocks were frozen together with columnar ice.
4. During 2021, the snow–ice content was found to be higher in the vicinity of the test channel, reaching 73%. In contrast, the snow ice content was lower in the ridges, measuring 58%, and significantly lower in the middle of the channel, where it was 21%. During winter 2023, the snow ice content was 54% in the vicinity of the main channel, 43% in the ridge, and 41% in the mid-channel. This indicates the melting of snow when submerged in the channel but also some melting of snow on the side ridges due to water being pushed sideways during the ship passages.
5. In winter 2021, the brash ice samples exhibited the lowest strength while the ridge samples from the MCh had the highest strength. Similarly, in winter 2023, the brash ice had the lowest compressive strength, while the level ice in the vicinity of the channel exhibited the highest compressive strength. This difference in strength can be attributed to the weaker bonds between brash pieces compared to the side ridges or level ice, which were exposed to a longer consolidation time. On average, the compressive strength of level ice and ridges ranged from 3 to 4 MPa for a strain rate of 10^{-3} s^{-1}, and from 3 and 5 MPa for a strain rate of 10^{-4} s^{-1}. Meanwhile, the brash ice cores had average compressive strengths of 1.8 and 3.4 MPa for strain rates of 10^{-3} and 10^{-4} s^{-1}, respectively.

Author Contributions: V.Z.: conceptualization, project administration, investigation, data curation, formal analysis, methodology, visualization, data analysis, writing—original draft. R.B.: conceptualization, project administration, investigation, funding acquisition, methodology, writing—review and editing. K.R.: conceptualization, project administration, methodology, supervision, writing—review and editing. J.N.: investigation, methodology, writing—review and editing. A.C.: funding acquisition, project administration, supervision. All authors have read and agreed to the published version of the manuscript.

Funding: This research was supported financially by TotalEnergies SE, Luleå University of Technology, Kolarctic CBC, and operational support by Luleå port.

Data Availability Statement: All data are available upon request.

Acknowledgments: The authors would like to acknowledge the financial support from Total Energies SE, Kolarctic CBC, and Luleå University of Technology. We are also thankful for the operational support we received from Luleå port and in particular the tug masters for maintaining two brash ice channels.

Conflicts of Interest: The authors declare no conflict of interest.

References

1. Juva, M.; Riska, K. On the power requirement in the Finnish-Swedish ice class rules. In *Winter Navigation Research Board 53*; Helsinki University of Technology: Helsinki, Finland, 2002.
2. Greisman, P. *Brash Ice Behavior*; Report No. USCG-D-30-81; U.S. Coast Guard Research and Development Center: New London, CT, USA, 1981.
3. Sandkvist, J. *Observed Growth of Brash Ice in Ships' Tracks*; Research Report; Series A No. 42; Division of Water Resources Engineering, Luleå University of Technology: Luleå, Sweden, 1980.
4. Sandkvist, J. Brash Ice Behaviour in Frequent Ship Channels. Licentiate Thesis, Series A No. 139. Division of Water Resources Engineering, Luleå University of Technology, Luleå, Sweden, 1986.
5. Kitazawa, T.; Ettema, R. Resistance to ship-hull motion through brash ice. *Cold Reg. Sci. Technol.* **1985**, *10*, 219–234. [CrossRef]
6. Ettema, R.; Schaefer, J.A.; Huang, H.P. Ice-tank data on brash-ice loads against barges. *J. Cold Reg. Sci. Technol.* **1998**, *12*, 153–161.
7. Bonath, V.; Zhaka, V.; Sand, B. Field Measurements on the Behavior of Brash Ice. In Proceedings of the International Conference on Port and Ocean Engineering Under Arctic Conditions, Delft, The Netherlands, 9–13 June 2019.
8. Zhaka, V.; Bonath, V.; Sand, B.; Cwirzen, A. Physical and Mechanical Properties of Ice from a Refrozen Ship Channel Ice in the Bay of Bothnia. In Proceedings of the 25th International Symposium on Ice, Trondheim, Norway, 23–25 November 2020.
9. Riska, K.; Bridges, R.; Shumovskiy, S.; Thomas, C.; Coche, E.; Bonath, V.; Tobie, A.; Chomatas, K.; Caloba Duarte de Oliveira, R. Brash ice growth model—Development and validation. *Cold Reg. Sci. Technol.* **2019**, *15*, 30–41. [CrossRef]
10. Zhaka, V.; Bridges, R.; Riska, K.; Cwirzen, A. A review of level ice and brash ice growth models. *J. Glaciol.* **2022**, *68*, 685–704. [CrossRef]
11. Andres, D.D.; Van der Vinne, P.G. Calibration of ice growth models for bare and snow covered conditions: A summary of experimental data from a small prairie pond. In Proceedings of the 11th Workshop on River Ice: River Ice Processes within a Changing Environment, CGU HS Committe on River Ice Proccesses and the Environment (CRIPE), Ottawa, ON, Canada, 14–16 May 2001.
12. Ager, B.H. Studies on the density of naturally and artificially formed fresh-water ice. *J. Glaciol.* **1962**, *4*, 207–214. [CrossRef]
13. Leppäranta, M. A growth model for black ice, snow ice and snow thickness in subarctic basins. *Hydrol. Res.* **1983**, *14*, 59–70. [CrossRef]
14. Leppäranta, M. A review of analytical models of sea-ice growth. *Atmos. Ocean* **1993**, *31*, 123–138. [CrossRef]
15. Saloranta, T.M. Modeling the evolution of snow, snow ice and ice in the Baltic Sea. Tellus A: Dyn. *Meteorol. Oceanogr.* **2000**, *52*, 93–108.
16. Cheng, B.; Vihma, T.; Rontu, L.; Kontu, A.; Pour, H.K.; Duguay, C.; Pulliainen, J. Evolution of snow and ice temperature, thickness and energy balance in Lake Orajärvi, northern Finland. Tellus A: Dyn. *Meteorol. Oceanogr.* **2014**, *66*, 21564.
17. Ashton, G.D. River and lake ice thickening, thinning, and snow ice formation. *Cold Reg. Sci. Technol.* **2011**, *68*, 3–19. [CrossRef]
18. Knight, C.A. Formation of slush on floating ice. *Cold Reg. Sci. Technol.* **1988**, *15*, 33–38. [CrossRef]
19. Coléou, C.; Xu, K.; Lesaffre, B.; Brzoska, J.B. Capillary rise in snow. *Hydrol. Process.* **1999**, *13*, 1721–1732. [CrossRef]
20. Colbeck, S.C. An analysis of water flow in dry snow. *Water Resour. Res.* **1976**, *12*, 523–527. [CrossRef]
21. Colbeck, S.C. The physical aspects of water flow through snow. *Adv. Hydrosci.* **1978**, *11*, 165–206.
22. Ashton, G.D. *River and Lake Ice Engineering*; Water Resources Publications: Littleon, CO, USA, 1986; U.S. Library of Congress Catalogue Number: 86–50681.
23. Leppäranta, M.; Kosloff, P. The structure and thickness of lake Pääjärvi ice. *Geophysica* **2000**, *36*, 233–248.
24. Ohata, Y.; Toyota, T.; Shiraiwa, T. Lake ice formation processes and thickness evolution at Lake Abashiri, Hokkaido, Japan. *J. Glaciol.* **2016**, *62*, 563–578. [CrossRef]

25. Light, B.; Maykut, G.A.; Grenfell, T.C. Effects of temperature on the microstructure of first-year Arctic sea ice. *J. Geophys. Res. Oceans* **2003**, *108*, 3051. [CrossRef]
26. Gow, A.J.; Langston, D. *Growth History of Lake Ice in Relation to Its Stratigraphic, Crystalline and Mechanical Structure (No. 77)*; Department of Defense, Army, Corps of Engineers, Cold Regions Research and Engineering Laboratory: Hanover, HN, USA, 1977; pp. 1–29.
27. Timco, G.W.; Frederking, R.M.W. A review of sea ice density. *Cold Reg. Sci. Technol.* **1996**, *24*, 1–6. [CrossRef]
28. Cox, G.; Weeks, W. Equations for Determining the Gas and Brine Volumes in Sea-Ice Samples. *J. Glaciol.* **1983**, *29*, 306–316. [CrossRef]
29. Bonath, V.; Edeskär, T.; Lintzén, N.; Fransson, L.; Cwirzen, A. Properties of ice from first-year ridges in the Barents Sea and Fram Strait. *Cold Reg. Sci. Technol.* **2019**, *168*, 102890. [CrossRef]
30. Maksym, T.; Jeffries, M.O. A one-dimensional percolation model of flooding and snow ice formation on Antarctic sea ice. *J. Geophys. Res. Ocean.* **2000**, *105*, 26313–26331. [CrossRef]
31. Jordan, R.E.; Hardy, J.P.; Perron, F.E., Jr.; Fisk, D.J. Air permeability and capillary rise as measures of the pore structure of snow: An experimental and theoretical study. *Hydrol. Process.* **1999**, *13*, 1733–1753. [CrossRef]
32. Hiemstra, C.A.; Liston, G.E.; Reiners, W.A. Observing, modelling, and validating snow redistribution by wind in a Wyoming upper treeline landscape. *Ecol. Modell.* **2006**, *197*, 35–51. [CrossRef]
33. Nicolaus, M.; Haas, C.; Bareiss, J. Observations of superimposed ice formation at melt-onset on fast ice on Kongsfjorden, Svalbard. *Phys. Chem. Earth* **2003**, *28*, 1241–1248. [CrossRef]
34. Granskog, M.A.; Vihma, T.; Pirazzini, R.; Cheng, B. Superimposed ice formation and surface energy fluxes on sea ice during the spring melt-freeze period in the Baltic Sea. *J. Glaciol.* **2006**, *52*, 119–127. [CrossRef]
35. Bonath, V.; Fransson, L. *Isförhållanden inom Luleå Hamnbassäng*; Research Report; Luleå University of Technology: Luleå, Sweden, 2013; p. 87, ISSN 1402-1528.
36. Chen, X.; Høyland, K.V. Laboratory work on heat transfer in submerged ice, theory, experimental setup and results. In Proceedings of the 23rd IAHR International Symposium on Ice, Ann Arbor, MI, USA, 31 May–3 June 2016.
37. Chen, X.; Høyland, K.V.; Ji, S. Laboratory tests to investigate the initial phase of ice ridge consolidation. *Cold Reg. Sci. Technol.* **2020**, *176*, 103093. [CrossRef]
38. Huang, H.P. Ice formation in frequently transited navigation channels. Ph.D. Thesis, The University of Iowa, Iowa City, IA, USA, 1988.
39. Ettema, R.; Huang, H.P. *Ice Formation in Frequently Transited Navigation Channels*; Special Report 90–40; U.S. Army Corps of Engineers, Cold Regions Research & Engineering Laboratory: Hanover, HN, USA, 1990; pp. 1–120.
40. Gow, A.J. Orientation textures in ice sheets of quietly frozen lakes. *J. Cryst. Growth* **1986**, *74*, 247–258. [CrossRef]
41. Høyland, K.V. Morphology and small-scale strength of ridges in the North-western Barents Sea. *Cold Reg. Sci. Technol.* **2007**, *48*, 169–187. [CrossRef]
42. Bonath, V.; Petrich, C.; Sand, B.; Fransson, L.; Cwirzen, A. Morphology, internal structure, and formation of ice ridges in the sea around Svalbard. *Cold Reg. Sci. Technol.* **2018**, *155*, 263–279. [CrossRef]
43. Cole, D.M. The microstructure of ice and its influence on mechanical properties. *Eng. Fract. Mech.* **2001**, *68*, 1797–1822. [CrossRef]
44. Ettema, R.; Schaefer, J. A Experiments on freeze-bonding between ice blocks in floating ice rubble. *J. Glaciol.* **1986**, *32*, 397–403. [CrossRef]
45. Urroz, G.E.; Ettema, R. Simple-shear box experiments with floating ice rubble. Simple-shear box experiments with floating ice rubble. *Cold Reg. Sci. Technol.* **1987**, *14*, 185–199. [CrossRef]
46. Ida Mari, B.; Høyland, K.V. Confined Compression Tests on Saline and Fresh Freeze-Bonds. In Proceedings of the International Conference on Port and Ocean Engineering under Arctic Conditions, Trondheim, Norway, 14–18 June 2015.
47. Boroojerdi, M.T.; Bailey, E.; Taylor, R. Experimental study of the effect of submersion time on the strength development of freeze bonds. *Cold Reg. Sci. Technol.* **2020**, *172*, 102986. [CrossRef]

water

MDPI

Article

Risk Evaluation of Ice Flood Disaster in the Upper Heilongjiang River Based on Catastrophe Theory

Yu Li [1], Hongwei Han [1,2,*], Yonghe Sun [3], Xingtao Xiao [3], Houchu Liao [3], Xingchao Liu [1] and Enliang Wang [1,2]

1 School of Water Conservancy and Civil Engineering, Northeast Agricultural University, Harbin 150030, China; 18846771809@163.com (Y.L.); dnlxc@neau.edu.cn (X.L.); hljwel@126.com (E.W.)
2 Heilongjiang Provincial Key Laboratory of Water Resources and Water Conservancy Engineering in Cold Region, Northeast Agricultural University, Harbin 150030, China
3 Heilongjiang Provincial Hydrology and Water Resources Center, Harbin 150001, China
* Correspondence: hanhongwei@neau.edu.cn

Abstract: The ice flood phenomenon frequently occurs in frigid locations of high latitude and high altitude, which triggers ice dam or ice jam flooding thus endangering personal and property safety. Hence, a scientific risk evaluation with enough consideration of each factor is a basic and necessary requirement for preventing ice flood disaster risks. This study establishes a risk evaluation system for ice flood disasters based on the catastrophe theory and utilizes the Pearson correlation coefficient to screen underlying indicators to evaluate the risk of ice flood in the upper Heilongjiang River region. Considering the correlation between different indicators, a hierarchical cluster analysis is invoked to simplify the indicator set and to select typical years. The results of the evaluation system indicate that the catastrophe membership values in the Mohe, Tahe, and Huma regions from 2000 to 2020 ranged from 0.86 to 0.93. Based on the membership values and the actual disaster situations, a four-level classification of risk ratings is conducted. The comparison between the results obtained from the catastrophe theory evaluation method and the fuzzy comprehensive evaluation method reveals similar risk levels, which verifies the effectiveness and practicality of the catastrophe theory applied to the ice flood risk evaluation and presents a novel method for the study of ice floods.

Keywords: ice flood disaster; catastrophe theory; indicator preference; risk evaluation; Heilongjiang River

Citation: Li, Y.; Han, H.; Sun, Y.; Xiao, X.; Liao, H.; Liu, X.; Wang, E. Risk Evaluation of Ice Flood Disaster in the Upper Heilongjiang River Based on Catastrophe Theory. *Water* **2023**, 15, 2724. https://doi.org/10.3390/w15152724

Academic Editor: Renato Morbidelli

Received: 29 June 2023
Revised: 20 July 2023
Accepted: 26 July 2023
Published: 28 July 2023

1. Introduction

Ice flooding is a unique hydrological phenomenon that occurs in frigid region rivers. It is mainly manifested in the flow and evolution of river ice during the winter and spring seasons, and due to a reduced overflow cross-section, ice jams and dams are formed, resulting in backwater staging and a high upstream water level, which can rapidly create a flooding hazard at short notice [1,2]. Due to the high backwater levels, extremely fast ice flow rates, and complex formation mechanisms of ice-induced floods, showing the relationship between floodwater levels and the probability of occurrence in hazardous watersheds becomes more difficult [3]. Therefore, it is significantly important to ensure that the risk evaluation of ice flood disasters is correct and has enough safety margin in regional disaster prevention and mitigation capabilities. Ice-jam floods often cause issues such as farmland submersion, building collapse, and embankment erosion. Additionally, these disasters can also affect the sedimentation and release of chemical substances in river water bodies [4,5]. Thus, ice floods are more likely than open-water floods to cause financial costs and damage to humans and their habitats [6]. In an ice flood risk evaluation, both the probability of occurrence, commonly quantified as the return period, and the damage consequences of the flood hazard need to be considered to assess the annual expected costs [7,8]. This is not only a necessary prerequisite for the prevention and control of natural disasters in frigid regions but also a crucial assurance for national sustainable development and the effective execution of significant programs.

Numerous elements, including environmental changes, water flow conditions, river borders, human activities, and socioeconomics, impact the evaluation of ice flood risk [8–13]. Due to its complicated hazard-inducing environment, various hazard factors, and fragile hazard-bearing body, it has been a challenging and popular subject of disaster research. While open-water flood hazard delineation and risk analysis are commonly used in traditional flood management approaches, methodologies for ice-related flood hazard delineation and risk analysis are not yet well established. This is due to the highly unpredictable and complex nature of ice-related flood events, which present significant challenges for accurate evaluation and assessment [14]. For this reason, some scholars have conducted a lot of research on the study of causes of ice floods, the forecasting of ice-jam and ice-dam floods, the delineation of flooding hazards, and flood risk calculation methods. The formation and evolution of ice floods are extremely complicated, and most scholars are currently studying the impact of various factors on the causes and disasters of flooding according to the geographical environment, hydrology, meteorology, and engineering of rivers [15]. The hydrotechnical approach proposed is the most appropriate method for assessing the risk of ice-jam floods in river systems when adequate historical and on-site data related to ice blockage are available [16]. However, the reliability of hydrometric gauge data can be compromised due to the extreme forces exerted by ice debris and blocks, which may lead to the unavailability and inaccuracy of the data [17,18]. Meanwhile, due to the harsh regional environments where severe ice floods occur, it is generally impossible to acquire real-time, complete, and precise data from field observations [19]. Beltaos [20] used the distributed-function method (DFM) to determine the frequencies and probabilities of ice-jam floods. Several studies [21,22] have utilized an ice-jam numerical model (RIVICE) to evaluate real-time ice-jam flood hazards along the Athabasca River at Fort McMurray, etc. These studies can establish a basis for real-time ice-jam flood risk analysis and improve our comprehension of the ice-jam flood risk of both property and inhabitants. Artificial intelligence techniques, such as neural networks and fuzzy logic systems [13,23], show promise in modeling the nonlinear processes underlying the formation of ice floods. In particular, combining multiple models to predict backwater levels can improve accuracy by 20–30%, albeit at a significant computational cost. Mahabir et al. [24–26] forecasted breakup water levels using multiple linear regression and based on this evaluated the application of soft computing in modeling the maximum water level during river breakout in flood and non-flood event years through fuzzy logic and artificial neural networks. Through the utilization of projection tracing, fuzzy clustering, and accelerated genetic algorithm, Wu [27] created a comprehensive evaluation model of the ice disaster risk that occurred in the Ning-Meng portion of the Yellow River. Luo [28] proposed the GM (1,1) evaluation model, which introduces three-parameter interval gray numbers, to simulate and anticipate the development trend of risk vectors. Numerous studies have already been conducted on flood hazard delineation and risk assessment in the context of managing the risk of ice floods [8,29–31].

In general, the process of nonlinear changes in ice-flooding behavior is a dynamic and irreversible evolutionary process under the influence of the realistic environment. Therefore, the ice flood disruption can be seen as an abrupt state catastrophe phenomenon driven by the energy of the river ice system. This study focuses on the disaster risk evaluation of ice floods using the catastrophe theory. Hazard-inducing environment, hazard factor, hazard-bearing body, and anti-icing capability are considered the criterion layer of the evaluation system, in order to research the indicators in the process of ice flood damage. The catastrophe evaluation method is combined with the Pearson correlation coefficient and hierarchical cluster analysis to solve the problem of index selection and optimization in the evaluation model. The actual ice flood situations in Mohe, Tahe, and Huma regions are taken as an example and are used in establishing the classification of the risk level of the ice flood in the regions. Meanwhile, a comparison is made between the results of the catastrophe theory evaluation method and the fuzzy comprehensive evaluation method, which verifies the efficacy and applicability of the catastrophe theory applied to the ice flood risk evaluation and introduces a novel concept for the study of ice floods.

2. Materials and Methods

2.1. Catastrophe Evaluation Method

Calculus, as a mathematical tool, is well-suited for modeling and problem-solving in natural phenomena characterized by continuity and differentiability. It enables the study of continuous, gradual, and smooth changes, allowing for a deeper understanding of such processes in nature. However, when a continuous development undergoes a transition from gradual and quantitative changes to abrupt and qualitative changes, calculus becomes inadequate for describing and addressing such phenomena. In order to solve the step change process, René Thom, a French mathematician, initially introduced the catastrophe theory in 1974 [32] to explore and research discontinuous changes and abrupt variations in phenomena. He discussed the three basic forms of the system and the mathematical principles of structural stability, singularity, and topology and described the transition from continuous asymptotic and quantitative changes to discontinuous jump mutations and qualitative changes using mathematical methods. The catastrophe theory utilizes potential functions to classify the critical points of a system and investigates the characteristics of discontinuous changes near each critical point. The properties of the discontinuous state that are located around the crucial points are uncovered in order to conduct a more in-depth investigation of the process underlying discontinuous occurrences [33]. There are seven different major forms of catastrophe theory [34], as determined by the geometry of the restriction criteria. Among them, different types of primary catastrophe models are used which include the cusp catastrophe type, swallowtail catastrophe type, butterfly catastrophe type, etc. The equilibrium surfaces and singular point sets associated with these models are shown in Table 1 and Figure 1.

Table 1. Normalization formulae for the catastrophe theory.

Category	Potential Function	Normalization Formula
Cusp	$V(x) = x^4 + ax^2 + bx$	$x_a = a^{1/2}, x_b = b^{1/3}$
Swallowtail	$V(x) = x^5 + ax^3 + bx^2 + cx$	$x_a = a^{1/2}, x_b = b^{1/3}, x_c = c^{1/4}$
Butterfly	$V(x) = x^6 + ax^4 + bx^3 + cx^2 + dx$	$x_a = a^{1/2}, x_b = b^{1/3}, x_c = c^{1/4}, x_d = d^{1/5}$

Note: The control variables a, b, c, and d can be viewed as factors influencing the system's behavior state and are decreasing in importance from a to d.

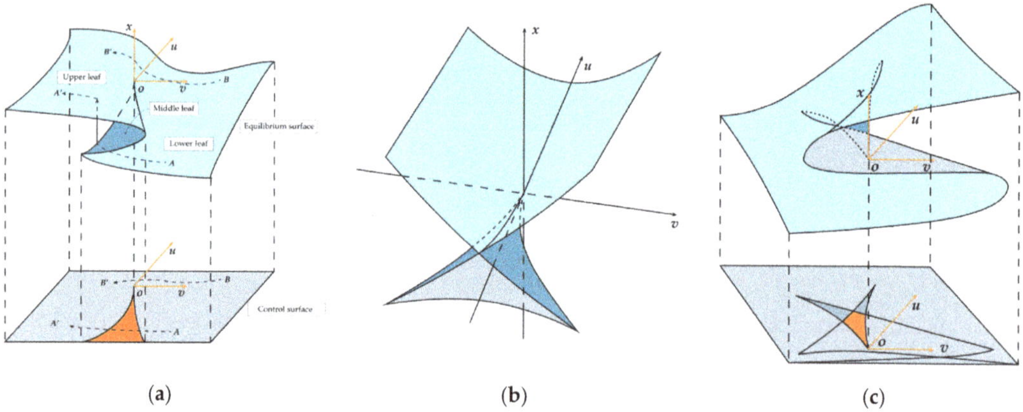

Figure 1. Equilibrium surfaces and singular point sets of catastrophe models. (**a**) Cusp catastrophe type; (**b**) Swallowtail catastrophe type; and (**c**) Butterfly catastrophe type.

It is worth noting that due to the four-dimensional and five-dimensional potential functions of the swallowtail and butterfly catastrophe types, the equilibrium surfaces in Figure 1b,c represent the projection of the original functions in three-dimensional space after constraining the variables.

In the process of risk assessment and judgment, it is important to consider the characteristics of each indicator layer and the actual situation. The choice should be made among the following three evaluation principles:

1. Non-complementary criterion.

In cases where the control variables of the system cannot be substituted for one another, the minimum value among the corresponding mutation values of the m control variables (a, b, …, m) is selected as x, as follows:

$$x = min(x_a, x_b, \ldots, x_m) \tag{1}$$

2. Complementary criterion.

If the variables of a system are mutually substitutable, the corresponding mutation values of each indicator should be calculated according to the catastrophe model of the system. Then, the average value of the variables should be computed as follows:

$$x = \frac{(x_a + x_b + \ldots + x_m)}{m} \tag{2}$$

3. Over-threshold complementary criterion.

The over-threshold complementarity criterion is based on the analysis of extreme events that exceed specific thresholds. It is used to estimate the tail probability or frequency of events surpassing the threshold. These tail probabilities or frequencies are typically low and are thus considered extreme events or exceptional circumstances. By analyzing these extreme events, we can gain a better understanding of the risk characteristics of the system and take appropriate measures for risk assessment and decision-making. The control variables of the system must reach a certain threshold before they can complement each other [33].

2.2. Data Preprocessing

The extreme value method is employed to standardize the preliminary selection indicators, aiming to mitigate the impact of diverse indicator data and their magnitudes on the indicator screening process and to enable effective comparison between indicators. Equation (3) is applicable in situations where larger values of the indicator are more advantageous for analysis. Equation (4) is applicable in situations where smaller values of the indicator are more advantageous for analysis. Assuming that the system indicators can be described by the state variables x_{ij}, the following extreme value normalization transformation formula can be used:

$$y_{ij} = \frac{x_{ij} - \min\limits_{1 \le i \le m}\{x_{ij}\}}{\max\limits_{1 \le i \le m}\{x_{ij}\} - \min\limits_{1 \le i \le m}\{x_{ij}\}} \tag{3}$$

$$y_{ij} = \frac{\max\limits_{1 \le i \le m}\{x_{ij}\} - x_{ij}}{\max\limits_{1 \le i \le m}\{x_{ij}\} - \min\limits_{1 \le i \le m}\{x_{ij}\}} \tag{4}$$

where y_{ij} represents the normalized value of the state variable x_{ij}, $\min\limits_{1 \le i \le m}\{x_{ij}\}$ represents the minimum value of x_{ij}, and $\max\limits_{1 \le i \le m}\{x_{ij}\}$ represents the maximum value of x_{ij}, i = 1, 2, 3, …, m (m is the designation of years); j = 1, 2, 3, …, n (n is the designation of indicators).

Then y_{ij} is the dimensionless data and between 0 and 1.

2.3. Pearson Correlation Coefficient

The Pearson correlation coefficient is a statistical measure that quantifies the strength and direction of the linear relationship between two continuous variables and is widely used in various fields of research and data analysis.

- STEP 1: Determine the constraints of Pearson correlation:
  - There is a linear relationship between the two variables;
  - The variables are continuous variables;
  - The variables are normally distributed, and the binary distribution is also normally distributed;
  - The two variables are independent.
- STEP 2: Calculate the Pearson correlation coefficient between X_i and Y_i. The Pearson correlation coefficient is represented by the symbol "r" and takes values between -1 and 1. The coefficient is calculated based on the covariance between the two variables and the product of their standard deviations. The formulation of the correlation coefficient can be described as follows:

$$r = \frac{\Sigma(X_i - \overline{X})(Y_i - \overline{Y})}{\sqrt{\Sigma_{i=1}^{n}(X_i - \overline{X})^2 \Sigma_{i=1}^{n}(Y_i - \overline{Y})^2}} \tag{5}$$

where X_i and Y_i are the individual data points in the two variables, $\overline{X}$ and $\overline{Y}$ are the means of the two variables, and Σ denotes summation over all data points.

The resulting value of "r" indicates the strength and direction of the linear relationship between the variables:

- If "r" is close to 1, it indicates a strong positive linear relationship, meaning that as one variable increases, the other variable also tends to increase;
- If "r" is close to -1, it indicates a strong negative linear relationship, meaning that as one variable increases, the other variable tends to decrease;
- If "r" is close to 0, it indicates a weak or no linear relationship between the variables.

It is noteworthy that in correlation analysis, we typically aim to determine whether the observed correlation coefficient is significantly different from zero, indicating the statistical significance of the correlation relationship. These significance tests are designed to evaluate whether the observed correlation coefficient is sufficiently large to reject the presence of correlation due to random sampling errors [35]. During significance testing, it is necessary to choose an appropriate significance level (typically 0.05) to determine whether to reject the null hypothesis. If the null hypothesis is rejected, it can be concluded that the observed correlation is significant.

2.4. Hierarchical Cluster Analysis

Cluster analysis is commonly used in scientific research to identify group associations and assess the affinity among different variables [36–39]. Hierarchical cluster analysis is a data analysis technique used to identify groups or clusters within a dataset based on their similarity or proximity. It is a form of unsupervised learning, as it does not rely on predefined class labels or target variables. Hierarchical cluster analysis, specifically, is used to determine associations between different parameters and ultimately identify the sources and processes related to them [40].

Ice flood disasters encompass various parameters from different disciplines, and there exists a significant correlation among these parameters. This correlation leads to a substantial increase in the time and computational resources required for data collection and processing.

In this study, the hierarchical cluster analysis method is employed to demonstrate the interrelationships among the variables under investigation [41]. Based on the criterion of the sum of distances, the typical years are selected by identifying the cluster centroids.

This approach guarantees the comprehensiveness of the evaluation index system while mitigating potential issues, such as result distortion caused by redundant indicators.

2.5. Fuzzy Comprehensive Evaluation Method

The fuzzy comprehensive evaluation method is a mathematical approach used to assess complex systems or phenomena that involve uncertainties and imprecise information [42,43], including the following steps:

- STEP 1: Assuming that there are n years to be evaluated to form a sample set, and based on the eigenvalues of m indicators, the eigenvalue matrix of ice flood risk to be evaluated can be expressed as Equation (6):

$$X = (x_{ij}) = \begin{bmatrix} x_{11} & x_{12} & \cdots & x_{1n} \\ x_{21} & x_{22} & \cdots & x_{2n} \\ \vdots & \vdots & \ddots & \vdots \\ x_{m1} & x_{m2} & \cdots & x_{mn} \end{bmatrix} \quad (6)$$

Upon applying data normalization using Equations (3) and (4), we derived the relative membership matrix $R = (r_{ij})$.

- STEP 2: Construct the index weight set.

In order to account for the varying importance of different factors in evaluating the objective, it is necessary to establish a set of indicator weights. Weighting techniques fall into two primary categories: statistical-based methods and participatory-based methods. The statistical-based methods analyze the indicator data to determine the weights, whereas the participatory-based methods involve incorporating expert or public opinions to determine the weights. In this study, the weight is determined using the entropy weight method, which falls under the statistical-based methods. The index weight set is constructed according to Equation (7).

$$A = (w_1, w_2, \ldots, w_m) \quad (7)$$

- STEP 3: Establishing the fuzzy comprehensive evaluation model.

To obtain the result of the fuzzy comprehensive evaluation for each sample, the evaluation matrix and the index weight set are quantified as shown in Equation (8). The synthesis operators used for fuzzy synthesis calculations include the dominant factor determining operator, the dominant factor prominent operator, and the weighted average operator, among others. Due to the interplay of factors in ice flood risk, this study employs the weighted average operator for fuzzy synthesis calculations.

$$B = A \circ R = (b_1, b_2, \ldots, b_n) \quad (8)$$

where B is the fuzzy comprehensive evaluation risk membership matrix of the assessed object.

3. Ice Flood Risk Evaluation

3.1. Study Area Overview

The main stream of the Heilongjiang River is located in the high latitude and cold region of the border between China and Russia. The river freezes for a long time and is prone to ice dam and jam disasters during the open river flow period. The Heilongjiang River originates from the Erguna River in the Mongolian Plateau, with a total length of approximately 4440 km and a drainage area of about 1,855,000 square kilometers. The Heilongjiang River flows through various terrains, including mountains, canyons, and plains. Within China's territory, the riverbed is rugged, forming a series of rapids and waterfalls. Upon entering Russia, the river becomes gentle and flows through vast plains.

This study was conducted along a specific section of the Heilongjiang River, spanning approximately 800 km from the Mohe region to the Huma region (Figure 2). The three regions of Mohe, Tahe, and Huma, situated between 50.9° N and 53.5° N, have historically experienced the highest occurrence of spring ice flood disasters and have been the most severely affected regions [44]. These three regions are part of the Greater Khingan Mountains region, characterized by a cold temperate continental monsoon climate. The average temperature in the region has been recorded at −2.1 °C over the years, while the annual average precipitation remains around 460 mm. Between 2000 and 2020, there were seven years marked by severe ice flood disasters, with no fixed occurrence location. The average duration of these events ranged from 2 to 3 days, while the longest recorded event lasted for 15 days. In the springs of 2000 and 2009, the upper sections of the Heilongjiang River experienced over seven ice dams and jams, resulting in the highest backwater heads reaching 7.58 and 9.23 m.

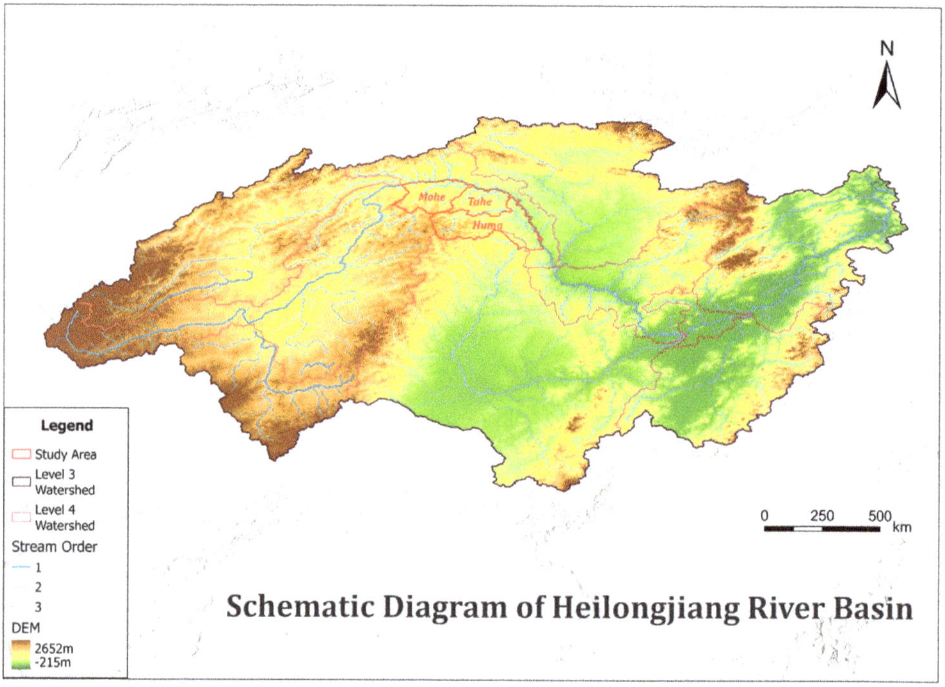

Figure 2. The geographic location of the study area.

3.2. Analysis of Ice Flood Risk

During the non-ice-blocked period of the ice-flowing period, the flow pattern of a river remains in a long-term linear relationship with hydraulic parameters such as channel characteristics, precipitation, snowmelt, and runoff. It exists in an equilibrium state. With changes in channel parameters, air temperature, and other conditions, ice floe may accumulate somewhere downstream of the channel or be pushed under the ice sheet, resulting in ice jams or dams. This disrupts the original balance of the water level in the river, resulting in a sudden change due to reduced flow velocity or backwater caused by ice blockages.

However, in the risk evaluation, it is also necessary to consider the differences in ice flood prevention and disaster resilience capacities in different regions. The above analysis indicates that the sudden accumulation of ice drains during the transition from stability to instability is the fundamental characteristic of potential disasters during the ice flood period. Therefore, it is possible to establish a risk evaluation index system and a model for evaluating the sudden changes in ice flood risks based on the catastrophe theory.

Given the challenges faced by certain methods, such as fuzzy evaluation methods which struggle with accurate weight determination and other issues involving factor analysis, complex calculations, and a substantial sample size requirement, opting for the catastrophe theory evaluation method is a reasonable choice to overcome these issues. The key advantage of this method is that it determines the importance of each indicator based on the inherent contradictions and mechanisms of various objectives within the normalization formula itself, without relying on indicator weights. As a result, the evaluation outcomes are objective in nature.

3.3. Data Acquisition and Processing

The research data used in this study were obtained from multiple sources, including the NOAA—National Centers for Environmental Information, ECMWF ERA5-Land monthly averaged data from 1981 to present, "China County Statistical Yearbook" (2000–2021), and the China Basic Geographic Information Sharing Website. The data underwent analysis and processing utilizing various tools such as ArcGIS Pro, SPSS 27.0, Origin 2022, and Excel 2021.

The risk of ice flood is influenced by multiple factors. This paper categorizes these factors into four guideline layers: hazard-inducing environment, hazard factors, hazard-bearing body, and anti-icing capability. Each subsystem plays a distinct role in the evolution of risk, with varying degrees of influence. Therefore, predicting the evolution of ice flood risk is a complex and uncertain task. To construct a risk evaluation index system that adheres to the principles of scientific, typical, comprehensive, systematic, and practical criteria, a four-level ice flood risk evaluation index system is developed. A total of 21 preliminary indicators are selected by integrating available information, as illustrated in Table 2.

Table 2. Preliminary selection of ice flood risk evaluation index system.

Criterion Layer	Index Layer	Indicator Nature	Unit
Hazard-inducing Environment	River length (X_{Q1})	(+)	km
	River gradient (X_{Q2})	(+)	–
	Meander coefficient (X_{Q3})	(+)	–
	Width-to-narrow ratio of sudden contraction in the river channel (X_{Q4})	(+)	–
Hazard Factor	Upstream average temperature from October to March (X_{P1})	(−)	°C
	Local average temperature from October to March (X_{P2})	(−)	°C
	Upstream cumulative precipitation from November to March (X_{P3})	(−)	mm
	Average temperature from April 1 to 20 (X_{P4})	(+)	°C
	Average high temperature from April 1 to 20 (X_{P5})	(+)	°C
	Upstream cumulative insolation from April 1 to 20 (X_{P6})	(+)	h
	Local cumulative insolation from April 1 to 20 (X_{P7})	(+)	h
	Snow depth on April 1(X_{P8})	(+)	mm
	Upstream average ice thickness in March (X_{P9})	(+)	m
	Local average ice thickness in March (X_{P10})	(+)	m
	Downstream average ice thickness in March (X_{P11})	(+)	m
Hazard-bearing Body	Population density (X_{R1})	(+)	people per km^2
	Primary industry value-added ratio (X_{R2})	(+)	–
	GDP per capita coefficient (X_{R3})	(+)	–
Anti-icing Capability	Number of hospital beds per capita (X_{S1})	(−)	sheet per people
	Resident deposit amount coefficient (X_{S2})	(−)	–
	Local fiscal general budget revenue coefficient (X_{S3})	(−)	–
Auxiliary Parameters	Ice flood hazard coefficient (X_M)	(+)	m·d
	Frequency of ice flood (X_N)	(+)	times

Note: "+" in the table represents that the indicator promotes the development of disaster risk in the ice flood and "−" represents the suppression of the development of disaster risk in the ice flood.

4. Results and Discussion

4.1. Construct the Ice Flood Risk Evaluation Index System

The risk of an ice flood is influenced by numerous factors, and there exist complex non-linear relationships among these factors. When these factors undergo changes and interact with each other, the risk of an ice flood disaster can experience sudden variations.

Figure 3a displays the correlation between hazard-inducing environmental factors and the frequency of ice floods in the Mohe, Tahe, and Huma regions. It indicates a low correlation between river length and the occurrence of ice floods. It is important to note that there is a strong negative correlation between X_{Q3} and X_N, which is due to the small size of the dataset. Previous studies [9,44–46] have already shown a positive correlation between the river meander coefficient and the frequency of ice floods. The correlation between the hazard factor and the ice flood hazard coefficient is depicted in Figure 3b. As the significant level values for indicators X_{P3}, X_{P6}, and X_{P7} are greater than 0.05, they are removed from the analysis. Additionally, there is a strong correlation between indicators X_{P4} and X_{P5}, as well as X_{P10} and X_{P11}, indicating redundancy in these indicators. Therefore, only X_{P4} and X_{P11}, which exhibit a higher correlation with X_M, are retained.

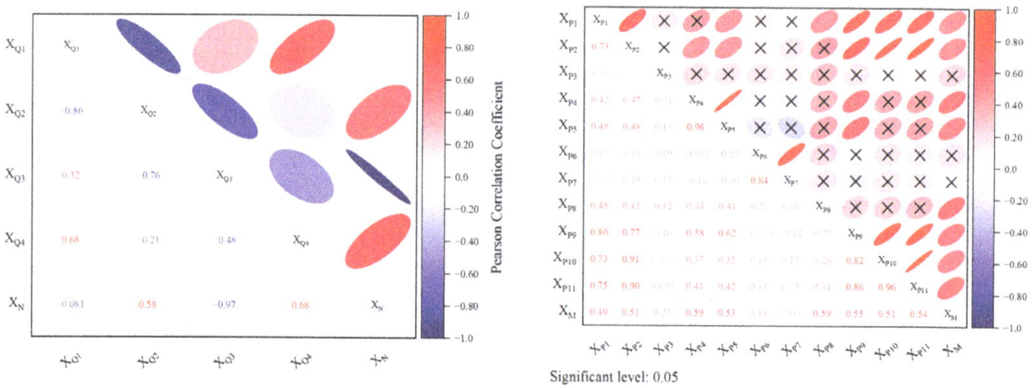

(a) (b)

Figure 3. Pearson correlation analysis for preliminary selection of indicators. (**a**) The correlation between hazard-inducing environmental factors and the frequency of ice flood. (**b**) The correlation between the hazard factor and the ice flood hazard coefficient.

The filtered variables X_{P1}, X_{P2}, X_{P4}, X_{P8}, X_{P9}, and X_{P11} undergo a hierarchical cluster analysis, resulting in the formation of the temperature element layer C_4 and the hydrological element layer C_5, as shown in Figure 4. It is important to highlight that when the clustering results of indicators lead to a distinct category, it signifies a crucial aspect of the evaluation system that was directly selected. Examples of such indicators include X_{R1}, X_{R2}, X_{R3}, X_{S1}, X_{S2}, and X_{S3}. These indicators hold significant value in the evaluation process.

Through the utilization of the Pearson correlation coefficient and hierarchical cluster analysis, a total of 15 indicators are selected as the final set of ice flood risk evaluation indicators. The priority ranking of the selected indicator set was determined based on previous studies [9], correlation coefficients, the vulnerability of the vulnerable entity, and the capacity for ice flood prevention. Please refer to Table 3 for more details on these indicators.

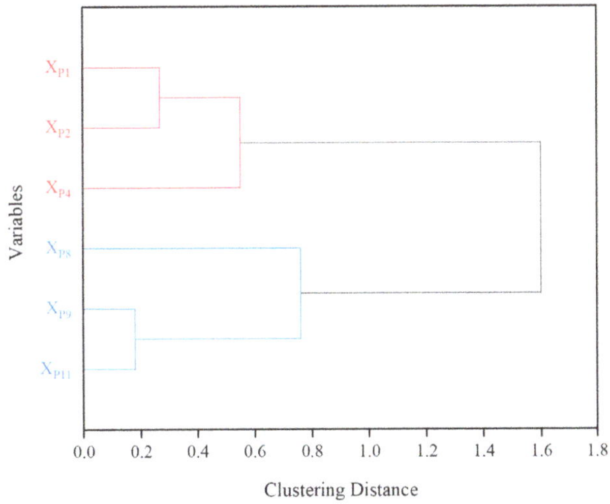

Figure 4. Hierarchical cluster analysis chart of hazard factors.

Table 3. Ice flood risk evaluation index system.

Target Layer	Criterion Layer		Index Layer	Pearson Correlation Coefficient: Correlation	Pearson Correlation Coefficient: Significant Level	Whether to Retain	Clustering Category	Indicator Layer
Comprehensive Risk Situation of Ice Flood (A)	Hazard-inducing Environment (B_1)		X_{Q1}	−0.08	0.949	N		
			X_{Q2}	0.58	0.609	Y	1	C_3
			X_{Q3}	−0.97	0.154	Y	1	C_2
			X_{Q4}	0.68	0.526	Y	1	C_1
	Hazard Factor (B_2)	Climatic Elements (C_4)	X_{P1}	0.49	0.025	Y	2	D_3
			X_{P2}	0.51	0.019	Y	2	D_2
			X_{P3}	0.23	0.307	N		
			X_{P4}	0.59	0.005	Y	2	D_1
			X_{P5}	0.53	0.013	N		
			X_{P6}	0.18	0.438	N		
			X_{P7}	0.03	0.893	N		
		Hydrological Elements (C_5)	X_{P8}	0.59	0.005	Y	3	D_4
			X_{P9}	0.55	0.010	Y	3	D_5
			X_{P10}	0.51	0.018	N		
			X_{P11}	0.54	0.011	Y	3	D_6
	Hazard-bearing Body (B_3)		X_{R1}			Y	4	C_6
			X_{R2}			Y	4	C_7
			X_{R3}			Y	4	C_8
	Anti-icing Capability (B_4)		X_{S1}			Y	5	C_9
			X_{S2}			Y	5	C_{10}
			X_{S3}			Y	5	C_{11}

Therefore, a four-level evaluation index system is established. The first layer is the target layer A, which is the evaluation of ice flood risk; the second layer is the criterion layer B, which is the hazard-inducing environment B_1, the hazard factor B_2, the hazard-bearing body B_3, and the anti-icing capability B_4; the third layer C is the indicators C_1 to C_{11} obtained after screening; and C_4 and C_5 consist of the bottom indicators D_1 to D_6, representing the influence of climatic elements and hydrological elements, as shown in Table 3.

4.2. Ice Flood Risk Evaluation Results and Grade Classification

In the criterion layer, the hazard-inducing environment, climatic elements, hydrological elements, hazard-bearing body, and anti-icing capability (B_1, C_4, C_5, B_3, and B_4) are composed of three variables, following a swallowtail catastrophe model. In addition, the hazard factor (B_2) in the criterion layer consists of two variables, following a sharp point catastrophe model. In the target layer, the ice flood hazard risk (A) is composed of four variables (B_1, B_2, B_3, and B_4), which follow a butterfly catastrophe model.

The data from the indicator layer of Mohe, Tahe, and Huma from 2000 to 2020 were incorporated into the mutation evaluation method. Over the 20-year period, ice dams and jams occurred in the upper Heilongjiang River during the ten years. The results of the ice flood risk evaluation and the classification of each region in the study year can be observed in Figure 5.

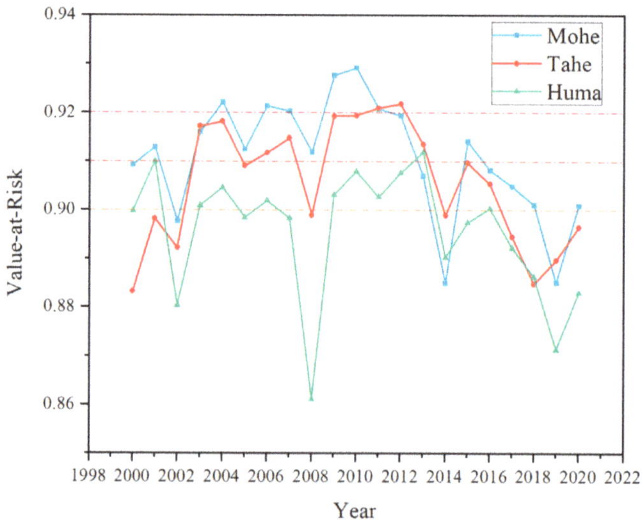

Figure 5. The results and classification of ice flood risk evaluation in the study year.

At present, there is no unified standard for the evaluation of ice flood risk, either domestically or internationally. Referring to the relevant studies [47,48], the evaluation criteria of their ice flood hazard evaluation index were determined by combining the actual situations of ice dam flooding in the upper reaches of the Heilongjiang River. Performing a hierarchical cluster analysis on the set of indicators for each year, as shown in Figure 6, we selected the years 2000, 2001, 2008, 2009, 2010, 2011, and 2015 as representative years based on the results of the clustering analysis. The clustered distances, value-at-risk, and realistic risk ratings for the typical years in the three regions are shown in Table 4.

The ice flood risk for each year and region was ranked according to the results of the catastrophe theory evaluation. Based on the historical ice flood data for the upstream region of the Heilongjiang River during the representative years, the evaluation results were divided into different levels. By following this approach, we determined the grade intervals corresponding to the risk levels of ice floods in the upper reaches of Heilongjiang Province. The specific grade intervals and corresponding details can be found in Table 5.

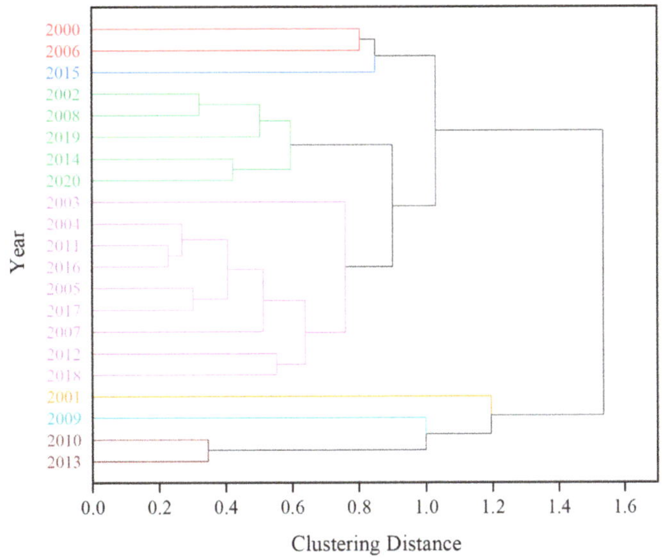

Figure 6. Clustering results of index sets for each year.

Table 4. The clustered distances, value-at-risk, and realistic risk ratings for the typical years.

Clustering Group	Year	Clustering Distance	Mohe		Tahe		Huma	
			Value	Rating	Value	Rating	Value	Rating
1	2000	0.804	0.909	Moderate	0.883	Low	0.900	Moderate
2	2015	0.850	0.914	High	0.910	High	0.897	Low
3	2008	0.322	0.912	High	0.899	Low	0.861	Low
4	2011	0.229	0.921	Critical	0.921	Critical	0.903	Moderate
5	2001	1.195	0.913	High	0.898	Low	0.910	High
6	2009	0.998	0.928	Critical	0.919	High	0.903	Moderate
7	2010	0.346	0.929	Critical	0.919	High	0.908	Moderate

Note: Selection method: sum of distances.

Table 5. The classification of ice flood risk.

Typical Years and Regions		Evaluation Results (*A*)	Grading Results	
Year	Region		Range	Grade
2010	Mohe	0.929	0.92~1	I
2011	Mohe	0.921		
2010	Tahe	0.920	0.91~0.92	II
2001	Huma	0.910		
2000	Mohe	0.909	0.90~0.91	III
2000	Huma	0.900		
2008	Tahe	0.899	0~0.90	IV
2008	Huma	0.861		

4.3. Results Analysis

A comparison of the average risk values in Mohe, Tahe, and Huma from a geographical perspective indicates that the evaluation level of ice flood risk in the upper main stream of the Heilongjiang River was highest in Mohe, followed by Tahe, and then Huma, throughout the period from 2000 to 2020. From the perspective of the hazard-inducing environment, the Mohe segment is characterized by the maximum river gradient and width-to-narrow ratio of sudden contraction in the river channel. The ratio compares the width of the original river section to the width of a narrowed section. A larger width-to-narrow ratio implies a more significant change in the width of the river channel. When a significant amount of ice flows through this section, the width of the channel decreases sharply, and ice debris is more likely to accumulate and cause congestion. As a result, the hazard risk of ice flood is higher in this particular segment of the river.

In addition, the Huma segment of the river gradient has a larger meander coefficient, indicating a steeper course compared to other sections. The river gradient is calculated as the ratio of the height difference between starting and ending points to the actual length of the river. Consequently, ice slush tends to accumulate in these bends, obstructing the water flow and quickly raising water levels. Together, the combination of a larger river gradient and a higher width-to-narrow ratio in the Mohe section increases the risk of ice floe accumulation, water flow blockage, and the subsequent occurrence of ice dams or jams.

From a temporal perspective, the risk levels in the Mohe, Tahe, and Huma areas follow a pattern of initial increase followed by a subsequent decrease. There was a notable increase in the overall risk index from 2000 to 2010. All three regions increased to the higher risk level, with the Tahe region showing the greatest increase at 4%. Regarding the factors contributing to these changes, the average temperature upstream during the October 2009 to March 2010 period was 3.1 °C higher compared to 2000. This period corresponds to the freeze-up phase of the Heilongjiang River, and the average temperature during this time directly affects the volume of ice and water in the river during the opening period. Furthermore, the local fiscal general budget revenue coefficient and the coefficient of resident savings deposits are generally lower compared to the year 2000. Additionally, the proportion of agricultural output value is significant. These factors contribute to an increased potential risk for the occurrence of ice floods.

In comparison to 2009, the risk levels of the three regions escalated in 2010. This can be attributed to the lower average temperatures experienced along the Heilongjiang River in April, with Mohe region being 4.68 °C lower than previous years, and Huma maintaining temperatures below zero. Temperature serves as a critical thermal condition for ice flood formation. In 2010, the Mohe and Huma sections failed to thaw due to the persistently low average temperatures. As a result, the upstream water carrying a substantial amount of floating ice exerted pressure on the downstream ice cover, leading to ice squeezing and accumulating. This scenario created favorable conditions for the formation of ice dams and jams.

The Heilongjiang River is primarily lined with villages, and a majority of rural residents rely on agriculture as their main source of livelihood. Consequently, during ice flood disasters, agriculture, in addition to the population, becomes the primary vulnerable entity. The severity of the consequences resulting from an ice flood disaster is directly proportional to the population density and per capita food possession. Enhancing the number of drainage structures and extending the length of embankments will enhance the region's capacity to mitigate ice flood disasters. As the proportion of total agricultural output in the regional GDP increases, the recovery process from the impact of ice flood disasters becomes more challenging. On the whole, the risk level of ice floods in the upper reaches of the Heilongjiang River demonstrates a decreasing trend, indicating a yearly improvement in the economic level and the ice flood prevention capabilities in the region.

4.4. Accuracy Evaluation

In order to examine the accuracy of the catastrophe evaluation method, the entropy weighting method was used to assign weights to the underlying indicators in the index system. The fuzzy comprehensive evaluation method was used to evaluate the risk of ice floods in Mohe, Tahe, and Huma regions from 2000 to 2020. Regression analysis of the underlying indicators and evaluation results showed that the correlation coefficient (R^2) based on the catastrophe evaluation method was 0.997 and the root mean square error

(*RMSE*) was 0.00898; the correlation coefficient (R^2) based on the fuzzy evaluation method was 0.995 and the root mean square error (*RMSE*) was 0.00002. The root mean square error of both algorithms is less than 0.05, and the correlation coefficients are significantly correlated, indicating that the catastrophe theory evaluation method has high accuracy in the application of ice flood hazard evaluation. The average risk values obtained from the fuzzy comprehensive evaluation method and the catastrophe theory evaluation method were compared for the Mohe region as presented in Figure 7 and Table 6. The results indicate that the two methods yielded relatively similar results, and the levels of risk values matched perfectly.

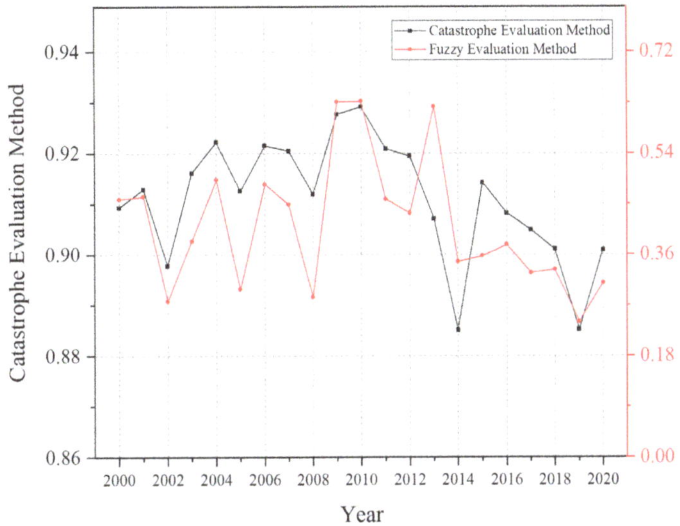

Figure 7. Comparison of the results of the ice flood risk evaluation.

Table 6. The level of value-at-risk for typical years.

Year	Catastrophe Evaluation Method		Fuzzy Evaluation Method	
	Value-at-Risk	Level	Value-at-Risk	Level
2010	0.929	1	0.632	1
2009	0.928	2	0.631	2
2011	0.921	3	0.458	3
2015	0.914	4	0.357	6
2001	0.913	5	0.463	4
2008	0.912	6	0.284	7
2000	0.909	7	0.458	5

5. Conclusions and Future Prospective

This study employed the catastrophe evaluation method to evaluate the risk of ice dam floods in the upper Heilongjiang River spanning from 2000 to 2020. The evaluation findings indicated that the Mohe section, characterized by an intricate and steep river topography, exhibits a higher comprehensive ranking of ice flood risk compared to the other two regions. Regarding the time series analysis, ice floods tend to occur more frequently during years with lower upstream temperatures between October and March, coupled with larger upstream and downstream ice thickness in March. Population, agriculture, economy, and other factors also affect the risk of the occurrence of floods, resulting in a trend where ice flood risk initially increases and then decreases. The results obtained through the catastrophe evaluation method exhibit a similar risk ranking as the fuzzy evaluation method. Furthermore, the catastrophe evaluation method offers the advantages of simplicity in calculation and reduced subjective factors. It eliminates the requirement for precise weighting of underlying indicators and results in a more rational overall distribution of risk values.

In addition, this study still has some limitations that need to be addressed and explored in future research:

1. Problems such as insufficient selection of indicators due to the difficulty of data accessibility may have some influence on the results of the ice flood disaster risk evaluation. However, as the construction and enhancement of the big data platform progress, it will be possible to include a wider range of indicators to enhance the ice flood disaster risk evaluation system. This improvement will contribute to more accurate and reliable results in the future.
2. Using the entropy weight method, in the fuzzy comprehensive evaluation method, to determine the weight of the index may result in distorted evaluation outcomes due to inaccuracies in some of the weights. In future research, we plan to explore alternative weighting techniques or enhanced fuzzy theory to obtain more robust and desirable conclusions. By doing so, we aim to address the limitations and potential distortions associated with the entropy weight method and improve the overall accuracy and reliability of our evaluations.

Author Contributions: Conceptualization, H.H. and E.W.; methodology, Y.L., H.H. and X.L.; formal analysis, H.H. and E.W.; investigation, Y.L., H.H., Y.S., X.X. and H.L.; data curation, Y.L., Y.S., X.X. and H.L.; writing—original draft preparation, Y.L. and H.H.; writing—review and editing, Y.L. and X.L.; funding acquisition, H.H. All authors have read and agreed to the published version of the manuscript.

Funding: This research was supported by the Major Scientific and Technological Projects of the Ministry of Water Resources of China (No. SKS-2022017), the Natural Science Foundation of Heilongjiang Province of China (No. LH2020E004), and the Project to Support the Development of Young Talent by Northeast Agricultural University.

Data Availability Statement: The data are available upon request.

Conflicts of Interest: The authors declare no conflict of interest.

References

1. Das, A.; Reed, M.; Lindenschmidt, K.-E. Sustainable ice-jam flood management for socio-economic and socio-ecological systems. *Water* **2018**, *10*, 135. [CrossRef]
2. Lindenschmidt, K.-E.; Rokaya, P. A stochastic hydraulic modelling approach to determining the probable maximum staging of ice-jam floods. *J. Environ. Inform.* **2019**, *34*, 45–54. [CrossRef]
3. Beltaos, S. *River Ice Jams*; Water Resources Publications, LLC.: Highlands Ranch, CO, USA, 1995; ISBN 978-0-918334-87-9.
4. Xu, X.; Xu, S.; Sun, G.; Ying, Y.; Li, Z. Experimental study on nitrobenzene distribution in cold river water. *Water Resour. Prot.* **2009**, *25*, 67–69+73. (In Chinese)
5. Nadeem, H.; Imran, M.; Saleem, S.; Rafiq, Z.; Batool, A.; Mehmood, K.; Pervaiz, M.; Hussain, S.; Saeed, Z.; Younas, U. GO-Ag-NPs as a promising agent for biomedical, catalytic, electrochemical detection and water treatment technologies; a comprehensive review. *Z. Phys. Chem.* **2023**. [CrossRef]
6. Beltaos, S. Progress in the study and management of river ice jams. *Cold Reg. Sci. Technol.* **2008**, *51*, 2–19. [CrossRef]
7. Williams, B.; Das, A.; Luo, B.; Lindenschmidt, K.-E. An ice jam flood hazard assessment of a lowland river and its terminus inland delta. *Nat. Hazards* **2021**, *105*, 2799–2817. [CrossRef]
8. Kovachis, N.; Burrell, B.C.; Huokuna, M.; Beltaos, S.; Turcotte, B.; Jasek, M. Ice-jam flood delineation: Challenges and research needs. *Can. Water Resour. J.* **2017**, *42*, 258–268. [CrossRef]
9. Song, C.; Zhu, X.; Han, H.; Lin, L.; Yao, Z. The influence of riverway characteristics on the generation and dissipation of ice dam in the upper reaches of Heilongjiang River. *J. Hydraul. Eng.* **2020**, *51*, 1256–1266. (In Chinese) [CrossRef]
10. Beltaos, S.; Prowse, T.; Bonsal, B.; MacKay, R.; Romolo, L.; Pietroniro, A.; Toth, B. Climatic effects on ice-jam flooding of the Peace-Athabasca Delta. *Hydrol. Process.* **2006**, *20*, 4031–4050. [CrossRef]
11. Turcotte, B.; Morse, B.; Pelchat, G. Impact of climate change on the frequency of dynamic breakup events and on the risk of ice-jam floods in Quebec, Canada. *Water* **2020**, *12*, 2891. [CrossRef]
12. Kusatov, K.I.; Ammosov, A.P.; Kornilova, Z.G.; Shpakova, R.N. Anthropogenic factor of ice jamming and spring breakup flooding on the Lena River. *Russ. Meteorol. Hydrol.* **2012**, *37*, 392–396. [CrossRef]

13. Das, A.; Lindenschmidt, K.-E. Modelling climatic impacts on ice-jam floods: A review of current models, modelling capabilities, challenges, and future prospects. *Environ. Rev.* **2021**, *29*, 378–390. [CrossRef]
14. Lindenschmidt, K.-E.; Huokuna, M.; Burrell, B.C.; Beltaos, S. Lessons learned from past ice-jam floods concerning the challenges of flood mapping. *Int. J. River Basin Manag.* **2018**, *16*, 457–468. [CrossRef]
15. Das, A.; Lindenschmidt, K.-E. Current status and advancement suggestions of ice-jam flood hazard and risk assessment. *Environ. Rev.* **2020**, *28*, 373–379. [CrossRef]
16. Burrell, B.C.; Huokuna, M.; Beltaos, S.; Kovachis, N.; Turcotte, B.; Jasek, M. Flood hazard and risk delineation of ice-related floods: Present status and outlook. In Proceedings of the CGU HS Committee on River Ice Processes and the Environment 18th Workshop on the Hydraulics of Ice Covered Rivers, Quebec City, QC, Canada, 18–20 August 2015.
17. Lindenschmidt, K.-E. *River Ice Processes and Ice Flood Forecasting: A Guide for Practitioners and Students*; Springer: New York, NY, USA, 2020; ISBN 978-3-030-28678-1.
18. Turcotte, B.; Alfredsen, K.; Beltaos, S.; Burrell, B.C. Ice-related floods and flood delineation along streams and small rivers. In Proceedings of the CGU HS Committee on River Ice Processes and the Environment 19th Workshop on the Hydraulics of Ice Covered Rivers, Whitehorse, YT, Canada, 10–12 July 2017.
19. Beltaos, S.; Burrell, B.C. Hydrotechnical advances in Canadian river ice science and engineering during the past 35 years. *Can. J. Civil Eng.* **2015**, *42*, 583–591. [CrossRef]
20. Beltaos, S. Distributed function analysis of ice jam flood frequency. *Cold Reg. Sci. Technol.* **2012**, *71*, 1–10. [CrossRef]
21. Lindenschmidt, K.-E. RIVICE-A non-proprietary, open-source, one-dimensional river-ice model. *Water* **2017**, *9*, 314. [CrossRef]
22. Warren, S.; Puestow, T.; Richard, M.; Khan, A.A.; Khayer, M.; Lindenschmidt, K.-E. Near Real-time ice-related flood hazard assessment of the Exploits River in Newfoundland, Canada. In Proceedings of the CGU HS Committee on River Ice Processes and the Environment 19th Workshop on the Hydraulics of Ice Covered Rivers, Whitehorse, YT, Canada, 10–12 July 2017.
23. Guo, X.; Wang, T.; Fu, H.; Guo, Y.; Li, J. Ice-jam forecasting during river breakup based on neural network theory. *J. Cold Reg. Eng.* **2018**, *32*, 04018010. [CrossRef]
24. Mahabir, C.; Hicks, F.E.; Robichaud, C.; Fayek, A.R. Forecasting breakup water levels at Fort McMurray, Alberta, using multiple linear regression. *Can. J. Civil Eng.* **2006**, *33*, 1227–1238. [CrossRef]
25. Mahabir, C.; Hicks, F.; Fayek, A.R. Neuro-fuzzy river ice breakup forecasting system. *Cold Reg. Sci. Technol.* **2006**, *46*, 100–112. [CrossRef]
26. Mahabir, C.; Hicks, F.E.; Fayek, A.R. Transferability of a neuro-fuzzy river ice jam flood forecasting model. *Cold Reg. Sci. Technol.* **2007**, *48*, 188–201. [CrossRef]
27. Wu, C.; Wei, Y.; Jin, J.; Huang, Q.; Zhou, Y.; Liu, L. Comprehensive evaluation of ice disaster risk of the Ningxia–Inner Mongolia Reach in the upper Yellow River. *Nat. Hazards* **2015**, *75*, 179–197. [CrossRef]
28. Luo, D. Risk evaluation of ice-jam disasters using gray systems theory: The case of Ningxia-Inner Mongolia reaches of the Yellow River. *Nat. Hazards* **2014**, *71*, 1419–1431. [CrossRef]
29. Lindenschmidt, K.-E.; Das, A.; Rokaya, P.; Chu, T. Ice-jam flood risk assessment and mapping. *Hydrol. Process.* **2016**, *30*, 3754–3769. [CrossRef]
30. Schanze, J.; Zeman, E.; Marsalek, J. *Flood Risk Management: Hazards, Vulnerability and Mitigation Measures*; Springer: New York, NY, USA, 2006; ISBN 1-4020-4597-2.
31. Carrivick, J.L.; Tweed, F.S. A global assessment of the societal impacts of glacier outburst floods. *Glob. Planet. Chang.* **2016**, *144*, 1–16. [CrossRef]
32. Thom, R. Stabilité structurelle et morphogenèse. *Poetics* **1974**, *3*, 7–19. (In French) [CrossRef]
33. Zhang, H.; Yu, J.; Du, C.; Xia, J.; Wang, X. Assessing risks from groundwater exploitation and utilization: Case study of the Shanghai megacity, China. *Water* **2019**, *11*, 1775. [CrossRef]
34. Singh, L.K.; Jha, M.K.; Chowdary, V.M. Application of catastrophe theory to spatial analysis of groundwater potential in a sub-humid tropical region: A hybrid approach. *Geocarto Int.* **2022**, *37*, 700–719. [CrossRef]
35. Rafi, M.N.; Imran, M.; Nadeem, H.A.; Abbas, A.; Pervaiz, M.; Khan, W.; Ullah, S.; Hussain, S.; Saeed, Z. Comparative influence of biochar and doped biochar with Si-NPs on the growth and anti-oxidant potential of *Brassica rapa* L. under cd toxicity. *Silicon* **2022**, *14*, 11699–11714. [CrossRef]
36. Reghunath, R.; Sreedhara Murthy, T.R.; Raghavan, B.R. The utility of multivariate statistical techniques in hydrogeochemical studies: An example from Karnataka, India. *Water Res.* **2002**, *36*, 2437–2442. [CrossRef] [PubMed]
37. Mahlknecht, J.; Steinich, B.; Navarro de León, I. Groundwater chemistry and mass transfers in the Independence aquifer, central Mexico, by using multivariate statistics and mass-balance models. *Environ. Geol.* **2004**, *45*, 781–795. [CrossRef]
38. Voudouris, K.; Panagopoulos, A.; Koumantakis, J. Multivariate statistical analysis in the assessment of hydrochemistry of the northern Korinthia prefecture alluvial aquifer system (Peloponnese, Greece). *Nat. Resour. Res.* **2000**, *9*, 135–146. [CrossRef]
39. Hussein, M.T. Hydrochemical evaluation of groundwater in the Blue Nile Basin, eastern Sudan, using conventional and multivariate techniques. *Hydrogeol. J.* **2004**, *12*, 144–158. [CrossRef]
40. Bonifacio, A.; Beleites, C.; Sergo, V. Application of R-mode analysis to Raman maps: A different way of looking at vibrational hyperspectral data. *Anal. Bioanal. Chem.* **2015**, *407*, 1089–1095. [CrossRef] [PubMed]
41. Suvedha, M.; Gurugnanam, B.; Suganya, M.; Vasudevan, S. Multivariate statistical analysis of geochemical data of groundwater in Veeranam catchment area, Tamil Nadu. *J. Geol. Soc. India* **2009**, *74*, 573–578. [CrossRef]

42. Yang, W.; Xu, K.; Lian, J.; Bin, L.; Ma, C. Multiple flood vulnerability assessment approach based on fuzzy comprehensive evaluation method and coordinated development degree model. *J. Environ. Manag.* **2018**, *213*, 440–450. [CrossRef]
43. Zou, Q.; Zhou, J.; Zhou, C.; Song, L.; Guo, J.; Yang, X. Flood disaster risk analysis based on variable fuzzy sets theory. *Trans. Chin. Soc. Agri. Eng.* **2012**, *28*, 126–132. (In Chinese)
44. Wang, T.; Liu, Z.; Guo, X.; Fu, H.; Liu, W. Prediction of breakup ice jam with Artificial Neural Networks. *J. Hydraul. Eng.* **2017**, *48*, 1355–1362. (In Chinese) [CrossRef]
45. Sun, W.; Trevor, B. Multiple model combination methods for annual maximum water level prediction during river ice breakup. *Hydrol. Process.* **2018**, *32*, 421–435. [CrossRef]
46. Lindenschmidt, K.-E.; Alfredsen, K.; Carstensen, D.; Choryński, A.; Gustafsson, D.; Halicki, M.; Hentschel, B.; Karjalainen, N.; Kögel, M.; Kolerski, T.; et al. Assessing and mitigating ice-jam flood hazards and risks: A European perspective. *Water* **2022**, *15*, 76. [CrossRef]
47. Wang, X.; Qu, Z.; Tian, F.; Wang, Y.; Yuan, X.; Xu, K. Ice-jam flood hazard risk assessment under simulated levee breaches using the random forest algorithm. *Nat. Hazards* **2023**, *115*, 331–355. [CrossRef]
48. Wang, T.; Guo, X.; Fu, H.; Guo, Y.; Peng, X.; Wu, Y.; Li, J.; Xia, Y. Effects of water depth and ice thickness on ice cover blasting for ice jam flood prevention: A case study on the Heilong River, China. *Water* **2018**, *10*, 700. [CrossRef]

water

MDPI

Article

Investigation of the Recent Ice Characteristics in the Bohai Sea in the Winters of 2005–2022 Using Multi-Source Data

Ge Li [1,2], Yan Jiao [1,2,*], Xue Chen [1,2,*], Yiding Zhao [1,2], Rui Li [1,2], Donglin Guo [1,2], Lei Ge [1,2], Qiaokun Hou [1,2] and Qingkai Wang [3]

1 North China Sea Marine Forecasting Center of State Oceanic Administration, Qingdao 266000, China; lige@ncs.mnr.gov.cn (G.L.); dingdingdingdin@126.com (Y.Z.); lirui@ncs.mnr.gov.cn (R.L.); gdlcici@163.com (D.G.); gexinlei@126.com (L.G.); qd_hdk@163.com (Q.H.)
2 Shandong Key Laboratory of Marine Ecological Environment and Disaster Prevention and Mitigation, Qingdao 266000, China
3 State Key Laboratory of Coastal and Offshore Engineering, Dalian University of Technology, Dalian 116024, China; wangqingkai@dlut.edu.cn
* Correspondence: jiaoyan87@163.com (Y.J.); chenxue@ncs.mnr.gov.cn (X.C.)

Abstract: The safety of winter activities in the Bohai Sea requires more detailed information on ice characteristics and a more refined ice zone division. In the present study, 1/12°-resolution sea ice characteristic data were obtained based on the NEMO-LIM2 ice–ocean coupling model that assimilated MODIS satellite sea ice observations from the years of 2005 to 2022 to acquire new sea ice hindcasting data. On this basis, the ice period, ice thickness, ice concentration, ice temperature, ice salinity, and design ice thickness for different return periods in the 1/4°-resolution refined zoning were analyzed, which were then compared with the sea ice characteristics in the previous 21-ice-zone standard. The distribution of ice temperature and ice salinity was closely related to the distribution of ice thickness. The results of ice period, ice thickness, and ice concentration, as well as design ice thickness for different return periods, and the comparison with the previous 21-ice-zone standards, showed that the ice condition on the west coast of the Bohai Sea has significantly reduced.

Keywords: sea ice; ice characteristics; ice zone division; winter; Bohai Sea

check for updates

Citation: Li, G.; Jiao, Y.; Chen, X.; Zhao, Y.; Li, R.; Guo, D.; Ge, L.; Hou, Q.; Wang, Q. Investigation of the Recent Ice Characteristics in the Bohai Sea in the Winters of 2005–2022 Using Multi-Source Data. *Water* **2024**, *16*, 290. https://doi.org/10.3390/w16020290

Academic Editor: Hung Tao Shen

Received: 19 November 2023
Revised: 3 January 2024
Accepted: 6 January 2024
Published: 15 January 2024

1. Introduction

The Bohai Sea, the inland sea in the northern China, has a cold winter climate and is the southern boundary of the frozen waters in the Northern Hemisphere [1]. In winter, sea ice in the Bohai Sea poses severe threats to port transportation, offshore oil and gas exploration, marine fisheries, etc. [2]. To ensure the security of human activities in the Bohai Sea during winter, it is necessary to know the characteristics of sea ice in the Bohai Sea in winter [3].

In order to facilitate distinguishing the differences in the ice characteristics in a large-scale sea area, the target sea area needs to be zoned according to various ice conditions [4]. Unlike the ice in the Arctic Ocean, all the ice in the Bohai Sea is first-year ice. Under the influence of local thermal and dynamic factors, the sea ice in the Bohai Sea is mainly drift ice, with significant interannual variations [5]. The engineering sea ice zone division in the Bohai Sea is primarily based on a series of sea ice parameters, such as ice concentration, ice thickness, ice type, ice period, etc. In 1984, the ARCTIC Corporation of the USA first divided the Bohai Sea into three ice zones: the Bohai Bay Ice Zone, the Bozhong Ice Zone, and the Laizhou Bay Ice Zone [6]. A few years later, Liu et al. divided Liaodong Bay into the northern and southern ice zones, considering the different sea ice properties and climatic conditions [7]. Afterward, Shen conducted an ocean survey to investigate China's sea ice intensity division in 1992 [6]. He divided the Bohai Sea into nine ice zones, considering the prospects for offshore oil and gas development. Entering the 21st century, Wu et al.

divided the nearshore areas of the Bohai Sea and the northern Yellow Sea into 21 ice zones based on the fundamental characteristics of the distribution and changes of the sea ice in China, considering the convenience of engineering design and production departments [8]. The 21-ice-zone scheme is currently the most commonly used standard [4,9].

The engineering ice-resistance designs are based on the sea ice characteristic parameters [10]. In recent years, the overall ice condition in the Bohai Sea has been milder, but extreme meteorological events, especially the strong cold waves, have gradually increased in winter [2]. Along the Bohai coast, the seasonal resources of wind energy [11], solar thermal energy [12], and wave energy [13] are abundant. With increased economic activity in the Bohai Sea in winter, potential sea ice disasters have become increasingly prominent. For the safe operation of industrial activities in Bohai, it is necessary to consider the impact of sea ice [1,3].

Ice engineering in the Bohai Sea has developed since the 1980s, and Chinese scholars and engineers have conducted many studies on sea ice problems [14,15]. Li et al. [16] proposed a division of ice engineering sub-areas based on the ice physical and mechanical parameter distributions in Bohai. Li et al. [2] studied the effect of porosity on the uniaxial compressive strength of Bohai sea ice. With global warming, the ice season shortens, the ice thickness decreases, and the drift ice occupies most of the existing sea ice in winter in the Bohai Sea [17,18]. The sea ice characteristic parameters and ice zone division obtained based on marine environmental survey data from decades ago cannot meet the current engineering ice-resistance designs. There is an urgent need for more detailed information on ice characteristics to ensure the safety of activities in the Bohai Sea in winter.

In the present study, ice characteristics with a resolution of $1/4^{\circ}$ in the Bohai Sea in the winters of 2005–2022 were evaluated using ice–ocean coupling model data that assimilated satellite data. Firstly, sea ice characteristic data with a resolution of $1/12^{\circ}$ were obtained based on the Nucleus for European Modelling of the Ocean, version 3.6 (NEMO3.6) and the large-scale Louvain-la-Neuve Sea Ice Model, version 2 (LIM2), which assimilated sea ice satellite remote sensing observations from the Moderate Resolution Imaging Spectroradiometer (MODIS) from 2005 to 2022 to acquire new sea ice hindcasting data. The hindcasting data were then compared and validated using long-term sea ice observations from coastal ocean stations and the oil platform in the Jinzhou area of Liaodong Bay from 2010 to 2018. Based on the long-time-series hindcasting data, the ice period, ice thickness, ice concentration, ice temperature, ice salinity, and design ice thickness for different return periods in the $1/4^{\circ}$-resolution refined zoning were analyzed and compared with the sea ice characteristics of 21 ice zones in the China National Offshore Oil Corporation (CNOOC) standard "Regulations for offshore ice condition and application in China sea (Q/HSn 3000-2002)" [19].

2. Materials and Methods

2.1. Study Area

The Bohai Sea is located in the mid-latitude monsoon region of the Northern Hemisphere (37°07′ N–41°00′ N; 117°35′ E–121°10′ E). From a geomorphological perspective, the Bohai Sea is a large bay that extends inland from the Yellow Sea. The Bohai Sea is mainly composed of five parts: the Liaodong Bay in the north, the Bohai Bay in the west, the Laizhou Bay in the south, the shallow sea basin in the center, and the Bohai Strait connected to the Yellow Sea. The Bohai Sea has a coastline length of approximately 3.80×10^{3} km, an area of approximately 8.00×10^{4} km^{2}, and an average depth of 18 m. In the Bohai Sea, it is controlled by the Asian continental high pressure and is dominated by northerly winds in winter. The changes in water temperature in the Bohai Sea are mainly influenced by the continental climate. Due to the shallow water in the Bohai Sea, the water temperature responds quickly to the air temperature. The water temperature in January is lower than in February, and the water temperature at the top of the three major bays is below 0 °C.

In general, the Bohai Sea freezes from north to south in late November or early December, starting from shallow water areas on the shore to deep water areas. In mid-to late-February of the following year, it melts from deep water areas to shallow water areas from south to north. The ice period is 3–4 months, and by mid-March, all sea ice disappears. The severe ice period in various sea areas generally occurs from mid-January to mid-February. Due to significant differences in ice period, ice thickness, ice drift speed and direction, ice cover range, and geographical environment among different sea areas, there are significant differences in ice conditions among different sea areas.

To distinguish the differences in the severity of ice conditions for marine engineering designs, the ice zone division in the Bohai Sea has been conducted in the past few decades [6–9]. The previously used ice zone division was proposed in the CNOOC standard "Regulations for offshore ice condition and application in China Sea (Q/Hsn 3000-2002)". Based on the actual situation in the Bohai Sea, the design and operation experience of 15 oil and gas fields in the Bohai Sea during 1968–1998 were summarized. The coastal place names were used as the division names, and the values of general sea ice conditions, such as ice days and ice thicknesses during ice periods, were set using the survey statistics of shore stations. The ice zone division in the standard is too rough, and the guidance and refinement of the zone for offshore platforms and other engineering projects need to be improved.

Based on completing the hindcasting data from 2005 to 2022, the present study provides a new division of the sea ice zones in the Bohai Sea area. Figure 1 depicts the new division scheme. The Bohai Sea (37.125° N–40.875° N and 117.625° E–122.125° E) is divided into a total of 139 ice zones with a spatial resolution of 1/4°.

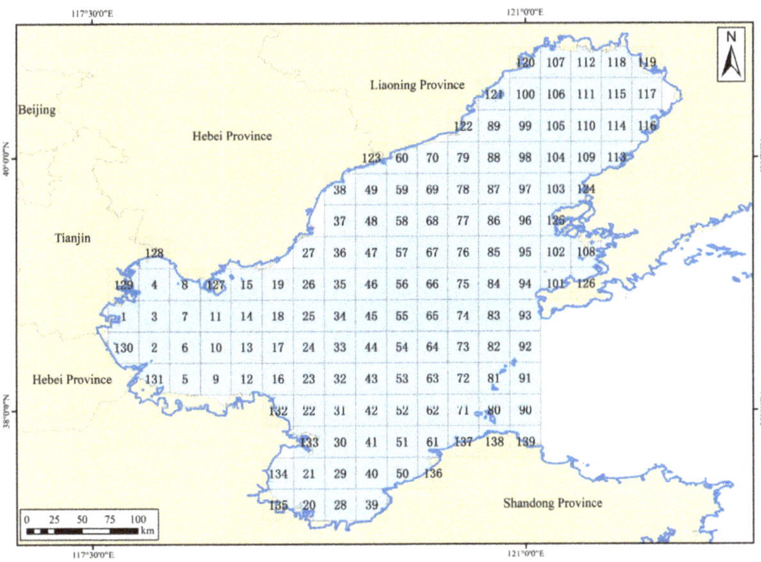

Figure 1. 1/4°-resolution sea ice zoning in the Bohai Sea.

2.2. Sea Ice Hindcasting Modelling

The sea ice hindcasting data used in the current study were acquired from a sea ice prediction model for the Bohai Sea established based on the NEMO3.6 ocean model in the ORCA configuration coupling the LIM2 sea ice model [20,21].

The model domain is located in a spatial range of 35° N–41° N, 117° E–127° E, covering the entire Bohai Sea and the Northern Yellow Sea with a spatial resolution of 1/12°. The topographic data in the NEMO model uses the ETOPO1 public data set, with a spatial resolution of 1 km, which is interpolated into the NEMO model grid through bilinear interpolation [22]. The ice–ocean coupling model can finely simulate the ocean and sea ice environments near complex islands and shorelines.

The ocean module is vertically divided into 36 layers along the *z*-axis, using the GLS framework and the k-ε turbulent closure scheme to calculate the vertical vortex viscosity and diffusion coefficient. The viscosity coefficient in the horizontal direction is set to 50 $m^2 \cdot s^{-1}$, and the side boundary is set to a non-sliding boundary. The bottom friction coefficient varies with space, with a minimum value of 5×10^{-4}. The sea ice module includes two calculation components: sea ice thermodynamics and dynamics, using the elastic-plastic viscosity (EVP) sea ice rheology calculation concept. The sea ice module includes a snow layer and two sea ice layers in the vertical direction. The sea ice module and ocean module perform real-time coupling calculations, exchanging data every 360 s. To accurately simulate the Bohai sea ice, the sea ice generation parameter is set to 0.3 m. The sea ice boundary is a non-sliding condition. The model is driven by wind speed, air temperature, relative humidity, longwave radiation, shortwave radiation, precipitation, and snow from the National Centers for Environmental Prediction (NCEP) Climate Forecast System Reanalysis (CFSR) reanalysis data. The initial field and open boundary conditions are interpolated using data such as seawater temperature, salinity, sea surface height, and flow velocity in each layer from the global forecast data of the European Copernicus Marine Environment Monitoring Service (CMEMS).

In order to improve the reliability of the sea ice hindcasting data, satellite remote sensing observations of sea ice concentration from the MODIS were assimilated into the NEMO3.6 ocean model using the Ensemble Adjustment Kalman Filter (EAKF) method. The EAKF method uses the dependency relationship and spatial correlation between variables in the model to estimate the optimal state variables of the model from a probabilistic and statistical perspective [23]. Figure 2 shows that the modeled sea ice concentration is closer to the satellite observation after EAKF assimilation. From the distribution map of sea ice concentration difference and assimilation adjustment amount (Figure 3), it can be seen that the assimilation adjustment of sea ice effectively compensates for the relative observation bias of the model, indicating that the sea ice assimilation method is correct and the assimilation effect is obvious.

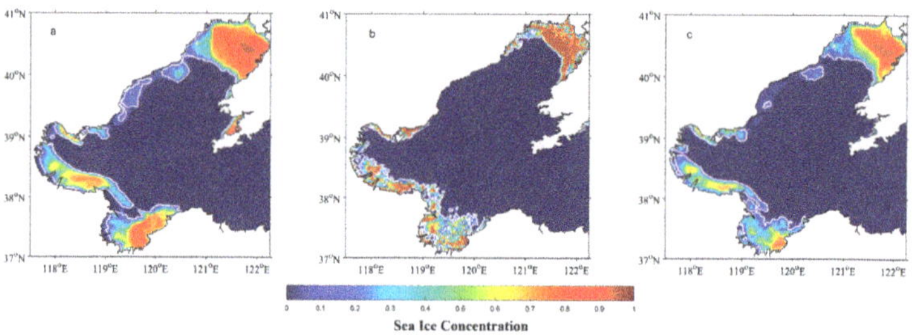

Figure 2. Distribution map of sea ice concentration from the (**a**) MODIS, (**b**) NEMO-LIM2 model, and (**c**) assimilated model data.

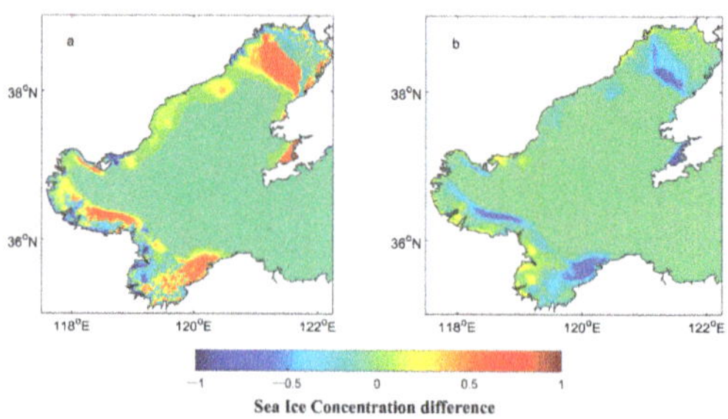

Figure 3. Distribution map of (**a**) sea ice concentration difference and (**b**) assimilation adjustment amount.

The assimilation frequency was once a week during the freezing period. Figure 4 compares errors in sea ice concentration before and after the assimilation of MODIS data into the NEMO3.6 ocean model in the Bohai Sea in the winter of 2021/2022. After assimilating the MODIS data, model prediction accuracy was significantly improved, with an average error of 6.7% in sea ice concentration.

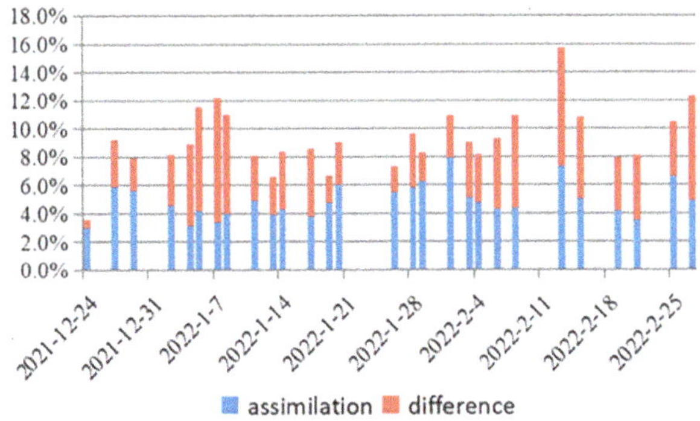

Figure 4. Errors in sea ice concentration before and after the assimilation of MODIS data into the NEMO3.6 ocean model in the Bohai Sea in the winter of 2021/2022. For each column, the blue bar represents the sea ice concentration error after the assimilation, and the red bar represents the difference between the sea ice concentration error before and after the assimilation.

In order to improve the accuracy of evaluation and reduce the impact of model errors on evaluation results, the adjacent grids are usually combined on the basis of subdivision grids, and several fine-grid data covered by coarse grids are evaluated by means of average or extreme value. In this paper, the sea ice data of nine 1/12° grids of the NEMO model covered by a 1/4° grid are used to evaluate the sea ice data of the 1/4° grid. That is, the maximum ice thickness of the nine 1/12° grids is selected to represent the ice thickness in the 1/4° grid.

2.3. Design Sea Ice Parameters

Sea ice temperature refers to the temperature inside the ice sheet. In actual natural ice sheets, the type of ice and its vertical temperature distribution are very complex, influenced by factors such as air temperature, water temperature, ice thickness, and ice heat transfer coefficient. In engineering designs, it is necessary to determine a single ice temperature design value under the most severe condition, i.e., the effective ice temperature. The effective ice temperature is the average temperature of the ice sheet under the lowest air temperature. Assuming no phase change occurs in the ice, the effective ice temperature can be calculated according to the following formula [10]:

$$T_{\mathrm{i}} = (T_{\mathrm{ia}} + T_{\mathrm{iw}})/2 \tag{1}$$

where T_{i} is the effective ice temperature (°C); T_{ia} is the upper surface temperature of the ice sheet (°C); and T_{iw} is the lower surface temperature of the ice sheet (°C). T_{ia} and T_{iw} are related to sea ice thickness, salinity, and air temperature.

Sea ice salinity mainly depends on the salinity and freezing speed of seawater before freezing. The effective salinity of sea ice can be calculated as follows [24]:

$$S_{\mathrm{i}} = 19.007H^{-0.387} \tag{2}$$

where S_{i} is the effective ice salinity (‰); H is the sea ice thickness (cm)

The determination of design ice thickness used the average of estimates by the Pearson-III (P-III) distribution and the Weibull distribution. When determining the design ice thickness, an excessively small estimate may lead to damage to marine structures, while an excessively large estimate can result in increased costs. The biggest advantage of the P-III distribution is its large elasticity. In most cases, it can be fitted well with the theoretical curve and empirical frequency points by repeatedly fitting the line or adjusting the coefficient of variation and mean appropriately.

The probability density function of P-III distribution is:

$$f(x) = \frac{\beta^{\alpha}}{\Gamma(\alpha)}(x - x_0)^{\alpha - 1}\exp[-\beta(x - x_0)] \quad (3)$$

where $\Gamma(\alpha)$ is the gamma function, and α, β as well as x_0 are the shape, scale and position coefficients, respectively, related to the statistical parameters of the random variables:

$$\alpha = \frac{4}{{C_s}^2} \quad (4)$$

$$\beta = \frac{2}{\overline{x}C_v C_s} \quad (5)$$

$$x_0 = \overline{x}(1 - \frac{2C_v}{C_s}) \quad (6)$$

where C_s is skewness coefficient, C_v is variation coefficient, and $\overline{x}$ is the average.

Therefore, the determination of the probability density function of the P-III distribution is transformed into the determination of the statistical parameters. In general, the observations of ice conditions are not long enough; the empirical distribution based on observations must be extended to determine the design ice thickness. The preliminary values of the statistical parameters can be obtained using the observation data, and an empirical distribution curve is depicted. Afterwards, the statistical parameters are adjusted until the corresponding empirical distribution curve fits well with the observations. The statistical parameters after adjustment are selected to determine the final P-III distribution, and the design ice thickness is determined from the corresponding cumulative frequency density curve according to the return period.

The probability density function of the Weibull distribution is:

$$f(x) = \frac{\beta}{\alpha}(\frac{x - \mu}{\alpha})^{\beta - 1}\exp[-(\frac{x - \mu}{\alpha})^{\beta}] \quad (7)$$

where α, β, and μ are scale, shape, and position coefficients, respectively.

The parameters of the Weibull distribution are determined using the maximum likelihood estimates. After determining the parameters, the design ice thickness is determined from the corresponding cumulative frequency density curve according to the return period.

3. Results

3.1. Model Verification

The JZ9-3 and JZ20-2 oil platforms are located in the northern waters of Liaodong Bay in the Bohai Sea, where ice conditions are severe in winter. Figure 5 shows the sea ice thickness monitoring data from the two oil platforms and the hindcasting sea ice thickness data from 2010 to 2018. At the location of the JZ9-3 platform, the correlation coefficient between the hindcasting sea ice thickness and observations is 0.64, with an absolute error of 3.15 cm. At the location of the JZ20-2 platform, the correlation coefficient between the hindcasting sea ice thickness and observation is 0.71, with an absolute error of 3.69 cm. The hindcasting data can better reflect different ice conditions between mild and severe ice years.

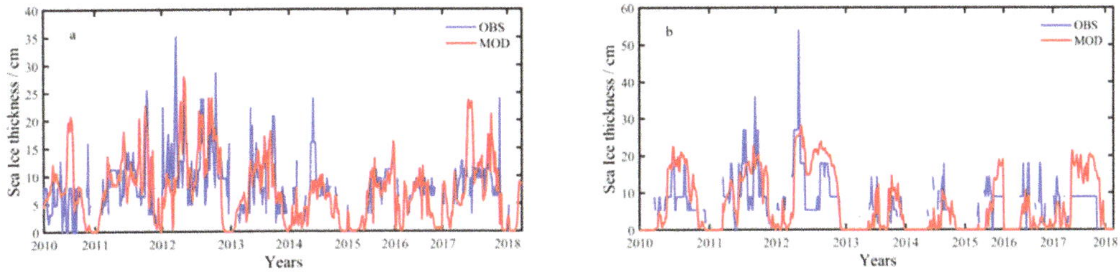

Figure 5. Comparisons of hindcasting sea ice thickness data and observations from (**a**) JZ9-3 and (**b**) JZ20-2 oil platforms in the winters of 2010–2018. The blue lines represent the observations, and the red lines represent the hindcasting sea ice thickness.

A comparative analysis of ice periods was conducted in Figure 6 using sea ice observations from three artificial observation stations along the coast of the Bohai Sea and hindcasting data from the nearest grid cells. The overall length of the monitoring ice period at each station is greater than that obtained by the numerical model. It mainly considers that the sea ice during the initial ice period mainly forms in the shallow shoals and semi-enclosed bays near the shore and eventually melts in these nearshore areas after warming up in spring. Therefore, the length of the artificial monitoring ice period conducted near the shore is generally slightly longer than that reported by the numerical model. At locations such as Bayuquan Ocean Station and Wentuozi Ocean Station, which are located to the north, data from ocean stations with longer ice periods are continuous, and the hindcasting results are consistent with the monitoring of ice periods. The hindcasting data can better reflect the distribution of ice periods in mild and severe ice years.

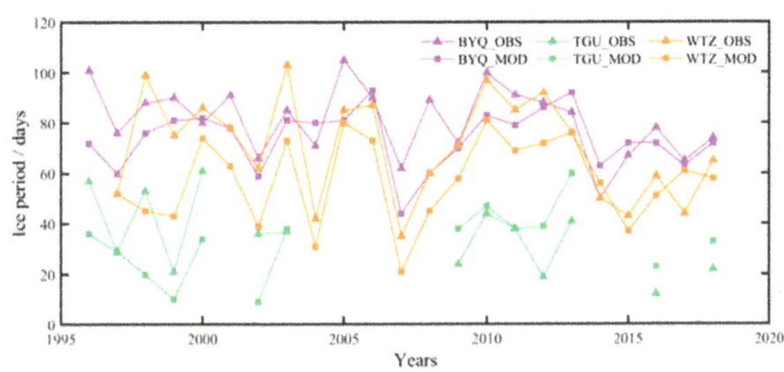

Figure 6. Comparisons of ice period from ocean stations and numerical model. The purple, green, and orange lines represent the Bayuquan (BYQ), Tanggu (TGU), and Wentuozi (WTZ) ocean stations, respectively. Lines with triangle and square marks represent observations from ocean stations and hindcasting data from the grid cell closest to the ocean station, respectively.

Sea ice temperature and salinity were not verified in this study because these two parameters are calculated by sea ice thickness and air temperature. Therefore, we focused our verification efforts on the ice thickness, which plays a key role in the estimation of sea ice temperature and salinity. According to the current standard used for sea ice observation in China, the sea ice concentration in the conventional shore station is mainly determined as the proportion of ice in the total observation area, which is different from the calculation of sea ice concentration in the numerical model (the proportion of ice units to the total number of units). Therefore, no validation was carried out for sea ice concentration. There is a good consistency between ice thickness and concentration, so the validation effect of ice thickness can also represent the accuracy of other sea ice parameters to a certain extent.

3.2. Ice Period, Thickness, and Concentration

Figure 7 plots the average severe ice period in the Bohai Sea from 2005 to 2022. The average severe ice period in Liaodong Bay was the longest, which is consistent with the phenomenon of sea ice accumulation on the east coast caused by the winter northwest monsoon. Moreover, influenced by the warm current of the Yellow Sea, the water temperature on the south coast of Bohai Bay is lower than that on the north coast, resulting in a longer severe ice period in the south of Bohai Bay compared to the north [25]. The average severe ice period in Laizhou Bay is shorter than that in Liaodong Bay and Bohai Bay, with only 10 to 20 days.

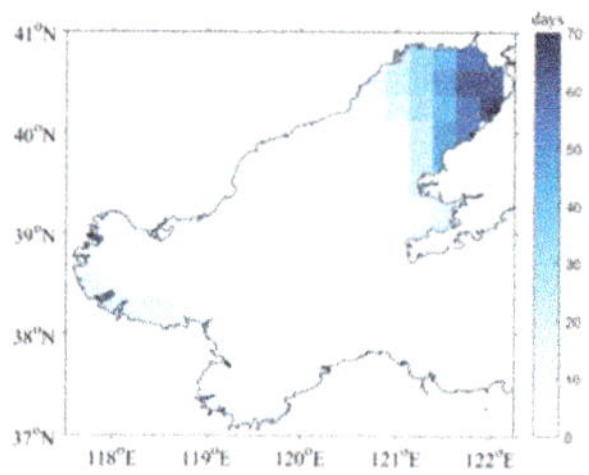

Figure 7. Average severe ice period in the Bohai Sea from 2005 to 2022.

Figure 8 shows the maximum sea ice thickness, average sea ice thickness, maximum sea ice concentration, and average sea ice concentration in the Bohai Sea from 2005 to 2022. The maximum and average ice thickness show relatively consistent spatial distribution characteristics. The sea ice thickness in the northeast of Liaodong Bay is the largest, and in the central area of the Bohai Sea it is the smallest. The sea ice thickness at the bottom of the bays is larger than that at the bay mouth. The sea ice thickness shows the characteristics of Liaodong Bay are large in the east and small in the west; those of Bohai Bay are large in the south and small in the north; and those of Laizhou Bay are large in the east and small in the west, which is consistent with the distribution of water depth.

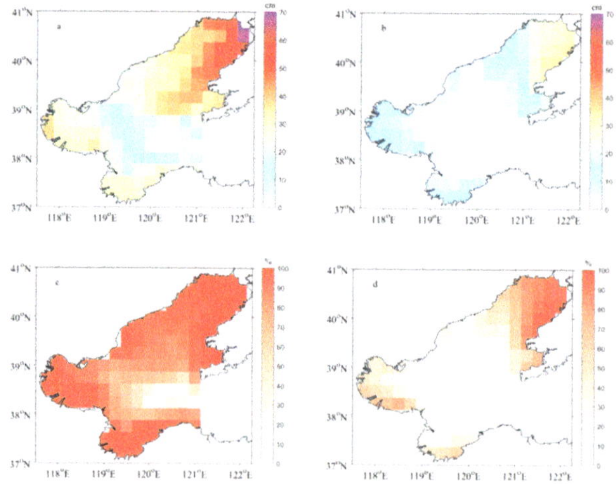

Figure 8. (**a**) Maximum sea ice thickness, (**b**) average sea ice thickness, (**c**) maximum sea ice concentration, and (**d**) average sea ice concentration in the Bohai Sea from 2005 to 2022.

The spatial distribution characteristics of sea ice concentration and thickness are relatively consistent. The sea ice in Liaodong Bay has the largest freezing range and slowest heat exchange with the outer sea. In addition, the thermal effect of the cold continent in winter is conducive to ice growth [26]. Under the influence of the northwest monsoon, sea ice accumulates towards the east. When the northward and eastward monsoons blow continuously in winter, under the action of clockwise reflux, drift ice can always exist in the areas of Bayuquan and Changxing Island [27].

Figure 9 plots the difference in the average severe ice period between the present study and the 21-ice-zone standard. Compared with the original 21-ice-zone standard, the latest average severe ice period in most areas of the Bohai Sea decreased, with the southern and western coast of the Bohai Sea experiencing more significant shortening, while the eastern part of Liaodong Bay increased slightly. Both the latest result and the old standard showed that there was no severe ice period in the central Bohai Sea.

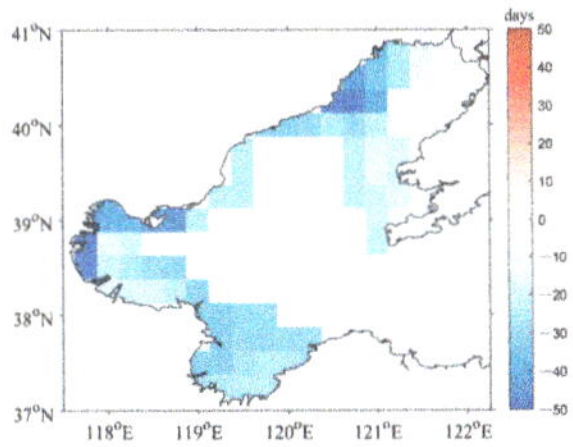

Figure 9. Differences in average severe ice period between the present study and the 21-ice-zone standard.

Figure 10 shows the differences in the sea ice thickness and concentration between the present study and the 21-ice-zone standard. Compared with the original standard values of the 21 ice zones, the maximum ice thickness in most areas of the Bohai Sea decreased, while the maximum ice thickness on the eastern side of Liaodong Bay increased, with an overall increase of about 10 cm. The decrease in sea ice thickness is most significant near the No. 43 zone in the central Bohai Sea. The average thickness and maximum thickness distribution of sea ice were similar, except for the eastern part of Liaodong Bay; there was also a slight increase in the central part of Bohai Bay.

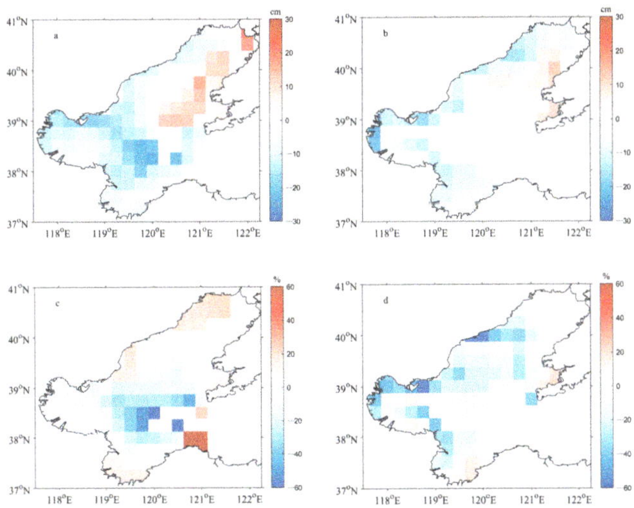

Figure 10. Differences in (**a**) maximum sea ice thickness, (**b**) average sea ice thickness, (**c**) maximum sea ice concentration, and (**d**) average sea ice concentration between the present study and the 21-ice-zone standard.

For the sea ice concentration, there were four significantly higher ice zones in the eastern area of Laizhou Bay, which had a maximum sea ice concentration of 18.5% in the

No. 20 ice zone in the original standard. By contrast, the maximum sea ice concentration of the four ice zones was between 80.3% and 97.2% in the latest results, with an average sea ice concentration of 4.8–9.5%. In extreme years, the ice condition is severe, and there is some accumulation of sea ice in the eastern part of Laizhou Bay, with a maximum sea ice concentration of 97.2% occurring in 1969.

3.3. Ice Temperature and Salinity

Figure 11 plots the results of the average ice temperature and salinity in the Bohai Sea from 2005 to 2022, calculated using Equations (1) and (2). The lower surface temperature of sea ice is equal to the freezing temperature of seawater. The difference in the lower surface temperature of sea ice in different ice zones is not significant, and the temperature of the upper surface of sea ice is mainly affected by air temperature. The average effective ice temperature of each ice zone in the Bohai Sea showed that the southern ice zone was higher than the northern ice zone, which is consistent with the air temperature distribution. The average effective ice temperature at the bottom of Liaodong Bay was the lowest. Because sea ice salinity is a function of sea ice thickness, the salinity of sea ice is lower in areas with thicker sea ice. The distribution of average effective salinity showed that the ice zones in the eastern part of Liaodong Bay, the southern part of Bohai Bay, and the southern part of Laizhou Bay were lower than other ice zones.

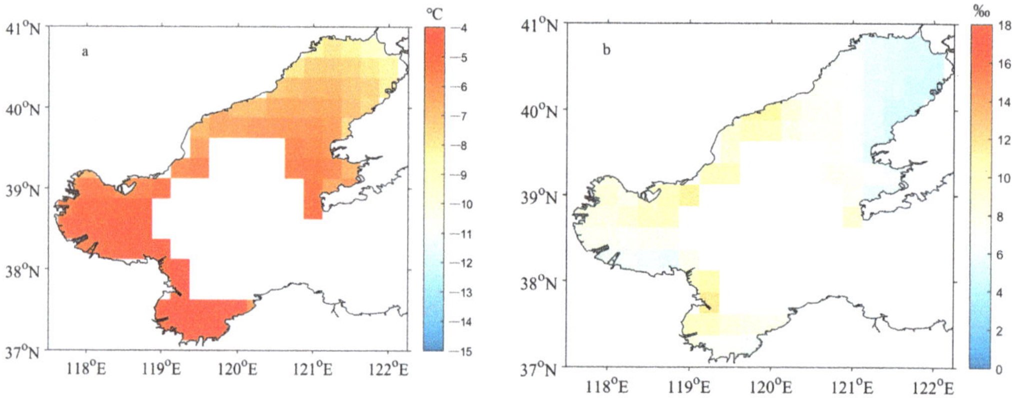

Figure 11. (**a**) Average ice temperature and (**b**) average ice salinity in the Bohai Sea from 2005 to 2022, calculated using Equations (1) and (2).

3.4. Design Ice Thickness

Figure 12 depicts the design ice thickness for different return periods in the Bohai Sea. The design ice thickness in Liaodong Bay was the thickest, followed by Bohai Bay and Laizhou Bay. There was no distribution of sea ice in the central area of the Bohai Sea. The largest design ice thickness for a 100-year return period in the Bohai Sea appeared in the No. 119 ice zone in the northeast of Liaodong Bay, reaching 62.5 cm. In Bohai Bay and Laizhou Bay, the largest design ice thicknesses were larger than 40 cm. For design ice thickness for a 10-year return period in the Bohai Sea, the largest value was larger than 50 cm in Liaodong Bay while smaller than 30 cm in Bohai Bay and Laizhou Bay.

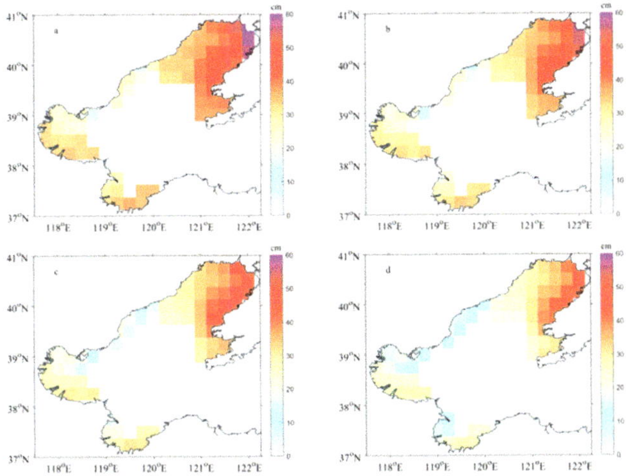

Figure 12. Design ice thickness for (**a**) 100-year, (**b**) 50-year, (**c**) 25-year, and (**d**) 10-year return periods in the Bohai Sea.

Figure 13 depicts differences in design ice thickness for different return periods between the present study and the 21-ice-zone standard. Compared with the design ice thickness for different return periods in the original 21-ice-zone standard, the latest results showed an overall decreasing trend in the Bohai Sea. The design ice thickness for 100-year and 50-year return periods significantly decreased along the west coast of the Bohai Sea compared to the original 21-ice-zone standard. The maximum decrease in coastal ice zones was 15 cm, and the zones with the largest decrease were mainly distributed in the coastal areas of Tianjin, Tangshan, Qinhuangdao, Jinzhou, and the west coast of Laizhou Bay.

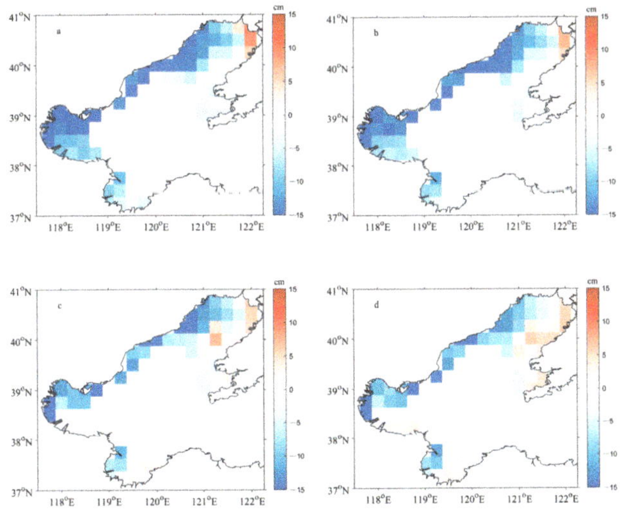

Figure 13. Differences in design ice thickness for (**a**) 100-year, (**b**) 50-year, (**c**) 25-year, and (**d**) 10-year return periods between the present study and the 21-ice-zone standard.

Compared with the design ice thickness for different return periods in the original 21-ice-zone standard, the eastern sea area of Liaodong Bay increased slightly. There are two main reasons for the increase in design ice thickness for different return periods in this area. Firstly, the increasing trend of distribution of sea ice in Liaodong Bay was mild in the west and severe in the east, which is caused by the drift and accumulation of sea ice from west to east [18]. Secondly, the ice zone division in the CNOOC standard often uses coastal ice conditions to represent the ice conditions of the entire ice zone, and the new refined zoning method can more prominently depict the changes and differences between coastal and offshore ice zones.

4. Discussion and Conclusions

In this study, the ice conditions predicted by the numerical modeling were compared with the information observed at the ocean stations along the coast. As shown in Figure 6, the agreements between hindcasting data and observations were good in general, but the ice periods provided by the hindcasting data were shorter than the observed ice periods. One reason is that, as mentioned earlier, sea ice in the shallow shoals forms earlier and melts later. The difference also reflects the low applicability of such hindcasting numerical models for predicting ice conditions in shallow coastal waters. For sea ice forecasts in the vicinity of ocean stations with long-term observations, the method of artificial intelligence analysis can be considered, which may be more suitable for modeling the development of sea ice in shallow areas with complex topography, depth, and tide.

It is also noteworthy from Figure 2 that the sea ice development on the west coast of the Bohai Sea provided by hindcasting data was later and lighter than the information drawn from satellite observations. The reason is that the west coast of the Bohai Sea is the first affected area after the invasion of cold air from Siberia into the Bohai Sea. Therefore, this area is prone to the growth of large areas of sea ice in the short term. It is necessary to further adjust the parameter setting of the numerical model on the west coast of the Bohai Sea in future research work.

Sea ice modeling research is very dependent on the observation of ice conditions for validation, especially ice thickness. While the accuracy of the sea ice thickness derived from the satellite images is difficult to validate. In future research, the EM-31 aboard the ship in ice will be used to accumulate observational data on ice thickness, which is finally used to improve the accuracy of hindcasting data.

This study applied sea ice observations and hindcast data from the NEMO-LIM2 ice–ocean coupling model to establish a refined sea ice zone division in the Bohai Sea and analyzed the ice condition characteristics, including ice period, thickness, concentration, temperature, and salinity, as well as the design ice thickness for different return periods constructed by the average values of the P-III and Weibull methods. According to the analysis, the average effective ice temperature of the Bohai Sea shows a distribution trend of warm southwest and cold northeast, and the average effective ice salinity shows a low salinity distribution at the bottom of Liaodong Bay, southern Bohai Bay, and southern Laizhou Bay. The distribution of ice temperature and ice salinity is closely related to the distribution of ice thickness; that is, there is consistency in the distribution of low ice temperature, low ice salinity, and large ice thickness areas. The results of ice period, ice thickness, and ice concentration, as well as design ice thickness for different return periods and the comparison with the current standards of CNOOC, show that the design ice thickness on the west coast of the Bohai Sea has significantly decreased. The calculated values of design ice thickness on the east coast of Liaodong Bay have increased, indicating that the fine ice zone division has a more refined evaluation conclusion than the coarse ice zone division. At the same time, it is also a manifestation of the distribution pattern of mild ice conditions in the west and severe ice conditions in the east of Liaodong Bay. The analysis methods adopted in this paper can also be utilized in Arctic Sea ice engineering.

Author Contributions: Conceptualization, G.L. and X.C.; methodology, Y.J.; software, Y.Z.; validation, R.L.; formal analysis, D.G.; investigation, L.G.; resources, Q.H.; data curation, R.L.; writing—original draft preparation, G.L. and X.C.; writing—review and editing, Y.J. and Q.W. All authors have read and agreed to the published version of the manuscript.

Funding: This research was funded by the National Key Research and Development Program of China (2022YFE0107000).

Data Availability Statement: Data are contained within the article.

Acknowledgments: The authors thank the editor and anonymous reviewers for their valuable comments and suggestions to this paper.

Conflicts of Interest: The authors declare no conflicts of interest. The funders had no role in the design of the study; in the collection, analyses, or interpretation of data; in the writing of the manuscript; or in the decision to publish the results.

References

1. Yang, G. Bohai sea ice conditions. *J. Cold Reg. Eng.* **2000**, *14*, 54–67. [CrossRef]
2. Li, Z.; Zhang, L.; Lu, P.; Leppäranta, M.; Li, G. Experimental study on the effect of porosity on the uniaxial compressive strength of sea ice in Bohai Sea. *Sci. China Technol. Sci.* **2011**, *54*, 2429–2436. [CrossRef]
3. Ouyang, L.; Hui, F.; Zhu, L.; Cheng, X.; Cheng, B.; Shokr, M.; Zhao, J.; Ding, M.; Zeng, T. The spatiotemporal patterns of sea ice in the Bohai Sea during the winter seasons of 2000-2016. *Int. J. Digit. Earth* **2019**, *12*, 893–909. [CrossRef]
4. Tao, S.; Dong, S.; Wang, Z.; Guedes Soares, C. Intensity division of the sea ice zones in China. *Cold Reg. Sci. Technol.* **2018**, *151*, 179–187. [CrossRef]
5. Li, Z.; Xiu, Y.; Wang, Q.; Li, G.; Lu, P.; Zhong, S.; Chen, X. On the method of design drift ice concentration and floe area for new-energy structures in ice-infested nearshore areas of the Bohai Sea, China. *Front. Energy Res.* **2022**, *10*, 947153. [CrossRef]
6. Shen, Y. Sea Ice Intensity Division Based on Cluster Analysis in the Bohai Sea and the Northern of Yellow Sea. Master' Thesis, Ocean University of China, Qingdao, China, 4 June 2013. (In Chinese).
7. Liu, D.; Li, T.; Zhang, T.; Wang, J. Probabilistic analysis of sea ice conditions in the Liaodong Bay in the Bohai Gulf. *J. Tianjin Univ.* **1987**, *4*, 48–56. (In Chinese)
8. Wu, H.; Yang, G.; Zhang, F.; Li, Y.; Zhang, M.; Li, H.; Li, H.; Wang, Z. Sea Ice Environment and Characteristics. In *Sea Ice Design Operating Conditions in the Bohai Sea*; China Ocean Press: Beijing, China, 2001; pp. 3–8. (In Chinese)
9. Zhang, X.; Zhang, Z.; Xu, Z.; Li, G.; Sun, Q.; Hou, X. Sea ice disaster and their impacts since 2010 in Laizhou Bay of Bohai Sea, China. *Nat. Hazards* **2013**, *65*, 27–40. [CrossRef]
10. Wang, A.; Tang, M.; Zhao, Q.; Liu, Y.; Li, B.; Shi, Y.; Sui, J. Analysis of sea ice parameters for the design of an offshore wind farm in the Bohai Sea. *Ocean Eng.* **2021**, *239*, 109902. [CrossRef]
11. Yu, J.; Fu, Y.; Yu, Y.; Wu, S.; Wu, Y.; You, M.; Li, M. Assessment of offshore wind characteristics and wind energy potential in Bohai Bay, China. *Energies* **2019**, *12*, 2879. [CrossRef]
12. Chao, J.; Gu, W.; Li, Y.; Xu, Y.; Zhang, H.; Tao, J. Temporal and spatial distribution characteristics of the effective wind and solar energy in the Bohai Bay coastal area. *J. Renew. Sustain. Energy* **2014**, *6*, 043133. [CrossRef]
13. Wang, Z.; Dong, S.; Li, X.; Guedes Soares, C. Assessments of wave energy in the Bohai Sea, China. *Renew. Energy* **2016**, *90*, 145–156. [CrossRef]
14. Li, Z.; Sui, J.; Dong, X.; Meng, G. Preliminary statistics of some sea ice conditions in Liaodong Gulf. *Ocean. Eng.* **1992**, *10*, 72–78. (In Chinese)
15. Li, Z. Field investigation of sea ice in the Liaodong Gulf. *Mar. Forecast.* **1999**, *16*, 48–56. (In Chinese)
16. Li, Z.; Lu, P.; Devinder, S.S. Ice engineering sub- areas in Bohai from ice physical and mechanical parameters. *Adv. Water Res.* **2004**, *15*, 598–602. (In Chinese)
17. Yan, Y.; Gu, W.; Xu, Y.; Li, Q. The in situ observation of modelled sea ice drift characteristics in the Bohai Sea. *Acta Oceanol. Sin.* **2019**, *38*, 17–25. [CrossRef]
18. Ma, Y.; Cheng, B.; Xu, N.; Yuan, S.; Shi, H.; Shi, W. Long-term ice conditions in Yingkou, a coastal region northeast of the Bohai Sea, between 1951/1952 and 2017/2018: Modeling and Observations. *Remote Sens.* **2022**, *14*, 182. [CrossRef]
19. Zhang, B. *Regulations for Offshore Ice Condition & Application in China Sea*; China National Offshore Oil Corporation: Beijing, China, 2002; Q/HSn 3000-2002.
20. Yan, Y.; Gu, W.; Gierisch, A.M.U.; Xu, Y.; Uotila, P. NEMO-Bohai 1.0: A high-resolution ocean and sea ice modelling system for the Bohai Sea, China. *Geosci. Model Dev.* **2022**, *15*, 1269–1288. [CrossRef]
21. Uotila, P.; Iovino, D.; Vancoppenolle, M.; Lensu, M.; Rousset, C. Comparing sea ice, hydrography and circulation between NEMO3.6 LIM3 and LIM2. *Geosci. Model Dev.* **2017**, *10*, 1009–1031. [CrossRef]
22. Chen, W.; Tenzer, R. Harmonic coefficients of the Earth's Spectral Crustal Model 180-ESCM180. *Earth Sci. Inform.* **2015**, *8*, 147–159. [CrossRef]
23. Toye, H.; Sanikommu, S.; Raboudi, N.F.; Hoteit, I. A hybrid ensemble adjustment Kalman filter based high-resolution data assimilation system for the Red Sea: Implementation and evaluation. *Q. J. R. Meteorol. Soc.* **2020**, *146*, 4108–4130. [CrossRef]
24. Xv, N.; Chen, W.; Yuan, S.; Yue, Q.; Li, Z.; Ji, S.; Liu, X.; Shi, W.; Zhang, D.; Wang, R. *The Specification for Engineering Sea Ice Techniques*; State Oceanic Administration: Beijing, China, 2016; HY/T 047-2016.

25. Gong, S.; Chen, W.; Zhang, C.; Yan, Q.; Hong, Y. Variability of sea ice from 2008 to 2019 in the Bohai and northern Huanghai Sea, China and the relationship with climatic factors. *J. Ocean Univ.* **2022**, *21*, 1189–1197. [CrossRef]
26. Ma, Y.; Guan, P.; Xu, N.; Xu, Y.; Yuan, S.; Liu, Y.; Yu, F. Determination of the sea ice parameters for the reliability design of the marine structures in Liaodong Bay. *Ocean. Eng.* **2019**, *37*, 136–142. (In Chinese) [CrossRef]
27. Tao, S.; Dong, S. Interval estimation of return sea ice thickness in the northern area of Bohai Sea based on maximum likelihood method. *Eng. Mech.* **2013**, *30*, 294–298. (In Chinese) [CrossRef]

water

MDPI

Article

Simulation and Key Physical Drivers of Primary Productivity in a Temperate Lake during the Ice-Covered Period: Based on the VGPM Model

Jie Zhang [1], Fei Xie [2], Haoming Song [1], Jingya Meng [1] and Yiwen Zhang [1,*]

[1] School of Ocean Science and Technology, Dalian University of Technology, Panjin 124221, China
[2] State Key Laboratory of Coastal and Offshore Engineering, Dalian University of Technology, Dalian 116024, China
* Correspondence: zhangyiwen@dlut.edu.cn

Abstract: The primary productivity of seasonal ice-covered water bodies is an important variable for understanding how temperate lake ecosystems are changing due to global warming. But there have been few studies on the complete change process of primary productivity during the ice-covered period, and the connection between ice physical and associated biological production has not been fully understood. In this study, a Vertically Generalized Production Model (VGPM) suitable for the ice-covered period was used to calculate the primary productivity of a temperate lake, and the key physical controlling factor was analyzed in the process of primary productivity change in the ice-covered period. The results showed that there was a high level of primary productivity, (189.1 ± 112.6) mg $C \cdot m^{-2} \cdot d^{-1}$, under the ice in the study site, Hanzhang Lake. The phytoplankton production under the ice was not as severely restricted by light as commonly thought. The water temperature played a more crucial role in the changes of primary productivity than the light beneath the ice. The study highlighted the variability in primary productivity covering the whole ice-covered age, and provided a better understanding of how the aquatic environment of lakes in seasonal ice-covered areas was affected by warmer temperatures.

Keywords: ice-covered period; primary productivity; temperate lakes; VGPM model; climate change

Citation: Zhang, J.; Xie, F.; Song, H.; Meng, J.; Zhang, Y. Simulation and Key Physical Drivers of Primary Productivity in a Temperate Lake during the Ice-Covered Period: Based on the VGPM Model. *Water* **2023**, *15*, 918. https://doi.org/10.3390/w15050918

Academic Editor: Karl-Erich Lindenschmidt

Received: 30 January 2023
Revised: 19 February 2023
Accepted: 20 February 2023
Published: 27 February 2023

1. Introduction

Temperate lakes with seasonal freezing are considered some of the most sensitive regions to global warming [1,2]. The warming of lakes affects the growth and melting of ice sheets [3]. As the ice cover weakens, the light increases and water temperatures rise, which lead to an increase in the activity of organisms living under the ice [4]. In recent years, researchers have found that the subglacial water bodies can be quite productive in winter months, despite the cold temperatures and lack of light, which breaks with the past [5–8]. Another reason why biological activity under the ice is crucial, is that the lake ecology will change with the alternation of the four seasons in temperate lakes. Each season has its own characteristics that can affect the lake's ecosystem, and there is a close relationship between seasons. For example, the changes in biological activity with the melting of ice in winter can affect the distribution, structure, and biomass of organisms in the following spring [9–11]. Therefore, the primary productivity beneath the ice of seasonal ice-covered water bodies is an important factor for a comprehensive understanding of how temperate lake ecosystems are changing under the background of global warming.

The model estimation method is one of the main ways of calculating primary productivity. This method involves using a mathematical or simulation model to estimate productivity based on various parameters, including temperature, light, and chlorophyll over a wide range of timescales. The concentration of chlorophyll on the water surface can be determined by remote sensing technology, and the Vertically Generalized Production

Model (VGPM) is a model used to estimate the primary productivity of the water bodies based on chlorophyll concentrations [12,13]. Lomas [14] confirmed that the VGPM model can accurately estimate primary productivity by comparing measurement of chlorophyll *a* (Chl.a) with radioactive ^{14}C isotope. The depth of the euphotic layer (which is the part of the water column where light can penetrate) is an input of the VGPM model, as it affects the amount of light that can reach the bottom layers of the water column. Generally, the depth at which 1% of the Photosynthetically Active Radiation (PAR) is taken on the water surface as the standard is typically used to determine the euphotic depths, or the photosynthetically active layer [15]. Yu calculated the primary productivity [16] below the ice by the amount of PAR that reaches the ice-water interface. The studies mentioned above have provided a foundation for the research into primary productivity in lakes with ice cover. But there have been relatively few studies on the complete change process of primary productivity during the ice-covered period. It is important to understand the primary productivity throughout the ice-covered period in order to inform our understanding of how aquatic ecosystems are responding to environmental changes. In addition, the connection between ice physical processes and ice ecosystem has not been fully established. For example, water temperature affects the vertical exchange of water [17], and the shortening of ice growth and melting lead to increased light availability under the ice, which enhance biological activity [18]. However, the specific effects of these physical factors on primary productivity remains to be studied. In this study, a VGPM model suitable for the ice-covered period was used to calculate the primary productivity and analyze key physical controlling factors, in order to establish the relationship between ice physics and ice ecology during the ice-covered period of a seasonal frozen temperate lake.

2. Materials and Methods

2.1. Study Site

Hanzhang Lake (40°40′ N–40°43′ N, 122°0′ E–122°08′ E) is located at the northern boundary of offshore China (Figure 1a), adjacent to Bohai Sea, with a salinity of 5–7 ppt. The lake has a surface area of 10 km^2 and it is generally shallow with an average depth of 6 m and a maximum depth of 10 m. The climate in this location is warm temperate continental semi-humid monsoon, with sufficient sunshine, and an average annual temperature of 10.5 °C. Hanzhang Lake generally enters the freeze period in December and melts in March of the next year, with a maximum ice thickness of 40 cm. The average winter temperature is approximately −6 °C, and the lowest temperature can reach −22 °C. The lake is eutrophic with an average total nitrogen (TN) concentration of 1.15 mg/L and an average total phosphorus (TP) concentration of 0.16 mg/L between 2019 to 2021.

2.2. Field Methods

The field study was conducted from 12 January to 14 March 2022, when the lake was covered by ice. The study was divided into two parts: floating remote observation platform and manual investigation. Ice thickness, PAR, water temperature, turbidity, and Chl.a were automatically monitored by sensors mounted on a floating remote observation platform (Figure 1b) [19]. The sensors used were: ultrasonic rangefinders with an accuracy of ±0.01 m and a monitoring frequency of every 1 min, solar radiation sensors with a monitoring frequency of every 30 min (two sets of sensors on the ice surface, measuring the incident light irradiance and the reflected light irradiance of the ice surface, respectively; one was set at 0.8 m below the ice), and YLS-ZDW chlorophyll and turbidity in situ monitoring sensors (the water temperature measurement range was −5 °C to 50 °C, with a resolution of 0.01 °C and an accuracy of ±0.15 °C; the turbidity range was 0 to 1000 FTU, with a resolution of 0.01 FTU and an accuracy of ± 2%; the range of Chl.a was 0 to 400 μg/L, with a resolution of 0.01 μg/L and an accuracy of ±5%; the monitoring frequency was once every 1 min; there were four sets at the water depth of 0.7 m, 1.5 m, 2.0 m, and 4.4 m, respectively). The transparency of water was measured by manual investigation with a Sayer's plate. Due to safety problems, the manual sampling time was based on the ice thickness, starting at 15 cm during the freezing period and ending at 15 cm during the melting period. The sampling time interval was 5 days.

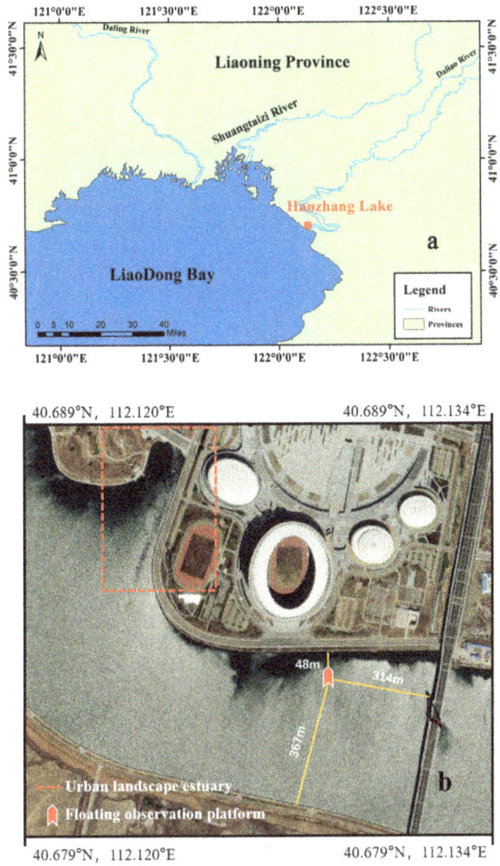

Figure 1. Study site and the floating observation platform ((**a**) Hanzhang Lake sampling point; (**b**), floating observation platform). The study sites are marked by solid lines.

2.3. Euphotic Depths

The calculation of euphotic depth (Z_{eu}) in this study refers to the following improved method [16]. Generally, non-icy water was considered to have uniform properties. According to Lambert-Beer law, the PAR of a beam of light passing through water will decrease exponentially with increasing depth. The euphotic depth was calculated by the Formula (1), the maximum depth at which light is still able to penetrate the water, allowing photosynthesis to occur:

$$Z_{eu} = \frac{2ln10}{K_d(PAR)} = \frac{4.605}{K_d(PAR)} \tag{1}$$

where K_d *(PAR)* was the diffuse attenuation coefficient of photosynthetically active radiation (m^{-1}). Holmes [20] defined the relationship between transparency and K_d *(PAR)* as:

$$K_d(PAR) = \frac{f}{SD} \tag{2}$$

where *SD* was the water transparency (in meters), and *f* was a constant that was determined by empirical data. Ma [21] investigated and analyzed data from 20 lakes in northern China and found that the relationship between *Kd(PAR)* and *SD* was relatively stable, with a generally accepted value of 1.4 for the constant *f*. It was appropriate to select an *f* value of 1.4 for this study.

In frozen lakes during the winter months, ice and snow covered the water bodies, which can absorb and scatter solar radiation, thus affecting the light radiation through water bodies and altering the calculated euphotic depth. Therefore, the extinction coefficient of the ice sheet was a necessary factor for calculating the euphotic depth of water bodies in the ice period. PAR decreased exponentially in the ice sheet, and when combined with the mirror reflection of light on the ice surface, the extinction coefficient of the ice sheet was expressed as:

$$K_i(PAR) = \frac{1}{h_i}\left[(1 - R_s)\frac{E_d(0, PAR)}{E_d(h_i, PAR)}\right] \quad (3)$$

In general, the euphotic depth was defined as the depth where underwater photosynthetically active radiation intensity was 14 μmol·m^{-2}·s^{-1}. The extinction coefficient and photosynthetically active diffuse attenuation coefficient of the ice sheet can be used to derive the calculation model of the euphotic depth of water in the ice-covered period:

$$Z_{eui} = \frac{SD}{f} ln\frac{(1 - R_s)E_d(0, PAR)}{14e^{k_i(PAR)h_i}} \quad (4)$$

where Z_{eui} was the depth of the euphotic layer (m), *f* was 1.4, *SD* was water transparency (m), h_i was ice thickness, K_i *(PAR)* was the extinction coefficient of the ice layer (m^{-1}), R_s was the reflectance of the surface mirror which was calculated by the ratio of surface reflection to surface incidence PAR, and E_d (0, *PAR*) was the ice surface *PAR* (μmol·m^{-2}·s^{-1}).

2.4. Primary Productivity

The core calculation formula of primary productivity used in this study was the VGPM model established by Behrenfeld and Falkowski [22], based on large-scale and long-term monitoring data.

$$PP_{eui} = 0.66125 \times P^B_{opt} \times D_{irr} \times \frac{E_i}{E_i + 4.1} \times Z_{eui} \times C_{opt} \quad (5)$$

In the formula, PP_{eui} was the primary productivity of the water body (measured by mgC·m^{-2}·d^{-1}) during the ice-covered period. P^B_{opt} was the maximum carbon sequestration rate of the water column (mgC·mg^{-1}Chl·h^{-1}) during the ice-covered period. D_{irr} is the illumination period (in hours) during the ice-covered period. E_i was *PAR* at the interface of ice–water mixing (mol·m^{-2}·d^{-1}). In this study, E_i was approximately replaced by *PAR* at 0.8 m water depth. C_{opt} was the concentration of Chl.a (μg·L^{-1}) at the depth of the euphotic layer. In this study, an average value of Chl.a concentration at the water depths of 0.7 m and 1.5 m was used as an approximate replacement of that depth of euphotic layer.

The maximum carbon sequestration rate in the water column was calculated based on the Equation (6), proposed by Behrenfeld and Falkowski [22]:

$$\begin{aligned} P^B_{opt} &= 1.2956 + 0.2749T + 0.0617T^2 - 0.0205T^3 + 2.462 \times 10^{-3}T^4 \\ &\quad - 1.348 \times 10^{-4}T^5 + 3.4132 \times 10^{-6}T^6 - 3.27 \times 10^{-8}T^7 \end{aligned} \quad (6)$$

where *T* was the water surface temperature (°C). There was a significant difference between the surface temperature in ice-covered and non-ice-covered water. In non-ice-covered water, the surface water temperature was usually the highest, and the maximum carbon sequestration rate of water column usually occurred at the surface water. However, for ice-covered waters, the water column temperature was lower, so the highest average temperature 2 m away from the water meter can be used as an approximate alternative.

The illumination period D_{irr} indicated the daily length, which was the amount of time that the site was illuminated each day. The value can be obtained by querying geographic information about the study site.

2.5. Statistical Analysis

Stepwise linear regression and principal component analysis were used to identify statistically significant trends in the potential drivers of changing productivity. Statistical analysis was conducted using SPSS 9.0 and Origin 18.0. Data were considered significant when the difference was set at $p < 0.05$.

3. Results

3.1. Ice Thickness, Water Temperature, PAR, Chl.a, and Transparency

The ice thickness, water temperature, incident PAR, ice-water interface PAR, Chl.a, and transparency were shown in Figure 2, which were detected by sensors on the floating observation platform. The ice thickness increased during the freezing period, and reached a maximum peak of 0.43 m on 1 February. The ice water interface PAR fell to its lowest point due to snow, although the ice incident PAR was still increasing. This prevented light from entering the water, resulting in a decrease in the ice water interface PAR. As the temperature rose after 28 February, the ice began to melt gradually. The water temperature increased sharply from 2 °C to 4 °C during the end of February, and the lake ice entered a rapid melting period, melting at a rate of 2–3 cm per day until it had completely melted on 14 March. The incident irradiance showed an obvious increasing trend during the observation period as a result of an increase in both the total radiation and light time, because the time of the experiment was after the winter solstice and the direct solar point gradually moved northward from the Tropic of Cancer.

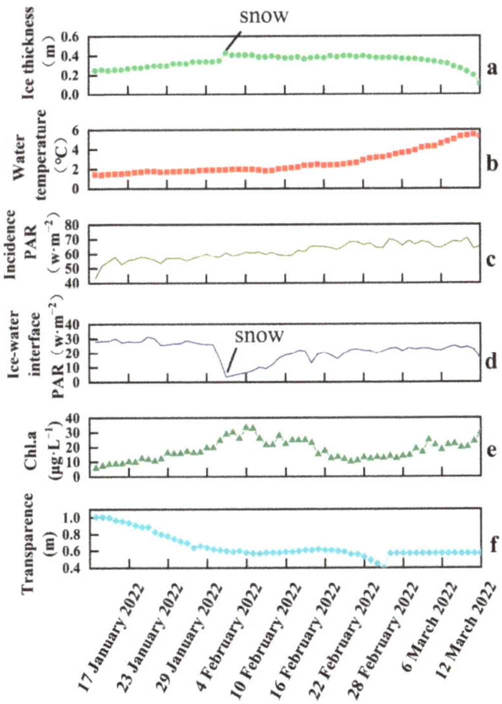

Figure 2. Changes of various environmental factors during the ice-covered period. ((**a**), Ice thickness; (**b**), water temperature; (**c**), ice incident PAR; (**d**), ice-water interface PAR; (**e**), Chl.a; (**f**), transparency).

The PAR at the ice-water interface was significantly lower than the incident irradiance on the ice surface and showed a trend of decreasing first before gradually increasing again. This was related to the formation and melting of ice. The ice thickness increased and enhanced the extinction effect during the freezing period. While the temperature rose and the solar radiation was further enhanced, the ice thinned, resulting in a compensation of the PAR at the ice-water interface in the melting period. The data showed that the average total PAR on ice surface was 61.73 $W \cdot m^{-2}$, and the average PAR at the ice-water interface was 23.83 $W \cdot m^{-2}$ without snow cover. The ratios of PAR at the ice-water interface and that at the ice surface were between 5.76% and 64.84%, respectively, with an average ratio of 35.69%.

During the whole period of ice cover, Chl.a showed a tendency to peak and then decrease. The average value of Chl.a was 17.89 $\mu g \cdot L^{-1}$. The transparency was decreasing throughout the ice-covered period, and showed a negative correlation with Chl.a, indicating a relatively high level of phytoplankton productivity during the ice-covered period.

3.2. Primary Productivity

Primary productivity was calculated based on the data of ice thickness, water temperature, PAR, Chl.a, and transparency using the VGPM model, as shown in Figure 3. The primary productivity of Hanzhang Lake during the ice-covered period showed a trend of fluctuation increase. The minimum primary productivity was 57.77 $mgC \cdot m^{-2} \cdot d^{-1}$, which appeared on 12 January. The maximum value of primary productivity was 666.9 $mgC \cdot m^{-2} \cdot d^{-1}$, which appeared on 13 March. The mean value was (189.1.3 $\pm$ 112.6) $mgC \cdot m^{-2} \cdot d^{-1}$ during the whole ice-covered period. In the last days of the ice-covered period, the primary productivity increased rapidly and reached its maximum value.

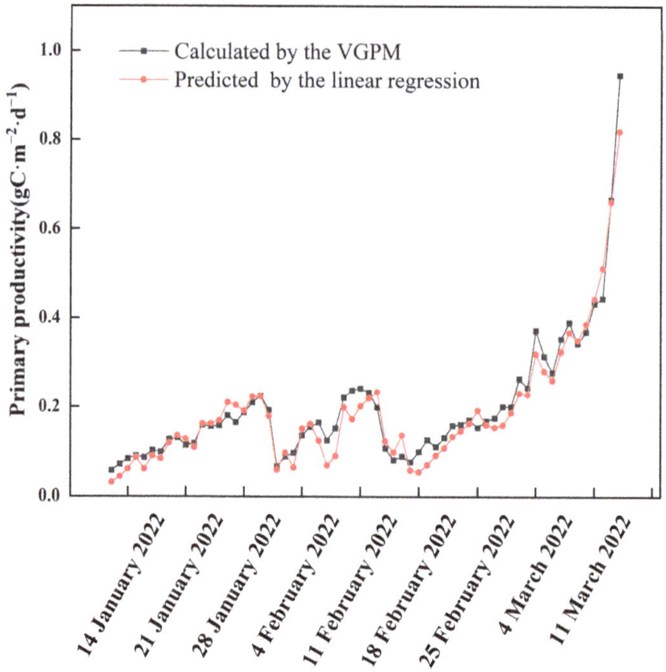

Figure 3. Primary productivity predicted by linear regression and calculated by the VGPM model.

A stepwise linear regression was used to figure out the importance of each calculation factor in the VGPM model, with transparency, ice thickness, ice surface incident PAR, ice-water interface PAR, water temperature, and Chl.a as independent variables, and primary productivity as the dependent variable. The results showed that predicted values of linear regression were in good agreement with the calculated values of the VGPM model ($R^2 = 0.957$, $p < 0.001$). The linear regression formula was primary productivity = 0.242–0.775 × ice thickness (m)—0.358 × transparency (m) + 0.01 × chlorophyll a ($\mu g \cdot L^{-1}$) + 0.009 × daily average ice water interface PAR ($W \cdot m^{-2}$) + 0.021 × water temperature (°C), indicating that ice thickness, transparency, Chl.a, ice water interface PAR, and water temperature have a significant impact on primary productivity. The process of ice growth and melting, such as ice thickness, affected the dynamics of primary productivity under the ice.

3.3. Key Physical Factors

In terms of non-frozen open water bodies, there were great differences in physical factors such as light and water temperature of frozen water bodies, which can have an impact on the water ecosystem under ice. The water temperature usually affects phytoplankton photosynthetic enzyme activity and primary productivity. Changes in euphotic depth and PAR will be impacted by the light intensity, which in turn affects the photosynthetic intensity of phytoplankton, resulting in a change in primary productivity. Principal component analysis was performed on a variety of parameters including ice thickness, water temperature, ice incident PAR, ice-water interface PAR, Chl.a, and transparency in the VGPM model, as shown in Figure 4. Water temperature had the highest correlation with primary productivity throughout the ice period ($p < 0.05$ in the Bartlett test), suggesting that water temperature was a key factor in affecting primary productivity. At the same time, the entire ice period was divided into three parts, which was consistent with freezing and melting periods according to the ice thickness.

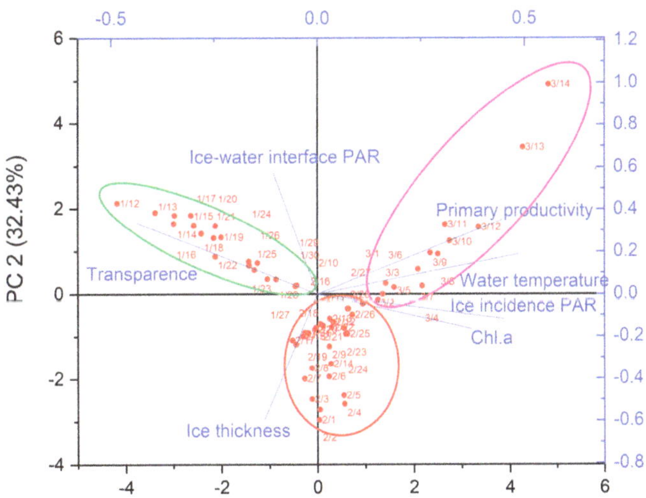

Figure 4. Principal component analysis of primary productivity and affecting factors in the VGPM model.

4. Discussion

The primary productivity of the Hanzhang Lake was 57.77–666.9 mg $C \cdot m^{-2} \cdot d^{-1}$ with an average of (189.1 ± 112.6) mg $C \cdot m^{-2} \cdot d^{-1}$. The result was comparable to that of other lakes or rivers calculated by the VGPM model (shown in Table 1), implying that there was a high level of primary productivity under the ice in Hanzhang Lake during the winter. The primary productivity of Hanzhang Lake was mainly attributed to phytoplankton. Hanzhang Lake was meso-eutrophic with an average Chl.a concentration of (18.4 ± 7.21) μg·L^{-1} and an average algae cell density of (5.85 ± 6.24) × 10^6 cells/L in water under the ice during the ice-covered period of 2022 [23]. The eutrophication of lakes changes the community structure of phytoplankton and benthic organisms significantly, which can lead to a shift in the water energy flow path from the bottom to the top [24,25]. As the phytoplankton was the main primary producer, the impact of environmental factors in the ice-covered period to the dynamics of phytoplankton was focused on in this study.

Table 1. Previous studies of primary productivity using the VGPM model.

Lakes and Rivers	Latitudes	Time	Primary Productivity mg C·m^{-2}·d^{-1}	References
Yenicaga Lake	40°47′ N	December	319	[26]
Pearl River Estuary	21°48′–22°27′	Winter	224.5	[12]
Taihu	30°55′40″–31°32′58″ N	Annual	207.67–2237.71	[27]
Tanganyika	3°20′–8°48′ S	Wet and Dry Seasons	110–1410	[28]
Cape Fear River	-	Annual	18–2580	[29]
Wuliangsuhai	40°36′–41°03′ N	Winter	86.34–96.34	[16]

The effective light radiation under the ice was the primary limiting factor for aquatic biological activities [30]. Both the thickness and the structure of ice contribute to the light attenuation [31,32]. However, some studies suggest that the maximum photosynthetic efficiency required by phytoplankton was only 30% of that of benthic organisms [33]. Two pieces of evidence in this study suggested that the phytoplankton under the ice were not limited by light in Hanzhang Lake. First, it was found that primary productivity did not decrease with the decrease of light under the ice in the freeing period of this study, even though ice thickness and snow cover decreased the depth of the euphotic layer. Observations showed that the lowest depth of the eukaryotic layer was 0.8 m, providing plenty of space for phytoplankton to undergo photosynthesis. Second, the average daily ice-water interface PAR of the Hanzhang lake was (21.7 ± 6.9) W·m^{-2}, which met the lighting requirement for primary production. The quantum irradiance required for primary productivity was approximately 25 μmol·m^{-2}·s^{-1}, which was equivalent to approximately 5 W·m^{-2} in terms of irradiance level [34]. Except for on overcast and snowfall days when light intensity decreased, primary productivity was observed to recover significantly with the melting of snow as PAR increased. This suggests that although PAR decreased due to snow and ice, it did not reach a level that limits primary production, possibly due to the geographic location of Hanzhang Lake.

This research suggests that water temperature is a key factor in determining the level of primary production in Hanzhang Lake, especially during periods of ice coverage. The data showed a statistically significant positive correlation between water temperature and concentration of chlorophyll a ($p < 0.05$), indicating that the warmer temperatures tend to lead to higher phytoplankton activities and primary production. The primary production of phytoplankton in the freezing period was mainly regulated by water temperature but not light. In addition, primary productivity increased quickly with the rise in temperature and improved light conditions during the period when the ice was melting. Water temperature and the duration of the ice cover were considered to be the main drivers of biological dynamics [35]. Aquatic conditions in winter played an essential role in the abundance and structure of phytoplankton communities, affecting the growth of phytoplankton during spring. Evidence showed change in phytoplankton in a long scale was linked to temperature rises, for example, phenology changes over a time period of 15 years in the Bassenthwaite Lake at the northern boundary of the English Lake District [1]. The dynamics and productivity of phytoplankton under the ice should be paid close attention to, in the background of global warming.

Convective mixing in the water layer beneath the ice played an important role in phytoplankton bioactivity, in addition to light and water temperature. This convective mixing can create favorable conditions for bioactivity by increasing nutrient supply and allowing for more efficient exchanges between the surface and lower layers of water. Among the stratified water bodies of ice-covered closed freshwater lakes, the convective mixing in vertical profiles was usually weak. The salty lakes had a typical salt drainage effect when they froze, which can lead to a concentration and diffusion of the salty water layer beneath the ice. This can enhance the vertical convection of the water body. Hanzhang Lake was adjacent to the Bohai Sea, with a salinity of 5 to 7 ppt. The salt of the ice sheet would be transferred downward, and the nutrients released by the sediments can be more easily exchanged to the upper water when vertical convection was enhanced [36]. Thus, the nutrients in the water were accumulated, and the phytoplankton moved freely between the water layers, which contributed to the growth of phytoplankton populations. The light limitation below the euphotic layer was weakened when vertical convection was enhanced [34], which can be related to the high chlorophyll concentration in Hanzhang Lake during the ice-covered period. There was a negative correlation ($p < 0.05$) between Chlorophyll *a* and ice-water interface PAR in this study, and the high-value area of chlorophyll did not always exist in surface water. The negative correlation of Chl.a and PAR in Hanzhang Lake was consistent with that in Antarctic Lake Bonney [37], which was due to the self-shading effect. The shelf-shading effect was that high chlorophyll concentrations attenuated light radiation.

The results from Hanzhang Lake indicated that the primary productivity was greatly affected by the ice-covered period. Phytoplankton was the main producer, and water temperature was the key physical factor driving the dynamics of phytoplankton. The phytoplankton under the ice was not as light limited as commonly thought. The primary production level of phytoplankton will not decrease significantly, even though the thick ice and snow cover can lead to a decrease in light radiation and a decrease in the depth of the euphotic layer. This study was meant to provide reference for the subsequent evolution of ice ecological environment.

5. Conclusions

The ice-covered period of Hanzhang Lake lasted approximately 3 months, with a maximum ice thickness of 0.43 m and an average ice thickness of 0.29 m during winter in 2021–2022. The primary productivity was 57.77–666.9 mg $C \cdot m^{-2} \cdot d^{-1}$ with an average of (189.1 $\pm$ 112.6) mg $C \cdot m^{-2} \cdot d^{-1}$ during the ice-covered period, using the VGPM model. The water temperature played a more critical role in the changes of primary productivity than light under the ice in Hanzhang Lake. The primary production level did not decrease when the thick ice and snow cover led to the decrease of light radiation and the depth of the euphotic layer in the freezing period. In addition, primary productivity increased quickly with the rise in temperature and improved light conditions in the melting period. It appears that in regions where there are seasonal changes of icy cover, warmer water temperatures may be more important for primary productivity than light conditions, providing a better understanding of the interaction between ice physicals and ecology.

Author Contributions: Investigation, analysis, and writing of the original draft, J.Z.; investigation and conceptualization, F.X.; analysis and visualization, H.S.; formal analysis, J.M.; writing review and editing, Y.Z. All authors have read and agreed to the published version of the manuscript.

Funding: This research was supported by the National Key Research and Development Program of China (Grant No. 2019YFE0197600) and the National Natural Science Foundation of China (Grant No. 42007150, 52211530038).

Institutional Review Board Statement: No human or animal studies involved in this study.

Data Availability Statement: Data from the VGPM model in this study can be downloaded from https://pan.baidu.com/s/1Lg0Hqa2aoyFI_OUyr3-BYw (code 1952); the meteorological data from the National Meteorological Data Center can be downloaded from http://data.cma.cn.

Acknowledgments: We are grateful to the editor and anonymous reviewers for their comments, which considerably improved this work.

Conflicts of Interest: The authors declare no conflict of interest.

References

1. Meis, S.; Thackeray, S.J.; Jones, I.D. Effects of recent climate change on phytoplankton phenology in a temperate lake. *Freshw. Biol.* **2009**, *54*, 1888–1898. [CrossRef]
2. Ladwig, R.; Appling, A.P.; Delany, A.; Dugan, H.A.; Gao, Q.T.; Lottig, N.; Stachelek, J.; Hanson, P.C. Long-term change in metabolism phenology in north temperate lakes. *Limnol. Oceanogr.* **2022**, *67*, 1502–1521. [CrossRef]
3. Woolway, R.I.; Kraemer, B.M.; Lenters, J.D.; Merchant, C.J.; O'Reilly, C.M.; Sharma, S. Global Lake responses to climate change. *Nat. Rev. Earth Environ.* **2020**, *1*, 388–403. [CrossRef]
4. Markensten, H.; Moore, K.; Persson, I. Simulated Lake phytoplankton composition shifts toward cyanobacteria dominance in a future warmer climate. *Ecol. Appl.* **2010**, *20*, 752–767. [CrossRef]
5. Kohlb Kohlbach, D.; Graeve, M.; Lange, B.A.; David, C.; Schaafsma, F.L.; van Franeker, J.A.; Vortkamp, M.; Brandt, A.; Flores, H. Dependency of Antarctic zooplankton species on ice algae-produced carbon suggests a sea ice-driven pelagic ecosystem during winter. *Glob. Chang. Biol.* **2018**, *24*, 4667–4681. [CrossRef]
6. Hampton, S.E.; Galloway, A.W.; Powers, S.M. Ecology under lake ice. *Ecol. Lett.* **2017**, *20*, 98–111. [CrossRef]
7. Bramburger, A.J.; Ozersky, T.; Silsbe, G.M.; Crawford, C.J.; Olmanson, L.G.; Shchapov, K. The not-so-dead of winter: Underwater light climate and primary productivity under snow and ice cover in inland lakes. *Inland Waters* **2022**. [CrossRef]
8. Song, S.; Li, C.Y.; Shi, X.H.; Zhao, S.N.; Tian, W.D.; Li, Z.J.; Bai, Y.L.; Cao, X.W.; Wang, Q.K.; Huotari, J.; et al. Under-ice metabolism in a shallow lake in a cold and arid climate. *Freshw. Biol.* **2019**, *64*, 1710–1720. [CrossRef]
9. Moss, B. Allied attack: Climate change and eutrophication. *Inland Waters* **2011**, *1*, 101–105. [CrossRef]
10. Yang, F.; Cen, R.; Feng, W.Y.; Zhu, Q.H.; Lepparanta, M.; Yang, Y.; Wang, X.H.; Liao, H.Q. Dynamic simulation of nutrient distribution in lakes during ice cover growth and ablation. *Chemosphere* **2021**, *281*, 130781. [CrossRef]
11. Yang, F.; Li, C.; Shi, X.; Zhao, S.; Hao, Y. Impact of seasonal ice structure characteristics on ice cover impurity distributions in Lake Ulansuhai. *J. Lake Sci.* **2016**, *28*, 455–462.
12. Ye, H.B.; Chen, C.Q.; Sun, Z.H.; Tang, S.L.; Song, X.Y.; Yang, C.Y.; Tian, L.Q.; Liu, F.F. Estimation of the Primary Productivity in Pearl River Estuary Using MODIS Data. *Estuar. Coasts* **2015**, *38*, 506–518. [CrossRef]
13. Zainuddin, M.; Kiyofuji, H.; Saitoh, K.; Saitoh, S.I. Using multi-sensor satellite remote sensing and catch data to detect ocean hot spots for albacore (*Thunnus alalunga*) in the northwestern North Pacific. *Deep-Sea Res. Pt. II* **2006**, *53*, 419–431. [CrossRef]
14. Lomas, M.W.; Moran, S.B.; Casey, J.R.; Bell, D.W.; Tiahlo, M.; Whitefield, J.; Kelly, R.P.; Mathis, J.T.; Cokelet, E.D. Spatial and seasonal variability of primary production on the Eastern Bering Sea shelf. *Deep-Sea Res. Pt. II* **2012**, *65–70*, 126–140. [CrossRef]
15. Huang, L.; Zhou, Y.; Zhou, Q.; Wang, W.; Dong, Y.; Li, K.; Chang, J. Temporal-spatial variation in phytoplankton primary production and its influencing factors in Lake Chenghai on the Yunnan Plateau, China. *J. Lake Sci.* **2019**, *31*, 1424–1436.
16. Yu, H.F.; Shi, X.H.; Zhao, S.N.; Sun, B.; Liu, Y.; Arvola, L.; Li, G.H.; Wang, Y.J.; Pan, X.R.; Wu, R.; et al. Primary productivity of phytoplankton and its influencing factors in cold and arid regions: A case study of Wuliangsuhai Lake, China. *Ecol. Indic.* **2022**, *144*, 109545. [CrossRef]
17. Kirillin, G.; Lepparanta, M.; Terzhevik, A.; Granin, N.; Bernhardt, J.; Engelhardt, C.; Efremova, T.; Golosov, S.; Palshin, N.; Sherstyankin, P.; et al. Physics of seasonally ice-covered lakes: A review. *Aquat. Sci.* **2012**, *74*, 659–682. [CrossRef]
18. Hintz, N.H.; Schulze, B.; Wacker, A.; Striebel, M. Ecological impacts of photosynthetic light harvesting in changing aquatic environments: A systematic literature map. *Ecol. Evol.* **2022**, *12*, e8753. [CrossRef]
19. Xie, F.; Lu, P.; Li, Z.J.; Wang, Q.K.; Zhang, H.; Zhang, Y.W. A floating remote observation system (FROS) for full seasonal lake ice evolution studies. *Cold Reg. Sci. Technol.* **2022**, *199*, 103557. [CrossRef]
20. Holmes, R.W. Secchi disk in turbid coastal waters. *Limnol. Oceanogr.* **1970**, *15*, 688–694. [CrossRef]
21. Ma, J.H. *Inversion of Kd (PAR) and Euphotic Zone Depth of Typical Water Bodys in Northeast China with Remote Imagery*; Northeast Institute of Geography and Agroecology: Changchun, China; Chinese Academy of Sciences: Beijing, China, 2016.
22. Behrenfeld, M.J.; Falkowski, P.G. Photosynthetic rates derived from satellite-based chlorophyll concentration. *Limnol. Oceanogr.* **1997**, *42*, 1–20. [CrossRef]
23. Wang, H.; Wang, H.; Li, Y.; Ma, S.; Yu, Q. The control of lake eutrophication: Focusing on phosphorus abatement, or reducing both phosphorus and nitrogen? *Acta Hydrobiol. Sin.* **2020**, *44*, 938–960.
24. Alexander, T.J.; Vonlanthen, P.; Seehausen, O. Does eutrophication-driven evolution change aquatic ecosystems? *Philos. Trans. R. Soc. B* **2017**, *372*, 20160041. [CrossRef] [PubMed]
25. Li, H.; Shen, H.; Li, S.; Liang, Y.; Lu, C.; Zhang, L. Effects of eutrophication on the benthic-pelagic coupling food web in Baiyangdian Lake. *Acta Ecol. Sin.* **2018**, *38*, 2017–2030.
26. Saygi-Basbug, Y.; Demirkalp, F.Y. Primary production in shallow eutrophic Yenicaga Lake (Bolu, Turkey). *Fresenius Environ. Bull.* **2004**, *13*, 98–104.
27. Wei, J.; Ji, X.N.; Hu, W. Characteristics of Phytoplankton Production in Wet and Dry Seasons in Hyper-Eutrophic Lake Taihu, China. *Sustainability* **2022**, *14*, 11216. [CrossRef]

28. Stenuite, S.; Pirlot, S.; Hardy, M.A.; Sarmento, H.; Tarbe, A.L.; Leporcq, B.; Descy, J.P. Phytoplankton production and growth rate in Lake Tanganyika: Evidence of a decline in primary productivity in recent decades. *Freshw. Biol.* **2007**, *52*, 2226–2239. [CrossRef]
29. Kennedy, J.T.; Whalen, S.C. Seasonality, and controls of phytoplankton productivity in the middle Cape Fear River, USA. *Hydrobiologia* **2008**, *598*, 203–217. [CrossRef]
30. Hampton, S.E.; Sharma, S.; Brousil, M.R.; Filazzola, A. Winter and summer storms modify chlorophyll relationships with nutrients in seasonally ice-covered lakes. *Ecosphere* **2022**, *13*, e4272. [CrossRef]
31. Xie, C.; Zhou, L.; Ding, S.F.; Liu, R.W.; Zheng, S.J. Experimental and numerical investigation on self-propulsion performance of polar merchant ship in brash ice channel. *Ocean Eng.* **2023**, *269*, 113424. [CrossRef]
32. Sun, Q.Y.; Zhang, M.; Zhou, L.; Garme, K.; Burman, M. A machine learning-based method for prediction of ship performance in ice: Part I. ice resistance. *Mar. Struct.* **2022**, *83*, 103181. [CrossRef]
33. Rautio, M.; Dufresne, F.; Laurion, I.; Bonilla, S.; Vincent, W.F.; Christoffersen, K.S. Shallow freshwater ecosystems of the circumpolar Arctic. *Ecoscience* **2011**, *18*, 204–222. [CrossRef]
34. Ozkundakci, D.; Gsell, A.S.; Hintze, T.; Tauscher, H.; Adrian, R. Winter severity determines functional trait composition of phytoplankton in seasonally ice-covered lakes. *Glob. Chang. Biol.* **2016**, *22*, 284–298. [CrossRef] [PubMed]
35. Katz, S.L.; Izmest'eva, L.R.; Hampton, S.E.; Ozersky, T.; Shchapov, K.; Moore, M.V.; Shimaraeva, S.V.; Silow, E.A. The "Melosira years" of Lake Baikal: Winter environmental conditions at ice onset predict under-ice algal blooms in spring. *Limnol. Oceanog.* **2015**, *60*, 1950–1964. [CrossRef]
36. Shatwell, T.; Thiery, W.; Kirillin, G. Future projections of temperature and mixing regime of European temperate lakes. *Hydrol. Earth Syst. Sci.* **2019**, *23*, 1533–1551. [CrossRef]
37. Obryk, M.K.; Doran, P.T.; Priscu, J.C. The permanent ice cover of Lake Bonney, Antarctica: The influence of thickness and sediment distribution on photosynthetically available radiation and chlorophyll-a distribution in the underlying water column. *J. Geophys. Res. Biogeosci.* **2014**, *119*, 1879–1891. [CrossRef]

MDPI
St. Alban-Anlage 66
4052 Basel
Switzerland
www.mdpi.com

Water Editorial Office
E-mail: water@mdpi.com
www.mdpi.com/journal/water

www.ingramcontent.com/pod-product-compliance
Lightning Source LLC
Chambersburg PA
CBHW040745250926
51850CB00007BB/1418
9783725806287